History of Earth Sciences

SCIENCES

History of Sciences, Field Director – Jean-Claude Dupont

Astronomy, Physics, Chemistry and Earth Sciences,
Subject Head – Christian Bracco

History of Earth Sciences

Coordinated by

Pierre Savaton

iSTE

WILEY

First published 2026 in Great Britain and the United States by ISTE Ltd and John Wiley & Sons, Inc.

ISTE Ltd
27-37 St George's Road
London SW19 4EU
UK

www.iste.co.uk

John Wiley & Sons, Inc.
111 River Street
Hoboken, NJ 07030
USA

www.wiley.com

Library of Congress Control Number: 2026937212

British Library Cataloguing-in-Publication Data
A CIP record for this book is available from the British Library
ISBN 978-1-78945-231-0

ERC code:
SH6 The Study of the Human Past
SH6_5 Archaeological science, bioarchaeology, environmental archaeology, geoarchaeology
SH6_14 History of science and technologies, environmental history

The manufacturer's authorized representative according to the EU General Product Safety Regulation is Wiley-VCH GmbH, Boschstr. 12, 69469 Weinheim, Germany, e-mail: Product_Safety@wiley.com.

Contents

Presentation of Field "History of Science"

Jean-Claude DUPONT
CHSSC, Université de Picardie Jules Verne, Amiens, France

The "History of Science" field of the *SCIENCES* collection is closely related to the sections devoted to scientific and technical disciplines, on the one hand, and to epistemology, on the other hand. In addition to providing a summary of current methods and issues in the history of science, all of the volumes offer an overview on the evolution of knowledge in mathematics, the physical and natural sciences, medicine, the humanities and technology, from the beginning to the present day, as well as in different cultural areas. This is aimed at non-specialists and scientists alike, providing them with a useful perspective on the origins and foundations of scientific knowledge, while keeping up with contemporary research in the history of science.

– Field Director: Jean-Claude Dupont.

– Subject heads of the English edition:

- "Issues, General Themes and Methods of History of Sciences": Stéphane Tirard.

- "Logic and Mathematics": Sabine Rommevaux-Tani.

- "Astronomy, Physics, Chemistry and Earth Sciences": Christian Bracco.

- "Life Sciences": Jean-Claude Dupont.

- "Humanities and Social Sciences": Alexandre Escudier.

History of Earth Sciences,
coordinated by Pierre SAVATON.

Introduction

Pierre SAVATON[1,2]
[1]*Université Caen Normandie, France*
[2]*Laboratoire CFV, Université de Nantes, France*

Translating into English a collective work on the history of the Earth sciences written in French is a first. Since *Histoire de la géologie* by Gabriel Gohau in 1987 (Gohau 1987), followed by François Ellenberger's (Ellenberger 1989, 1994) two-volume work in 1989 and 1994, publishers have provided other excellent studies in this field of research, though with less general scope. The proposal of the ISTE SCIENCES collection to produce a volume specifically devoted to the history of Earth sciences is therefore an opportunity to renew this ambition of disseminating research to a wider audience than that of readers of scientific articles, and particularly to students.

In 1976, François Ellenberger, professor of structural geology at the Sorbonne since 1962, president of the Geological Society of France in 1972, and a tectonic specialist recognized both nationally and internationally, founded with a handful of geologists and historians the *Comité français d'histoire de la géologie* (French Committee for the History of Geology, COFRHIGEO[1]) to contribute collectively to the development and dissemination of specific research on the history of geological sciences. The committee, which will celebrate its 50th anniversary in 2026, represents France at the International Commission on the History of Geology (INHIGEO), one of the commissions of the International Union of Geological Sciences (IUGS[2]). In a

1 See: https://hal.science/COFRHIGEO.

2 See: https://inhigeo.com/.

landscape of institutional research where the history of science was only beginning to emerge, the creation of COFRHIGEO reflected the interest of several geologists in questions of history and philosophy within their own discipline, alongside their university work. While the committee seeks to unite and encourage geologists' interest in the history of their field, it is also open to historians and epistemologists, which quickly led to a diversity of approaches and perspectives in a context where debates between "internalist" and "externalist" approaches were gradually fading. Its original disciplinary grounding, however, remains in 2024 one of its defining characteristics. The research presented during its sessions gives rise to an annual publication, Les *Travaux du COFRHIGEO*, which over the years has become both a showcase for French research in the history of geology and a resource for French and foreign scholars.

In the 1970s, books on the history of science gave little attention to the history of geology. One had to turn to Frank Dawson Adams's (Adams 1954) *The Birth and Development of the Geological Sciences* or to Helmut Hölder's (Hölder 1960) *Geologie und Paläontologie in Texten und ihrer Geschichte*, or even earlier to Karl Alfred von Zittel's (Zittel 1899, 1901) *Geschichte der Geologie und Paläontologie* (English translation in 1901), or to Sir Archibald Geikie's 1897 *The Founders of Geology*, in its revised and expanded 1905 edition reissued in 1960 (Geikie 1905, 1960). Adams focused mainly on the books of the 16th and 17th centuries, and Zittel focused on the 19th century; Geikie was sometimes imprecise and reductive; Hölder provided access to early texts, but only for those able to read German.

For the study of medieval texts dealing with geology, one must turn to volume IX of Pierre Duhem's *Système du monde* (Duhem 1959). Access to Greek and Latin texts concerning geology is often an obstacle. Ellenberger, who read both languages, praised the 1976 dissertation of Bouillet-Roy (1976), *La géologie dynamique chez les anciens Grecs et Latins d'après les textes*, which contained 875 excerpts from original sources, though unfortunately this book was never published. Finally, let us note Roy Porter's 1977 book (Porter 1977) *The Making of Geology*, a marker of a new historiography, though limited to the history of Earth sciences in Great Britain from 1660 to 1815. The field of the history of geology was still to be cultivated. A few French researchers then oriented their studies in this direction, but most often outside institutional research and teaching positions, or on the margins of their main activities. This institutional situation has hardly changed since, but the accumulation of individual books and the major contributions of a few have, over 50 years, built a French historiography of geology, both identified and recognized internationally. François Ellenberger (1915–2000) and Gabriel Gohau (1934–2023) have left, and continue to leave through their publications, a lasting mark on this history and its dissemination.

In 1987, Gohau published *Histoire de la géologie*, a 259-page book that retraces a history of discourses about the Earth from Antiquity to the formulation of plate tectonics theory in 1968 (Gohau 1987). The book was reissued 3 years later by Éditions du Seuil as a pocket edition in the "Points Sciences" collection and translated into English in 1991 as *A History of Geology* by Albert V. Carozzi and Marguerite Carozzi, distributed in the United States by Rutgers University Press. The history of geology in it truly begins with the first theories of the Earth in the 17th century, the "act 1", but only after a presentation of the debates between continuity and discontinuity that opposed the Stoics and the disciples of Aristotle in Antiquity. This "prehistory" of geology is, for the author, a necessary prelude for situating his book within a philosophical history of the sciences. There is no question in this book of studying the history of mining practices and techniques, which nonetheless likely founded the earliest discourses about the Earth, nor of tracing the development of mineralogical discourses. Gohau justified his choice. He wished to offer his readers a history of history itself, a history of ideas, a history of historical geology – the one that has aimed to reconstruct the history of events that have succeeded one another on the surface of the globe since its creation. A declared student of Georges Canguilhem and of the ideas of Gaston Bachelard, Gohau develops an epistemological history of the Earth sciences. The geological sciences have the particularity of combining questions that aim to reconstruct the history of transformations of the Earth's surface since its creation with questions that seek to identify and characterize the causes of those transformations. These two aspects correspond to the often-made distinction between historical geology and physical or causal geology. The historical dimension, palaetiological (term coined by William Whewell in 1837), is for Gohau the one that determines the emergence of a geological science, beyond the ancient mineralogical discourses and theories of the Earth – the "act 2" of the history of geology. This first general, synthetic book on the history of geology aimed to make this science known to an audience of geology teachers, students and curious readers. Beyond its substance, the form and style contributed to the success of the book, which now appears in all general French bibliographies of the history of geology.

The *Histoire de la géologie* in two volumes[3], published by Ellenberger in 1988 and then 1994, aimed both "to offer historians of science material that could serve as

3 Respectively 352 pages and 382 pages.

a basis for their own research" (Ellenberger 1988, p. 1) and to serve "all those who, teaching Geology, will wish to illuminate their teaching by recalling the historical stages of the development of this discipline". One finds in these stated intentions the same purpose that motivated the COFRHIGEO, of which Ellenberger remained president until 1995. This pedagogical concern also reflects the author's desire for the history of science to be taken into account in science education itself to awaken a critical spirit that he felt was increasingly lacking among his contemporaries, students and researchers alike (Ellenberger 1988, p. 2). Ellenberger's book is that of a scholar who accesses source texts often in their original language and makes them available to his readers while tirelessly urging them to make the effort to return to the sources. The author lets us read by presenting numerous excerpts from texts he regards as those of predecessors, "brothers in geology" and friends "who have preceded me on the eternal road of time". As a geologist, Ellenberger sought to place himself in the position of these early authors studying nature; he accompanied them in the field: "Together we have contemplated, scrutinized the face of the Earth, and all those curious objects yielded by its subsoil. I did not smile upon hearing them propose naïve, fragmentary, rudimentary solutions" (Ellenberger 1988, p. 5). But being a geologist is neither necessary nor sufficient; one must first be a historian, and that requires a few methodological rules worth recalling. Ellenberger asserts: "The first duty is to read attentively what the authors actually wrote: there, and there only, lies the solid foundation of any history of science. Scholarly dissertations must come only afterward. Everything begins with the texts" (Ellenberger 1988, p. 3). He insisted: "There is an absolutely imperative duty: it is always to go back to the source, to the literal meaning of the texts. One cannot emphasize this point enough. Doing everything possible to obtain an exact copy of the original texts is the historian's first duty in the study of a science" (Ellenberger 1988, p. 3).

In the absence of university teaching in the history of geology, Ellenberger also hoped to make his book a first manual for apprentice historians of his discipline. The first volume of his Histoire de la géologie (*Des Anciens à la première moitié du XVIIe siècle* [From the Ancients to the First Half of the 17th Century]) traces 2,000 years of a "fragmented history", of discourses about the surface of the globe, its objects and its phenomena. This does not constitute a science in the strict sense, but if one seeks to reconstruct a history of the observations and interpretations of the objects and phenomena studied by our modern geological sciences, these early fragmentary and scattered discourses can in no way be excluded from that history. The author, "as a practicing geologist, considers himself fully qualified to declare that during those twenty centuries, genuine geology was indeed practiced: under other names, within other frameworks, in a punctual, fragmented, dispersed way". "It was a geology before Geology", and it fully belongs to that history (Ellenberger 1988, p. 6). This volume 1 ends with a very long development devoted to the contributions of Nicolas Steno (1638–1686) and more specifically to his short 78-page book that laid the axiomatic foundations of geology to come and can be considered a "first birth of geology". *De solido intra solidum naturaliter contento dissertationis prodromus* (*Prodrome d'une dissertation sur un solide naturellement contenu dans un solide*), published in 1669, announced a more substantial book that was never written.

But this *Prodrome* already contains the statement of the first principles of stratigraphy and marks a break with its predecessors, while offering the framework needed to connect scattered observations and ideas into a founding discourse. Historians of geology agree in considering this text as a founding book of geology, and Steno and his writings have been the subject of numerous publications, so it is unnecessary to discuss them further in this general introduction. Ellenberger titled his volume 2 *La grande éclosion et ses prémices, 1660–1810* (The Great Emergence and Its Precursors, 1660–1810) and concluded it by discussing "the founding revolution", which, in less than 20 years at the beginning of the 19th century, brought forth a modern geology already resembling that which would later develop and which we now call Geology. "The present-day geologist, if he happens to leaf through books of the 18th century, feels as though in a foreign world. Then, when he comes to the writings of the 1830–1840 decade (such as the first volumes of the *Bulletin* of the young Société géologique de France), he no longer feels out of place. He easily recognizes his own discipline in its beginnings" (Ellenberger 1994, p. 318). Ellenberger left it to his readers to write the history of the rise of this new science from the 1830s onward. Roy Porter defined this pivotal period as spanning from 1660 to 1815, Rachel Laudan from 1650 to 1830 (Laudan 1987), and Rhoda Rappaport divided it into two phases, emphasizing the first, from 1665 to 1750. The diversity of writings and the large number of authors who marked the 150 years separating the first theories of the Earth from the emergence of a modern geology led Ellenberger to choose a sequence of studies on the major figures of this nascent geology, from Athanasius Kircher (1602–1680) to Abraham Gottlob Werner (1749–1817) and James Hutton (1726–1797). A community took shape during the 18th century, and the books circulated. Ideas seemed to stagnate, reinvent themselves, or sometimes ignore one another, yet they contributed to building the framework of a shared theory of the Earth. Ellenberger tirelessly sought to share his readings of the texts and authors that he deemed essential for understanding this period and the conditions of its emergence.

Since the 1970s and these two general books on the history of Earth sciences, numerous books in French have shed light more specifically on the historical development of particular geological subjects, historical periods, authors or institutions. The publication of articles in the *Travaux du COFRHIGEO* and in other generalist journals of the history of science and technology has complemented these syntheses and now provides readers with greatly enriched material and perspectives. The development of university research within teams in ancient or medieval history, literature or philology, history and philosophy of science, epistemology, sociology of science, and other fields has greatly broadened the range of study approaches, expanded and renewed the initial questions, diversified and facilitated access to sources. The digitization of printed sources, especially journal collections, their online availability, and the online publication of archival and collection inventories have multiplied opportunities for access to materials for the history of science and technology. This broadening of interest in the history of Earth sciences has also resulted in a dispersion and distancing of researchers into distinct communities with few opportunities to meet. Occasions to bring these contributions together and enrich the writing of a history of geology are rare, and the writing of this book is only a modest contribution to such rapprochements.

In presenting this book, we wished to give an account of the French research carried out since the momentum of the 1970s and to make it visible to a wider audience than that of the dispersed communities of philologists, historians, philosophers, epistemologists and geologists.

The richness and diversity of books on the history of geology in English are a major resource for any historian of the geological sciences, especially for the modern and contemporary periods. Our book in English does not intend to overlook them, and we can only invite the reader to continue this reading with the many reference books published over the past 20 years. Three examples are provided. For those interested in the moment of the emergence of geological science, in the precursors and the great flourishing as Ellenberger wrote, in the "act 2" described by Gohau, the essential readings are the monumental books of Martin J. S. Rudwick, *Bursting the Limits of Time* and *Worlds Before Adam* (Rudwick 2005, 2008). For "act 3", the construction of the plate tectonics model, the reference is the four-volume book of Henry R. Frankel, *The Continental Drift Controversy* (Frankel 2012). The history of endogenous petrology, absent from our book, is magnificently presented since 2003 in Davis A. Young's (Young 2003) *Mind over Magma*, which has no equivalent in French.

This volume on the history of Earth sciences is a collective work bringing together contributions aimed at providing a succession of insights on periods, subjects and themes. It results from choices and constraints. Adopting for ourselves the idea that the history of geology begins before the establishment of a geological science characterized by shared principles, common theses, well-defined objects, results discussed and subjected to the approval of a specialized collective, institutions, publications and teaching, our book begins with the study of the earliest discourses about the Earth in Greco-Roman Antiquity.

Chapter 1, "What Earth Sciences in Greco-Roman Antiquity?", was written by Frédéric Le Blay. He presents Aristotelian meteorology, a body of discourses on the sublunar world encompassing many objects and phenomena that fall within geology in the modern sense. The author draws on original texts, which he translates and discusses, making available to the uninitiated reader primary sources that are always difficult to access, both because of their language and the concepts on which they are based, often far from our current representations and liable to generate confusion and misunderstanding. He precisely presents the ancient interpretations of earthquakes, representations of the subterranean world and notably the hypothesis of subterranean fire. Subsurface materials used in the production of glass or metals, or in construction, are also the subject of discourses and theses. Variations in sea level, the idea of a deluge and more broadly the transformations of the Earth's surface also nourished ancient theories.

Chapter 2, "Studying and Knowing the Earth in the Middle Ages", was written by Joëlle Ducos and Fleur Vigneron. It provides an overview of discourses about the Earth and stones, scattered among books often unnoticed by historians of geology, because they belong to fields of knowledge as diverse as geography, medicine, mining, architecture or agronomy. The authors present the lapidaries – those descriptive texts about stones – inheritors of both a naturalistic tradition and a magical, astrological, even symbolic and spiritual one. One of the difficulties in describing the state of knowledge about the Earth and stones is directly linked to the consulted sources, which show that scholarly knowledge and secondary didactic knowledge, or even practically oriented knowledge, could coexist and differ greatly. The diversity of knowledge, writings and sources constitutes so many difficulties to overcome for anyone seeking to establish a state of knowledge. The Middle Ages was moreover a time of reappropriations and transformations, inheritances and reinterpretations. Tracing the evolution of discourses about the Earth during this period is complex. The authors then present the ideas and discourses about the formation and evolution of the Earth and mountains, the generation of rocks, minerals and fossils, about earthquakes and volcanoes, about mines, grottos and caves, about the circulation of water, and the nature and evolution of soils. Three authors stand out: Albert the Great, Jean Buridan and Pietro de' Crescenzi.

Chapter 3, "Geology from the Renaissance to the Enlightenment", was written by Gaston Godard. Since the beginnings of the great flourishing – the passage from act 1 to act 2 – have already been the subject of numerous studies, the author chose to revisit this period through several emblematic or, on the contrary, lesser studied subjects and authors. The chapter first revisits some well-known or little-known discourses about the origin of fossils and "figured stones", about the supposed fossil giant humans, fossilized wood and the origin of stones, minerals and crystals that filled the earliest cabinets of curiosities. From discourses about objects, we move to discourses about phenomena, reexamining the theory of the forces of nature or that of the Deluge, and following those about volcanoes and earthquakes. This period is also the one that builds the foundations of a common framework for speaking about the Earth, drawing from the diversity of theories of the Earth from Descartes to Hutton. The bases of stratigraphy laid down by Steno in 1667 are only briefly mentioned, as they are studied in this volume in the chapter devoted to the history of stratigraphy.

The first three chapters of this volume have examined, in a synthetic way and through selected perspectives, the history of scholarly discourses about the Earth, its objects and its phenomena over the 20 centuries and more that preceded the emergence of a modern geological discourse. From 1750 to 1810 or 1830, depending on the markers favored by historians of geology, a shared, reasoned discourse was built around the history of the Earth's surface. It reexamined objects and reinterpreted them in connection with questions about their origin and their place in a history of the Earth that gradually made room for a long timescale. To the speculations of the theories of the Earth, the second half of the 18th century substituted the accumulation of field observations and later those made in mineralogical cabinets. Minerals and crystals came to be viewed differently. People sought to classify them, and to do so, they described them by their appearance, their regularity and soon their composition. Fossils, recognized as traces, as mineralized remains of living organisms, were viewed with a naturalist eye, where the question of species definition led at the beginning of the 19th century to making them a key element in reconstructing the history of the Earth's surface. The next three chapters have chosen, within this fertile moment, sometimes called a founding revolution, during this period when geological science seemed to emerge suddenly, fully formed, to follow the construction of three specialties or components: crystallography, cartography and stratigraphy.

Chapter 4, “Crystallography: 17th–19th Centuries”, was written by Bernard Maitte. The 18th century inherited the conjectures of previous centuries about rocks, minerals and crystals but marked a turning point: the light passing through crystals attracted the attention of physicists. Double refraction (birefringence), observed by Bartholin in 1669, raised questions both about the properties of light and about those of crystals. Huygens developed a wave theory of light; Newton rejected it. Crystals surpassed the mere curiosity of naturalists; chemists and mathematicians became interested in them in turn during the 18th century. Mineralogy, which undertook to classify its objects in the manner of botanists and zoologists, questioned its species. Romé de l'Isle, in seeking to classify, asserted a constancy of angles and defined six classes according to the angles between faces. But mineralogists hesitated and debated over which criteria were best suited for classification: form or chemical composition? Haüy adopted Romé's geometric criteria but showed no interest in crystallization processes (the cause, in his view, being necessarily divine!); only the description of the object mattered. As his thought evolved, his *Traité de minéralogie* came to classify using both geometry and chemistry, before his *Traité de cristallographie* refocused solely on geometry. The early 19th century revisited the question of birefringence. Fresnel theorized the polarization of light. Weiss introduced the notion of crystalline symmetry. Classification was refined, and a geometric notation for crystal faces became established in Germany and England, though not in France, where Haüy's followers emphasized lattice structures. They later converged after some cross-border disputes. Physical, chemical and geometric properties offered a new perspective on mineral species, whose number continued to increase throughout the 19th century, without ever leading to a unified classification. Throughout this chapter, we follow these intersecting approaches, which extend far beyond geological concerns. The history of mineralogy and the history of crystallography also show us how much these objects resist categorization.

Chapter 5, “Geology and Visual Discourse: The Birth of the Geological Map”, was written by Pierre Savaton. It traces the intellectual construction of the geological map from the first projects of soil maps or fossil mining maps that appeared at the end of the 17th century, through the first mineralogical maps produced by transferring statistical inventories onto a geographical base map from the middle of the 18th century, then to the invention of the geognostic map at the end of the century, and finally to the invention of the geological map at the beginning of the 19th century. To maps that merely indicate the location of deposits succeed maps that attempt to reconstruct the continuity between two deposits, then maps that attempt to account for the spatial arrangement of the various rocks and rock groupings. The Neptunist framework, the view that all layers of rock originated from an initial deposit left by a global ocean in constant retreat, led to conceiving of the Earth's surface as structured by a stack of successive layers. It was therefore necessary to reconstruct this stack through field observation of the geometric arrangements. But how could one correlate at a distance rock formations that are sometimes so similar without risking mistakes and disturbing the chronological order? The question was resolved at the very end of the 18th century and the beginning of the 19th by increasing attention to the diversity of fossils contained in most of these layers. Fossils vary over time: their chronological succession made it possible to use them like medals to date the layers relative to one another. By the beginning of the 19th century, rocks had become archives of the history of the Earth's surface. A new science, geology, now claimed to reconstruct that history by making those rocks speak. The geological map was now capable of producing that historical discourse. The emerging geology had developed a visual language just as the construction of the map produced a new discourse.

Chapter 6, "The Birth and Development of Stratigraphy", was written by Philippe Grandchamp. The history of the geological map is inseparable from that of stratigraphy, and these two consecutive chapters intersect frequently. Their order could even have been reversed. Nicolas Steno's *Prodrome*, the act 1 of geological discourse, is also the birth act of stratigraphy, but of a stratigraphy in potential, separated by a century and a half from the biostratigraphy introduced by Cuvier and Brongniart. In the meantime, the study of rock layers, guided by a Neptunist vision, led first to focusing on their arrangement, on their initial stacking. Understanding the history of stratigraphy involves studying models that sometimes had to be abandoned in order to progress, such was the case for geognosy, theorized by Abraham Gottlob Werner. It also fundamentally involves a reexamination of the history of discourses about fossils, reflecting on how they were used and how they could be used. Making fossils the best indicators for distinguishing similar lithologies, as William Smith did, was not enough to make stratigraphy a science capable of recording a history. It was necessary to make the fossil an indicator of time, a bearer of history. The author explains in more detail how, at the beginning of the 19th century, a decisive breakthrough led to the emergence of paleontological stratigraphy, the determination of new classifications of terrains and the invention of the concept of the stage. The fossil-based scale of geological time became established.

The transformation of ideas between 1750 and 1810 was so profound and creative that it sometimes gives the impression that geology imposed itself as a matter of course between 1810 and 1830, already with its specializations. In the crucible of geognostic empiricism, elements of theories of the Earth, echoes of controversies between Neptunists and Vulcanists (the latter defended the idea that there existed ancient volcanic rocks), experimental attempts to demonstrate the origin of basaltic and granitic rocks, a statistical desire to inventory mineral wealth, a rereading of fossils and their origin, questions about the nature of species, and a decline in the intensity of debates between Wernerians and Huttonians would have produced this geological science with shared objects and methods. The narrative imposes itself, but it is complex, as it always is when one looks more closely at the texts and broadens the community of actors. These insights into the history of crystallography, cartography and stratigraphy leave in the shadows the parts of the picture that could have been provided by a history of tectonics, petrology or paleontology. The history of ideas about fossils, present since the first chapter of this book, still nourishes that of stratigraphy. At the beginning of the 19th century, the encounter between the traces of species preserved in rocks and living species built a new science that flourished in the 19th century: paleontology. Between the heritage of the history of ideas about fossils, the debate on the transformation of species and the nascent history of humankind, paleontology was from the outset at the intersection of the biological and geological sciences. A single chapter in this volume would not have sufficed to convey this complexity. The history of ideas about the formation of mountain chains and the arrangement of terrains on the Earth's surface, frequently cited in previous chapters, has already been very well studied in French, allowing us to refer the reader to the books of Gohau (1990, 2003, 2010) and Balan (2011, 2018). As for the emergence of petrology, surpassing the lithology of the 18th century and the heritage of mineralogy, a synthesis remains to be written, though we already have, in English, the book on magmatic petrology by Davis A. Young, already cited (Young 2003).

The principles of geology clearly established at the beginning of the 19th century nourished a science that developed and specialized very rapidly. The object Earth, examined by naturalists, was also studied by chemists and physicists. The latter, in particular, by studying gravity, magnetism, seismicity, geodesy, the melting of rocks, the cooling of the Earth, its deformations, its age, etc., constructed a much broader perspective than that of early geology and asserted new experimental and modeling approaches. One may profitably read the history of these ideas, for example, in *Voyage à l'intérieur de la Terre* by Deparis and Legros (2000), *L'âge du monde : à la découverte de l'immensité du temps* by Richet (1999) and the recent book by Richard J. Howarth, *The Emergence of Geophysics* (Howarth 2024). The Earth, its structure and its history now concern a large number of sciences, and this broadening of studies and methods has led to speaking more generally of Earth sciences rather than of geology.

This book concludes with a final chapter on the mobilist model developed by Alfred Wegener at the beginning of the 20th century. Chapter 7, "History of Mobilist Ideas: Wegener's Model", was written by Philippe Le Vigouroux. To understand this mobilist idea, one must first grasp the fixist view that preceded it, and for that, one must revisit the physical ideas about the globe mentioned earlier. The study of gravity led in the United States to the concept of isostasy and the permanence of the shape of the Earth's surface. The idea of the cooling of an initially molten Earth gave rise in Europe to a contractionist model, in which the central idea of vertical movements curiously coexisted with the description of traction nappes. To these two fixist models, Alfred Wegener opposed a theory of continental drift, asserting the relative movement of these landmasses at least since the Carboniferous. The conception of the globe had to be reconsidered. To think about the Earth's surface, Wegener, a climatologist by profession, drew as much on naturalist observations as on geophysical modeling: he sought to bring together scientific specialties that at the time ignored one another. The history presented here of the construction and later rejection of this mobilist model, the history, in fact, of a first attempt at synthesizing the dynamic history of the globe, places us directly within the geological sciences of the 20th century, with their diversity, their specialization and their sometimes mutually incompatible methodologies. The study of the controversy sparked by this model immerses us in the relationships between sciences and scientific communities, but also between national communities during the period of the First World War.

The history of the Earth sciences, presented in this book through seven chapters written by eight different authors, should be seen both as a collective contribution to a historiography of the Earth sciences, grounded in the research conducted by these authors, and as an invitation to readers to continue their exploration of the history of the Earth sciences through the extensive bibliography provided. This history does not end with Chapter 7; act 3, or the construction of the plate tectonics model, which would be presented in 1968, was then only in its infancy. The comprehensive synthesis by Frankel (2012) is essential for anyone wishing to study in detail this major moment in the history of the Earth sciences. However, much remains to be done to disseminate this complex model and its history to a wider audience beyond that of Earth scientists and historians of the Earth sciences.

References

Adams, F.D. (1954). *The Birth and Development of the Geological Sciences*. Dover Publications, New York.

Balan, B. (2011). *L'évolution des idées en Géologie. Des cosmogonies à la physique du globe*. Vrin, Paris.

Balan, B. (2018). *Montagnes et continents : la tectonique*. Vrin, Paris.

Bouillet-Roy, G. (1976). La géologie dynamique chez les anciens Grecs et Latins d'après les textes. PhD Thesis, Université Pierre et Marie Curie, Paris.

Deparis, V. and Legros, H. (2000). *Voyage à l'intérieur de la Terre. De la géographie antique à la géophysique moderne. Une histoire des idées*. CNRS Éditions, Paris.

Duhem, P. (1959). *Le système du monde. Histoire des doctrines cosmologiques de Platon à Copernic*. Hermann, Paris.

Ellenberger, F. (1988). *Histoire de la géologie 1 : des Anciens à la première moitié du XVIIe siècle*. Tec & Doc, Paris.

Ellenberger, F. (1994). *Histoire de la géologie 2 : la grande éclosion et ses prémices, 1660-1810*. Tec & Doc, Paris.

Frankel, H.R. (2012). *The Continental Drift Controversy. T1. Wegener and the Early Debate. T2. Paleomagnetism and Confirmation of Drift. T3. Introduction of Sea-floor Spreading. T4. Evolution into Plate Tectonics*. Cambridge University Press, Cambridge.

Geikie, A. (1962). *The Founders of Geology*. Dover Publications, New York.

Gohau, G. (1987). *Histoire de la géologie*. La Découverte, Paris.

Gohau, G. (1990). *Les sciences de la Terre aux XVIIe et XVIIIe siècles. Naissance de la géologie*. Albin Michel, Paris.

Gohau, G. (2003). *Naissance de la géologie historique. La Terre, des "théories" à l'histoire*. Vuibert, Paris.

Gohau, G. (2010). *Histoire de la tectonique, des spéculations sur les montagnes à la tectonique des plaques*. Vuibert, Paris.

Hölder, H. (1960). *Geologie und Paläontologie in Texten und ihrer Geschichte*. Karl Alber, Freiburg.

Howarth, R.J. (2024). *The Emergence of Geophysics: A Journey into the Twentieth Century*. Geological Society, London.

Laudan, R. (1987). *From Mineralogy to Geology. The Foundations of a Science, 1650-1830*. University of Chicago Press, Chicago.

Porter, R. (1977). *The Making of Geology. Earth Science in Britain, 1660-1815*. Cambridge University Press, Cambridge.

Quenet, G. (2005). *Les tremblements de Terre aux XVIIe et XVIIIe siècles. La naissance d'un risque*. Champ Vallon, Seyssel.

Rappaport, R. (1997). *When Geologists Were Historians, 1665-1750*. Cornell University Press, Ithaca.

Richet, P. (1999). *L'âge du monde : à la découverte de l'immensité du temps*. Le Seuil, Paris.

Rudwick, M.J.S. (2005). *Bursting the Limits of Time. The Reconstruction of Geohistory in the Age of Revolution*. University of Chicago Press, Chicago.

Rudwick, M.J.S. (2008). *Worlds Before Adam. The Reconstruction of Geohistory in the Age of Reform*. University of Chicago Press, Chicago.

Young, D.A. (2003). *Mind over Magma. The Story of Igneous Petrology*. Princeton University Press, Princeton.

von Zittel, K.A. (1899). *Geschichte der Geologie und Paläontologie*. Scribner, London.

1

What Earth Sciences Existed in Greco-Roman Antiquity?

Frédéric LE BLAY
Laboratoire CFV, Université de Nantes, France

Writing a history of Western sciences by tracing back to what we consider their origins, that is, the first theories formulated in the Greek language, is always a challenge. This challenge is as historical as it is epistemological. The difficulty of the task, beyond the always delicate access to sources, lies in the fact that our current definitions of science, disciplines or scientific fields do not really overlap with the formal and intellectual frameworks of ancient thought. Western science may have been born in Greece, as a rational way of thinking about the world; our academic sciences, moreover, still generally bear the Greek and Latin legacy. This observation can be qualified depending on the fields under consideration: mathematics or medicine are not in the same position as physics or chemistry. As for the Earth sciences as we define them today, this journey back in time seems, in many respects, doomed to fail, which does not mean that the Ancients were uninterested in the history of the Earth or in understanding its structure.

Georgia L. Irby recently oversaw the writing of a two-volume work presenting a synthesis of the state of science and technology in Greco-Roman Antiquity (Irby 2016b). For the sake of clarity, its main sections are organized according to our scientific classifications. A section of the first volume is thus devoted to the Earth sciences; it includes five chapters dedicated to hydrology (Irby 2016a), geology and mining (Craddock 2016), Greco-Egyptian and Byzantine alchemy (Martelli 2016), meteorology (Taub 2016) and geography (Roller 2016). This breakdown immediately reveals the nature of the problem; it would be pointless to suggest any carelessness on the part of the editor, who knows her subject perfectly well. The first two fields seem to fit easily within what we call "Earth sciences", even though hydrology/hydrography may open onto areas that extend beyond this framework. However, it will be seen that geology does not exist as such in the ancient world. The title of the second chapter, therefore, introduces an anachronistic concept[1]. Alchemy, meteorology, and geography appear to us to belong to different fields. Alchemy would be an early form of chemistry. Geography is indeed, strictly speaking, an Earth science, but it stops at the surface, leaving the exploration of the depths to others. We shall see, however, that Strabo's *Geographica* is not devoid of considerations likely to interest our Earth sciences. We shall show that, to trace the beginnings of the latter, it is toward meteorology that we must turn. Contemporary and ancient terminologies do not refer to the same realities.

1.1. Defining the field: Aristotelian meteorology

In the ancient context, meteorology must be understood in a much broader sense than it has today:

> It is the science of the phenomena of the sublunary world, the imperfect world of change and corruption, the world of the mixture of the four elements, a science that encompasses both phenomena we now place in the astronomical domain and terrestrial events such as earthquakes. (Cusset 2003, p. 8)

Aside from its etiological framework, this definition would hold until the time of Descartes who, in the fourth part *De terra* of his *Principles of Philosophy* (1644), as well as in his *Meteors* published together with the *Discourse on Method* (1637), takes up the program of Aristotle's *Meteorologica*.

Aristotle plays the role of a founder in a twofold sense. He is first the one who defines the outlines of a unified field of inquiry. But it is above all with him that begins the literary tradition concerning this part of the investigation of nature destined to become, for a long period, meteorology. His *Meteorologica* (*Μετεωρολογικά*), in four books, logically follows the great early cosmological texts such as the *Metaphysics*, the *Physics* and the treatise *On Generation and Corruption*.

1 It is interesting to note that many designations of current scientific disciplines, although borrowed from the Greek language, do not belong to the vocabulary of the Ancients. Thus, with geology. Another interesting case is the term "cosmology", which is recent.

They precede the writing of the biological and zoological treatises. The first chapter sets forth a program and gives a clear definition of this new territory of speculation:

> We have treated first of the primary causes of nature and of all that concerns natural motion, of the arrangement of the stars in their movement in the upper regions, of the corporeal elements, their number and their qualities, of the transformations of one into another as well as of generation and corruption in general (γενέσεως καὶ φθορᾶς τῆς κοινῆς).
>
> It remains, within this investigation, to establish a theory of what all our predecessors have called the study of meteors (μετεωρολογίαν). It concerns all phenomena which, though subject to the principles of nature, have a nature less ordered than that of the first element of bodies and are situated in the region nearest to the movement of the stars, such as the Milky Way, comets, fiery and mobile appearances, all phenomena which we may suppose occur in the space common to air and water. It also concerns the earth, all its kinds, its parts, and the affections[2] of these parts (ἔτι δὲ γῆς ὅσα εἴδη καὶ μέρη καὶ πάθη τῶν μερῶν), from which we may establish a theory of the causes of winds and earthquakes (περί τε πνευμάτων καὶ σεισμῶν θεωρήσαιμεν ἂν τὰς αἰτίας), as well as of all the phenomena produced by their movements, among which some escape our explanation, while others can be grasped to some extent. Finally, it concerns lightning, typhoons, the combustion of the air, and all other periodic phenomena that occur through the solidification[3] of these same elements[4].

It will therefore be a matter of studying all the observable phenomena occurring in the part of the world that no longer obeys the regularity characteristic of the higher celestial spheres. Within this cosmology, meteorology is thus the science of the phenomena taking place in the sublunar part of the world. This domain is the one involving the four elements, whose nature is subject to the principle of generation and corruption. The meteors are not characterized by the immutability of the laws

2 Here, we give πάθη its proper meaning, insofar as it designates, specific to an element and its place, the earth, different events or accidents that can be "suffered".

3 The elements in question are again air and water, which can experience, as the author will show, different processes of solidification. One may speak sometimes of freezing, sometimes of condensation.

4 I, 1, 338a 20-339a 5. We translated from the text established by Pierre Louis (collection des Universités de France, Les Belles Lettres, Paris, 1982).

governing them, unlike the stars, yet they still obey a certain rationality that it is the philosopher's task to uncover.

In this partition of the cosmos established by Aristotle, the following phenomena are called meteors, listed in the order in which they are treated:

– Book I: celestial flames, shooting stars, various colorations of the sky, comets, Milky Way; clouds and fog; dew, frost/white frost, rain, snow, hail; formation of winds (mistaken opinions); rivers and waterways; cycles of inversion of lands and seas.

– Book II: the sea, its salinity, its currents, its place and role; winds; earthquakes and tsunamis; lightning, thunder.

– Book III: storm wind/hurricane, typhoon, whirlwind of fire, thunderbolt; photometeors/halo, rainbow; sun rays and parhelia; effects of exhalation in the depths of the Earth, formation of non-metallic rocks (*τὰ ὀρυκτά*) and metals (*τὰ μεταλλευτά*).

– Book IV: study of the different transformations of matter[5].

This plan clearly shows that ancient meteorology covered a very broad field encompassing several of our current disciplines. One may naturally be surprised to see at the forefront the study of shooting stars, comets and the Milky Way, which we now classify within astronomy or astrophysics. The phenomena in question are in fact understood as kinds of combustion; they take place in the sphere of fire, situated in the sublunar part of the cosmos. The supralunar part of our world is indeed composed of a single unalterable element, "ether", while four corruptible elements (fire, air, water, earth) share the sublunar world, the first of the four occupying the upper part of this space of meteors. One may also note that, for the observer, such phenomena do not display the predictable and regular character of the motion of the astral spheres. The theory on which Aristotle bases his unified explanation of meteors allows causal links to be established between phenomena such as comets, winds and earthquakes – a connection that contemporary science can no longer maintain, assigning each of these phenomena to its own specific domain or field. One may readily see in this program an attempt to theorize an Earth System in which all phenomena are linked with one another, all parts or regions interacting with each other. In fact, the etiological framework adopted by Aristotle allows for

5 The authenticity of this latter part of the treatise, often described as the first treatise of Western "chemistry", has been much debated. Commentators now seem no longer to doubt its integration into Aristotle's original project. This debate dates back to Alexander of Aphrodisias (2nd century CE).

the interrelation of all meteors instead of dividing them into entities governed by specific causalities. Such is the true tour de force of this enterprise, which explains the influence it would exert over the long term[6].

It would, however, be wrong to claim that Aristotle was the first to take an interest in the phenomena he groups in his meteorological inquiry. In this respect, he is the heir of the first natural philosophers (*φυσιολόγοι*) of Greece, whom he cites, comments on and critiques extensively. The prologue we have quoted acknowledges this heritage by recalling the antiquity of this study. But the hazards of the transmission of texts mean that it is essentially thanks to the Aristotelian treatise that we have preserved a trace of these early theoretical attempts. Thus, although the inquiry itself does not begin with Aristotle, our knowledge of its earliest outlines begins with him.

The unifying etiological framework introduced by Aristotle rests on a principle designated as *ἀναθυμίασις*. The term is not easy to translate. English translators have generally opted for the term "exhalation", which most commentators also adopt. The term is introduced as early as Chapter 3 of the first book, at the same time that Aristotle revisits the natural places of the four elements:

> The body that lies beneath the upper region of the revolution of the stars is like a matter that is potentially hot, cold, dry, and moist, with all the resulting states, but becomes or is this or that[7] only through the effect of motion or the absence of motion, whose cause and principle[8] we have previously indicated. At the center and around the center lies what is heaviest and coldest, in distinct places, that is, earth and water (γῆ καὶ ὕδωρ). Around and contiguous to them are air and what we habitually call fire, though it is not really fire (ἀήρ τε καὶ ὃ διὰ συνήθειαν καλοῦμεν πῦρ, οὐκ ἔστι δὲ πῦρ): for fire is an excess of heat and a form of incandescence. As for what we designate as "air" (τοῦ λεγομένου ὑφ' ἡμῶν ἀέρος), we must understand that the part near the Earth is moist and warm, by being vaporous and containing the exhalation coming from the Earth (διὰ τὸ ἀτμίζειν τε καὶ

6 One must return to Bachelard's (1938) essay, which identifies as a major epistemological obstacle of the first state of inquiries into nature the search for a unifying agent. Any etiological system allowing the unification of all phenomena gains an explanatory power that ensures its fortune.

7 That is, a hot matter or a cold matter or a dry matter or a humid matter.

8 This concerns the preceding chapter, where the author posits that the sublunar world is under the influence of the movement of the celestial spheres (339a 21–24).

> ἀναθυμίασιν ἔχειν γῆς), while the part situated above it is warm and dry. For the nature of vapor is moist and cold, that of exhalation warm and dry. Vapor is potentially like water, exhalation is potentially like fire (καὶ ἔστιν ἀτμὶς μὲν δυνάμει οἷον ὕδωρ, ἀναθυμίασις δὲ δυνάμει οἷον πῦρ)[9].

This passage clearly and concisely restates the distribution of the four elements around the center of the Earth while introducing at the same time the theory of exhalation. It is immediately clear that Aristotle distinguished two exhalations, one moist and cold, the other warm and dry. He specified, through the term designating vapor, the nature of the first and reserved the generic term "exhalation" for the second. This theory of "double exhalation" is developed in Chapter 2[10]. The principle of exhalation is not an Aristotelian invention, for the author reproached certain predecessors for having conceived a single exhalation, whereas he maintained that two must be posited[11]. It is on the basis of this principle that Aristotle succeeds in explaining seismic and volcanic phenomena.

1.2. The movements of the Earth

The understanding of earthquakes and related phenomena is indeed a key part of the study of meteors. The strong seismicity of the Mediterranean Basin has led populations to learn to live with this hazard (Compatangelo-Soussignan et al. 2022) despite its often-disastorous effects. Like any other natural phenomenon, the movements of the Earth have given rise to a great variety of myths and superstitions, but the Ancients rather early attempted theory, which was itself marked by the diversity of explanatory models. The question is addressed by Aristotle in the second book of the *Meteorologica* in Chapters 7 and 8 (365a 14–369a 7). In accordance with a method characteristic of the Aristotelian way of approaching a subject, the reader is first presented with a summary of the main theories put forward by his predecessors on the matter. These are then systematically refuted. Thus, we learn that Anaxagoras based his seismic etiology on the movement of air trapped beneath the surface of the Earth: since its natural motion is an upward one, it can only seek to rise again if it finds itself confined in the lower parts of the world. In Aristotle's critique of this causality, it is not so much the element of air that is questioned as the assumption about the orientation of the world that this theory

9 340b 14–29.

10 I, 4, 341b 4–24.

11 This critique may target Plato's *Timaeus* (56d).

implies[12]. Democritus is said to have held that water was trapped beneath the surface of the Earth. We know from other sources that Thales, whom Aristotle does not cite, also based his system on the element of water (Le Blay 2022). Anaximenes is said to have maintained that the earth, naturally imbued with water, could dry out and break apart as it desiccated. Of all its rival theories, this one comes closest to an empirical observation: indeed, the fissures that appear on the ground's surface during an earthquake may quite naturally suggest, on a larger scale, the cracks that appear in soil subjected to extreme dryness. The last theory summarized is attributed to Anaximander, who invoked the action of fire present beneath the earth.

Aristotle invokes the dry exhalation, in the same way as for the phenomenon of winds. The term *πνεῦμα* may therefore designate both what common usage calls "winds" and this exhalation that can shake the subterranean parts of the world. From an etiological point of view, the two phenomena are of the same nature. The comparison with winds had already been explicitly stated in the introduction of the treatise, announcing in the same passage the examination of the causes of winds and earthquakes[13]. In fact, to put it more simply, earthquakes are caused by "winds" rushing into the ground[14]. The surface of the Earth thus more generally plays the role of a zone of exchange, exchanges that may be hindered by its density and its structure:

> But since it is evident that an exhalation necessarily arises either from the moist or from the dry, as we have said before, it is necessary that earthquakes (τοὺς σεισμούς) occur when they do arise. For the earth itself happens to be dry. But, as a result of rains, it contains much moisture, so that when it becomes heated by the sun and by the fire within itself, a great deal of breath (πνεῦμα) is generated both outside and inside (πολὺ μὲν ἔξω πολὺ δ' ἐντὸς γίγνεσθαι τὸ πνεῦμα). This

12 The notions of "high" and "low" indeed presuppose an orientation. The Aristotelian cosmological system, however, is not devoid of such intention, since the geocentric representation on which it rests presupposes an orientation from a center. In fact, one of the fundamental postulates of the *Meteorologica* is that the four elements of the sublunar world have a natural place: the upper part of the world is the natural place of fire, then, approaching the center of the Earth, one finds the natural places of air, water and earth. But orientation relative to a center, in the mode of cosmic circularity, is not the same as the definition of a bottom and a top, which rests on the principle of horizontality of a plane. This methodological difficulty of the orientation of the cosmos and the place of bodies is raised in *Physics*, III, 5, 205b 25–30.

13 338b 25–26.

14 Etiology developed in Book II, Chapter 8.

> breath flows continuously either outward or inward, or sometimes divides between the two. [...] Therefore neither water nor earth can be the cause of the motion (τῆς κινήσεως), but the breath (πνεῦμα), when that which is exhaled outward happens to flow back inward (ὅταν εἴσω τύχῃ ῥυὲν τὸ ἔξω ἀναθυμιώμενον). This is why most earthquakes, and the most violent ones, occur in the absence of wind. For the exhalation, being continuous, for the most part follows its initial impulse, rushing entirely either inward or outward. That some earthquakes may occur in the presence of wind is not strange: we sometimes see several winds[15] blowing at once; if one of them rushes into the earth, the earthquake will occur in the presence of wind[16]. But these earthquakes are of lesser intensity because their origin and cause are divided[17].

The argument then continues by assessing the frequency and strength of earthquakes according to the seasons (some being windier than others), the configuration of places (proximity to the sea, which reinforces the effect of the breath through its combination with water) and the nature of the soil, more or less spongy and rich in cavities[18]. The question of the porous nature of the ground thus became central and debated from Aristotle onward.

This etiology makes it possible to provide a simple explanation for the fire observed in certain cases – that is, what we call a volcanic eruption: it is nothing other than the result of the friction of the breath within the narrow passages through which it moves, Aristotle considering that the dry exhalation is of a nature close to fire. We may then return to the case of comets and consider that there exists an

---

15 The term designates winds in their geographical definition, that is, oriented according to a precise direction. The same place can be subjected to opposing winds. If *πνεῦμα* designates the wind as a generic phenomenon, as soon as it is a question of referring back to winds known and felt as meteorological phenomena, the author prefers to use other, more common terms.

16 The explanation is simple: if an earthquake is caused by the passage of the wind beneath the earth's surface, one does not feel wind on the surface. But if several winds blow in the same place, the disappearance of one of them beneath the surface does not make the phenomenon on the surface disappear. For that to happen, all the winds would have to rush in, which the author does not conceive.

17 II, 8, 365b 21-366a 12.

18 366a 25.

etiological kinship within the triad they form with winds and earthquakes[19]. The comet is a phenomenon of fiery nature, as is the eruption of a volcano. Nevertheless, the treatment of volcanoes remains brief, and from this standpoint, the reading of the *Meteorologica* proves disappointing. Only one passage seems to refer to a volcanic eruption, concerning the appearance of the island of Hiera:

> In this island, indeed, a part of the earth had swollen and was noisily rising like a mound; it finally burst, and a great breath came out, which hurled high the embers and the ashes, covering the whole city of Lipari, which is not far away, with ashes, and reaching even some of the cities of Italy; and even now one can see where this eruption (τὸ ἀναφύσημα τοῦτο)[20] took place.

That is very little for the founding text of meteorological science. Few commentators have noted this relative silence of Aristotle on the question of volcanoes[21]. Indeed, this mention provides an opportunity for an explanation[22]. An eruption is a fire produced by the collision of air particles trapped underground, whose escape is obstructed. Thus, the sea striking certain coasts under the effect of violent winds can form an obstacle. Combined with the opposing push of the sea, the outward thrust of the underground breath becomes violent. The argument followed by Aristotle suggests that volcanic eruptions occur mainly in coastal areas, where the sea meets the land. A map of active volcanoes in the Mediterranean clearly shows the coastal nature of volcanic zones. But this point alone would suffice to invalidate the Aristotelian hypothesis if applied to other regions of the world, despite the high concentration of volcanoes near coasts – when they are not islands – due to their position along the fault lines between continental plates. It is undeniable, however, that every natural philosopher sees in nature only what he knows of it, and that his environment, the landscapes and places in which he lives and conceives the world, strongly determines his way of reasoning. Aristotle could have only spoken of the volcanoes he knew.

---

19 Aristotle noted this similarity twice: I, 6, 344a 35-343b 7, where he reported a comet appearing concomitantly with an earthquake; I, 7, 344b 18-345a 10, where he observed that the appearance of comets could announce periods of wind and drought.

20 II, 8, 367a 3–9.

21 We are not at liberty here to develop this aspect or to try to understand this relative silence on one of the most impressive natural phenomena, familiar above all to Aristotle's contemporaries. This is partly the subject of our recent monograph, from which we partially reuse the material (Le Blay 2023).

22 II, 8, 367a 9–17.

This etiology that grants subterranean winds or breaths a crucial role in triggering earthquakes and eruptions stands as the explanation to which all authors will adhere, even those least inclined to support Aristotelian theses. One may then be tempted to summarize this theory of exhalation as the "outline of a general pneumatics" (Gruet 2004, pp. 207–209). Its strength lies in integrating all meteors into a coherent system, but this systematic dimension may also be seen conversely as a weakness of the argument. The first critiques soon arose; they are already present in the approach of the disciple Theophrastus. His treatise *On Winds* (*De ventis/Περὶ ἀνέμων*) offers a commentary on the concept of *ἀναθυμίασις* developed by the master and does not hide the criticisms that can be made of it (Bakker 2016).

One of the main purposes of Epicurean physics was the critique of the Aristotelian system; Epicurus sought to propose an alternative to the cosmology of the Peripatos. Of this great naturalistic philosophy that was Epicureanism, only residual traces remain, except for the scientific poem of Lucretius. Yet there is no doubt that questions related to meteors particularly attracted the attention of Epicurus and his disciples. The *Letter to Pythocles* attributed to the master bears witness to this. As a compendium of cosmology and meteorology as well as a discourse on method that presents itself both as a treatise on semiology and epistemology, it questions the conditions for properly reading the signs (*σημεῖα*) that natural phenomena represent and defends the thesis of the plurality of explanations in physics. It is thus indeed a polemical response to Aristotelian dogmatism. Two paragraphs concern seismic activity, attributing its cause to subterranean winds[23]. On this point, there thus seems to be agreement between Epicurus – or one of his followers – and the one he seeks to contradict. To be fully fair in tracing these convergences and influences, one must not lose sight of the *physiologoi* of the Presocratic period, some of whom had already laid the groundwork for a pneumatic physics. According to the preserved fragments, Anaximander had already developed a pneumatic explanation of meteors[24], whereas Aristotle attributed to him rather an explanation based on the fiery element. Book VI of Lucretius's *De rerum natura*, which is itself a complete abridgment of meteorology[25], addresses earthquakes and

23 §§ 23–24 Bollack and Laks = 105-106 Arrighetti.

24 D-K. A 23.

25 The plan of this canto is explicit regarding its object: vv. 1–95: praise of Epicurus; vv. 96–422: thunder, lightning, and lightning bolt; vv. 423–450: presters (hurricanes/tornadoes mixing air and fire, also studied by Aristotle); vv. 451–534: clouds and rain; vv. 535–607: earthquakes; vv. 608–638: sea and tides; vv. 639-711: Etna; vv. 712–737: Nile floods; vv. 738–839: Avernes (places or lakes from which toxic vapors emanate); vv. 840–905: prodigious springs and fountains; vv. 906–1089: the magnet; vv. 1090–1137: origin and cause of epidemics; vv. 1138–1286: the plague of Athens.

volcanoes after the master, reduced to the canonical example of Etna. The proposed explanation is indeed pneumatic, though it does not coincide in every respect with that of the founder of the Lyceum[26].

This agreement on pneumatic etiology is clearly noted by Seneca, the continuator of Aristotle in the definition of meteors. Through his *Natural Questions* (*Naturales quaestiones*), he resumed Aristotle's doxographical and critical project (Waiblinger 1977; Gross 1989; Beretta et al. 2012) and proposed to add the presentation of later theories to defend his own arguments, which conform to the physical doctrine of the Stoics, of which he considered himself a faithful disciple. When defining the scope of his investigation, he thus declared:

> *Omnis de uniuerso quaestio in caelestia, sublimia, terrena diuiditur.*
>
> Every inquiry concerning the universe is divided into celestial phenomena, phenomena of the intermediate region, and terrestrial phenomena[27].

The term *sublimia*, distinct from *caelestia*, is not easy to translate. Seneca explained his tripartition: the first part corresponds without doubt to astronomy. The second part concerns, he said, "what is found between heaven and earth" (*inter caelum terramque uersantia*), which he further specified as[28] *nubila* (clouds), *imbres* (rains) and *nives* (snow). A further clarification is given later by the author himself:

> *Quomodo, inquis, de terrarum motu quaestionem eo posuisti loco quo de tonitribus fulguribusque dicturus es? – Quia, cum motus spiritu fiat, spiritus autem aer sit agitatus, etiamsi subit terras, non ibi spectandus est; cogitetur in ea sede in qua illum natura disposuit.*
>
> You ask me why I placed the investigation of the movements of the earth in the section where I will speak of thunder and lightning? It is because movement is produced by a breath, and a breath is agitated air. Even when it penetrates beneath the surface of the earth, that is not where it must be observed: it must be studied in the place that nature has assigned to it[29].

---

26 vv. 680–702.

27 II, 1, 1. For this excerpt as for those that follow, we offer our translation from Hine's (1996) edition.

28 II, 1, 2.

29 II, 1, 3.

With this programmatic justification, we find again the Aristotelian distribution of elements in the sublunar space, which the Stoic author does not seek to overturn. It is precisely the necessity for subterranean air to regain its natural place that creates the agitation at the origin of earthquakes. Hence, the possibility of classifying earthquakes among the *sublimia*. What then of the *terrena*? The text describes them as the domain in which one must discuss the following subjects: *de aquis* (waters), *de terris* (lands), *de arbustis* (trees), *de satis* (plants) and finally *de omnibus quae solo continentur* (everything that is connected to the soil[30]). In a certain sense, geography, but also the very objects of meteorology as defined by Aristotle. Where Aristotle established a clear and strict bipartition of the cosmos, Seneca introduced a tripartition with much less distinct boundaries.

From the doctrinal point of view of a Stoic, the obviousness of a pneumatic origin does not seem to present any problem. Indeed, his fellow Stoics likely found in this unanimity a confirmation of the foundations of their physics. For the Stoics, the cosmos is traversed by a pneuma – translated into Latin as *spiritus* – which ensures its cohesion (*ἕξις*) through an alternating and perpetual movement of expansion and contraction. This material entity is both an element (described in the most precise sources as a mixture of air and fire) and a universal agent. Its essential property is tension (*τόνος*), which it exerts on all physical entities. One of the main purposes of Seneca's meteorological treatise is precisely to demonstrate the action of this agent as the primary cause of most natural phenomena. While Seneca recalled that Epicurus himself favored the pneumatic etiology of seismic phenomena, while affirming the plurality of possible causes[31], he nevertheless radically opposed Epicurean physics by rejecting the principle of the porosity of matter: according to him, matter is subject to the action of the *spiritus* insofar as it is held together by its dynamic cohesion (Le Blay 2010).

The role assigned to the *spiritus* by Seneca undoubtedly owes much to one of his main models, Poseidonius of Apamea. A reference by Diogenes Laërtius is explicit on this point:

> A whirlwind of fire is a cloud torn apart by fire accompanied by breath. Earthquakes occur when breath rushes into the cavities of the earth or when it is trapped there, as Poseidonius said in his eighth

30 II, 1, 2. The last member cannot include underground phenomena because the expression is borrowed, specifies Seneca, from the vocabulary of jurists. It therefore concerns everything that touches the surface of the soil, the place of all legal attention.

31 VI, 20, 5–7.

> book [devoted to physics]. One distinguishes tremors, tremors with the opening of faults, landslides, and shocks[32].

The name of the Stoic Poseidonius of Apamea indeed appears in most studies dealing with the sources of Latin-language meteorology. This great thinker seems to sum up by himself the science of the Stoics. He was, with his master Panaetius of Rhodes, the most eminent scholarch of his time. The considerable body of work he left can only be compared in scope to that of Aristotle, and few fields seem to have escaped his reflection, as his writings covered nearly all areas of knowledge: history, geography, ethnography, cosmology and physics, meteorology. This work, unfortunately, has come down to us only in the form of quotations. The geographer Strabo, in particular, was a great reader of Poseidonius, whose theories he used on meteorological phenomena, on the question of tides, and on that of volcanoes; thanks to him, we know in particular that Poseidonius was the first to provide a complete explanation of the phenomenon of tides, developed from field observations (Compatangelo-Soussignan 2015). Several of Strabo's quotations make it possible to reconstruct the main outlines of a volcanological investigation, which seems to have advanced reflection much further than most other known ancient sources[33]. He is credited with a treatise on meteorology in seven books, also summarized in the form of *Elements of Meteorology*, a treatise *On the Ocean* mentioned by Strabo[34] and a treatise *On the World* comprising at least two books, of which we have the testimony of Diogenes Laërtius[35].

To conclude this section, it must again be emphasized that there is an almost total absence of discussion of volcanoes in Seneca's treatise, following the example of his model Aristotle. Commentators have not failed to question the reasons that may have led the two most important surviving sources to avoid this issue, even though volcanoes, Etna first among them, are otherwise very prominent in Greek and Latin literature. We do, however, retain a crucial witness in the form of a didactic poem in Latin entirely devoted to Etna and to the explanation of volcanism in general; its author is unknown, despite numerous attributions proposed throughout the history of its transmission, none of which can be confirmed with certainty. The dating we have proposed places its composition between 40 and 79 CE (Le Blay 2023). This *terminus ante quem* appears certain, for there is no mention in the text of the catastrophic eruption of Vesuvius, although this event was significant enough to be mentioned many times in later literature. The etiology

32 VII, 154 (Marcovitch editions; we translated).

33 We refer to the fragments edited by Edelstein & Kidd.

34 I, 2, 1.

35 VII, 142.

proposed is consistent with other sources, even if the poet's discussion dwells at some length on the description of volcanic rocks and matter, whose properties sustain Etna's activity.

1.3. Probing the bowels of the Earth: networks and cavities

If meteorological texts concentrate their attention on seismic phenomena, it is obviously because these are experienced as phenomena of a particularly striking nature, capable of dramatically affecting human activity, but also because the question relates to a rich imaginary of the Earth's depths, a hidden and unfathomable world that invites both reverie and speculation. The underground world is the ideal receptacle for all the elements that can exert their power over the inhabited world. In Greco-Roman culture, this imaginary remains inseparable from a symbolic tripartition of the world: Zeus/Jupiter governs the sky, Poseidon/Neptune the seas[36] and Hades/Pluto the underground world that shelters the Underworld.

Seneca's *Natural Questions* brings this dream of the depths to life and reveals an undeniable fascination with this invisible part of the world, which the philosopher did not hesitate to describe with its landscapes, its reliefs and its seas. He imagined that the oceans of the depths may be the sites of hidden life, a kind of mirror world with its own fauna[37]. Aristotle had already mentioned the existence of these aquatic spaces beneath the Earth's surface, in his critique of Plato's *Phaedo*:

> What is written in the *Phaedo* concerning the rivers and the sea is impossible. It is said there that all things can communicate with one another under the earth and that the principle and source of all waters is what is called Tartarus, a kind of mass of water situated at the center, from which all waters come, both those that flow and those that do not[38].

For Aristotle, it was essential to maintain that the elements each had their proper place, and that the place of water was the part of the world located above the earth, on the surface. If streams can sink into the ground, they eventually reappear at

36 It is also the deity that causes earthquakes, the part of the world inhabited by mortals, that is, the surface of the Earth, being a kind of interface between sky, sea and underground world, hence the fact that it is the theater of all divine powers and angers.

37 III, 8; III, 16, 4–5.

38 *Meteorologica*, II, 2, 355b 32-356a 2. The reference to the *Phaedo* is 111c-113c, where Plato cites the rivers of the Underworld.

another point on the Earth, for water, like all elements, seeks to regain its natural place, materialized by the marine space:

> In fact, it seems that all rivers end in the sea, except those that flow into one another; none end in the earth, and when they sink into it, they reappear elsewhere[39].

Nevertheless, the question of rivers that can disappear underground, like that of the springs that emerge from it, inevitably arouses curiosity, and numerous discussions can be found in the surviving sources. It is an observable fact that naturally invites inquiry into the nature of the Earth's depths and their capacity to contain the same elements found on its surface, whether air or water. The texts moreover cite many examples of rivers flowing alternately on the surface and underground (Irby 2016a, p. 192). The most famous of these, the river Alpheus, born in the Peloponnese near Olympia, was said to flow beneath the sea to Sicily, where it would emerge at Ortygia to mingle its waters with the fountain of Arethusa; this marvel of nature gave rise to a fable celebrating the river's desire for the nymph[40]. The myth was supported by reports claiming that objects thrown into the river's waters in Greece later reappeared in the waters of the spring[41].

Such springs emerging from the depths are quite common in the Mediterranean landscape. The aqueducts, which collect and carry water to the cities, make the existence of vast underground reservoirs of water likely. From an anthropological point of view, it is known that the sacred or magical dimension of springs lies in their seeming to arise miraculously from the bowels of the world. One cannot, of course, refrain from thinking here of the rivers of fire that form part of the infernal topography in mythological literature. Traditional representations of the Underworld are probably partly responsible for this ancient imagery (Craddock 2016).

Seneca reviewed various hypotheses to explain the massive presence of water beneath the Earth's surface[42]: the depths of the Earth would constantly produce and expel moisture; air would be converted into water under the effect of cold and darkness (one element can be born from another element according to Aristotelian and Stoic physics); the element earth could also be transformed into water. The bowels of the Earth are thus conceived as a place of transformation of elements and matter.

39 *Meteorologica*, II, 2, 356b 22–25.

40 Ovid, *Metamorphoses*, V, 599–641.

41 Pliny, *Natural History*, II, 225; Pausanias, *Description of Greece*, VII, 2.

42 *NQ*, III, 7–10.

As for the "circulation" of air, Strabo's *Geography* repeatedly refers to the existence of underground networks interconnecting volcanoes[43], a network with which he associated Vesuvius, proof that the Ancients indeed regarded it as akin to the volcanoes known to be active before its awakening in 79 CE. A thinker such as Diogenes of Apollonia was famous for the interest he is said to have taken in this subterranean world, to the point of having been called the "master of conduits" (Lloyd 2006). According to Simplicius, this same man wrote *Meteorologica*[44]; the various surviving testimonies about his book agree on the central role he assigned to air, which he regarded as the principle of nature; he is also said to have compared the stars to pumice stones and emphasized the porous nature of the Earth's structure[45]. The conception that emerges fairly early is that of a subsoil allowing the movement and circulation of the elements, whether liquid or aerial, an Earth whose structure is traversed by channels, networks and scattered with cavities more or less vast. Two conceptions, however, oppose each other: one emphasizing the porosity of this structure (advanced by the Epicureans in the philosophical tradition) and the other rejecting this porosity in principle, preferring to posit more open natural pathways specific to each fluid (defended by a Stoic like Seneca, precisely against Epicurean atomic physics and its corpuscular representations). Behind the meteorological developments, two biological models are in fact opposed, and one notes that physiological analogy often serves as a reference for demonstration. Indeed, the progress of anatomy made possible by Alexandrian medicine encouraged imagining the Earth's depths, inaccessible to observation, in the manner of bodily entrails (Le Blay 2005, 2010, 2011).

Book VI of Seneca's *Natural Questions*, the one dealing specifically with earthquakes, illustrates the theme of the power of the underground breath (*spiritus*) by presenting it as the agent and cause of telluric shocks. The question of the origin of this air present in the subterranean regions nevertheless had to be resolved. It was a physiological analogy that served as the foundation for the argument:

> That air is the cause of this evil, I also admit. But what I will discuss is the manner in which it penetrates beneath the earth: is it through thin pores invisible to our eyes, or through larger and wider passages? Does it come from the depths or enter from the surface? In our body, the skin does not reject air either, and it can enter only through the channels by which it is inhaled. And even when it has entered inside, it can remain only in a part of our body that is relatively open. It

43 *NQ*, I, 3, 16 ; V ; 4, 9 ; X, 1, 19.

44 D.-K. A 4.

45 This thesis is recalled by Seneca, *NQ*, IVa, 2, 28–30.

> therefore does not stop among the nerves and muscles, but in the viscera and the large inner cavity[46].

The idea of breathing through the pores of the skin is refuted; the air inhaled must pass through specific channels and organs. This question was debated very early among physicians, dividing those who supported respiration through organs dedicated to this function and those who supported respiration through the skin or perspiration (Le Blay 2010). The Stoic's theoretical position on this point is absolutely clear:

> "On the contrary," he declares, "a trembling occurs when we shiver from the effect of cold; it is also a breath coming from the outside that shakes the earth. That is absolutely impossible. The earth would have to be capable of feeling cold for it to experience what happens to us when an external cause makes us shiver. What happens to the earth is indeed comparable to the condition that affects us, but I would not go so far as to say that the causes are comparable. It is necessarily an internal and deep affliction that shakes the earth[47]".

The hypothesis of air coming from the outside is therefore rejected. The hypothesis in question most likely refers to Poseidonios[48]. One also thinks of the famous passage from the *Meteorologica* in which Aristotle establishes an analogy between earthquakes and the tremblings of the body in order to distinguish between the different forms of tremors, which may have inspired the Stoic Poseidonios:

> For he represents that, just as in our body the enclosed power of the breath (ἡ τοῦ πνεύματος ἐναπολαμβανομένη δύναμις) is the cause both of shivering and of pulsations (τρόμων καὶ σφυ γμῶν), so too in the earth the breath produces similar effects, and that one kind of tremor resembles a shiver (τρόμον), another a pulsation (σφυγμόν), and that, just as often happens after one has urinated (for a shiver runs through the body when the breath moves continuously from the outside inward), such things also occur in the earth. How great the power of the breath is must be observed not only from what happens in the air (for there one might suppose that it is by its volume that it can produce such effects), but also in the bodies of animals: for contractions and spasms are movements of the breath (οἵ τε γὰρ

46 VI, 24, 1–2.

47 VI, 24, 4.

48 VI, 14, 3.

τέτανοι καὶ οἱ σπασμοὶ πνεύματος μέν εἰσι κινήσεις), and they display such force that several people trying at once to restrain the motion of those who suffer from them cannot succeed. One must therefore imagine that the same thing happens in the earth, comparing the greater with the lesser[49].

1.4. The subterranean fire hypothesis

Among the elements present in the Earth's depths, some authors wished to place fire. Such a hypothesis is unthinkable in Aristotelian cosmology, but even if that system came to dominate at a certain point in the history of thought, it is far from sufficient to summarize the views of the Ancients.

The Pythagoreans believed in the existence of an eternal fire at the center of the Earth, and according to Plato, the Pyriphlegethon, the famous river of the Underworld in mythology, would be the source of the lava spewed by volcanoes[50]; it is this myth that Aristotle addresses in the criticism mentioned earlier. The founder of the Academy, moreover, gives one of the most astonishing descriptions of the geography of the depths, with its rivers and springs, accompanied by an explicit reference to the eruptions of Etna. Empedocles, the Sicilian from Agrigento whom legend says died in the crater of Etna, had also described torrents of subterranean fire[51]. In his explanation of hot springs, Seneca indeed referred to the Agrigentine and to his theory of fires hidden beneath the earth, and he summarized it through a comparison with the devices used to heat water in thermal buildings, consisting of conduits passing above or within a furnace[52]. It is not easy to know whether this technological comparison is due to the Latin philosopher or whether it was already present in Empedocles. We lean a priori toward the first hypothesis. Another passage from *Natural Questions* states unambiguously the fiery nature of the Earth's depths:

> *Adice nunc ignes, non tantum manufactos et certos, sed opertos terris, quorum aliqui erupuerunt, innumerabiles in obscuro et condito flagrant semper.*

---

49 II, 8, 366b 14–30 (we translated). The Aristotelian distinctions closely follow medical definitions, some of which are posterior in the state of the sources.

50 *Phaedo* 113b.

51 In his *Commentary on Plato's Timaeus*, Proclus attributes to him the existence of torrents of fire under the earth (D.-K. B 52). See also fragments A 68 and 69.

52 *NQ*, III, 24, 1–3. Also VI, 9, 2–3 and VI, 11.

> Let us add finally the fires, not only those produced by the hand of man and visible, but those hidden beneath the earth, some of which have erupted, and countless others that always burn, concealed in darkness[53].

This passage, which concerns the variability of air under the effect of temperature changes, is striking. It in fact presents a theory of subterranean fires that seems difficult to reconcile with the pneumatic etiology otherwise accepted. It also appears to allude to volcanic eruptions. There could therefore exist in Seneca a unified theory of volcanism involving subterranean fires, though the surviving presentation is a collection of disparate observations. Indeed, it seems inconceivable to ignore the fiery element when observing an eruption. According to Pliny, volcanoes would be signs announcing the final conflagration of the world[54]. The phenomenon would thus confirm the Stoic scenario of *ekpyrosis*. These apocalyptic visions, which are not merely products of our imagination, have continued to fuel the fascination we feel in the face of the unleashing of telluric forces[55].

Thus, Seneca introduced an idea that departs from the Aristotelian framework. It is difficult to identify the source from which the Stoic philosopher could have drawn this thesis: noting that air is warmer near the earth, he posits the principle of the calorific nature of the metals enclosed within it[56]. Supporting this thesis, in particular, is the empirical observation of a decrease in ambient temperature as one ascends in altitude, hence the presence of persistent snows at the tops of mountains, contrary to the view held by certain authors, whom he did not name, who claimed that greater proximity to the sun caused a rise in temperature with altitude. Certain bodies would be produced by the action of subterranean heat, all the more perceptible the deeper one goes beneath the Earth's surface, such as metals and volcanic rocks. The meteorological tradition following Aristotle did not adopt this theory, although Theophrastus seems in his writings, including the *Meteorologica*, to have been very receptive to this etiology by subterranean fires, within an

53 II, 10, 4.

54 II, 236.

55 In his *De igne subterraneo physica prolusio* published in Florence in 1641, Alsario hypothesized the presence of an underground fire capable of generating most known geological phenomena but also the tides. He referred to the movement of atoms in a clearly Lucretian inspiration.

56 *NQ*, IVb 9. Halleux reprised the presentation of this theory while limiting the scope of Seneca's demonstration (Halleux 1974, pp. 76–81).

exposition accepting a plurality of causes[57] (Régnier 2022). For Aristotle, the principle of the natural places of the elements forbids any presence of fire at the center of the world: of the four elements of the sublunar cosmos, fire occupies the highest zone, adjacent to the first astral sphere, that of the Moon. Yet we read several times in the *Meteorologica* that the Earth is warm and contains heat. One passage in particular may cause confusion, in the discussion devoted to the origin and formation of winds from dry exhalation:

> There is in the earth much fire and heat, and the Sun not only draws up the moisture that lies on the surface of the Earth, but also dries, by heating, the earth itself[58].

It is not so much the use of the term we translate as "heat" that poses a problem, but that of the term we translate as "fire"[59]. The word designating "earth" or "Earth" is itself always difficult to render, depending on whether one chooses to translate it with or without a capital letter. Following other translators, we alternate the spelling, for the sake of coherence. One may consider the Earth as a part of the world, which lies at the center, or earth as one of the elements present in the sublunar world, that of the meteors. Aristotle's text includes both meanings. The sense of the argument is clear: the Sun acts to warm the element earth or the Earth. Thus, Aristotle can explain rainfall: when the Sun approaches the surface of the Earth, the heat of this star produces a strong evaporation of the moisture present on and within the Earth, resulting in a great humid exhalation. When the Sun recedes, this humid but warm exhalation cools and transforms into rain, hence the fact that rains are more frequent during the cold season[60]. The Earth therefore contains heat, which is conferred upon it by the Sun's rotation. All the heat present in the world of meteors derives from this principle of the rotation of the celestial spheres. This also explains why the fiery element is the one neighboring these spheres. One must return to one of the initial definitions in the *Meteorologica*.

> The body that lies beneath the upper region of the revolution of the stars is like a matter that is in potentiality both hot, cold, dry, and wet, with all the qualities that follow from this, but which becomes or is one or another only through the effect of movement or the absence of

57 Halleux (1974) shows that the idea was mainly taken up by medieval alchemists (metals produced by the central fire).

58 II, 3, 360a 5–8.

59 One could opt for "igneous element" but the problem remains identical.

60 We summarize the entire passage.

> movement, whose cause and principle we have previously indicated. At the center and around the center lies what is heaviest and coldest, in separate places, that is, earth and water. Around and adjacent to them are air and what we habitually call fire, which, however, is not truly fire, for fire is an excess of heat and a form of incandescence[61].

The last part of this passage must help us understand the different meanings of the term "fire" in Aristotle's exposition. It sometimes designates the element itself and sometimes the nature that characterizes that element, the excess of heat and incandescence. If one can say that the earth contains much fire and heat, it is in the sense that the earth contains a great heat which, when excessive, can produce incandescence. The origin of this "fire" is, however, not to be found within the Earth itself. There is therefore no subterranean fire in Aristotle.

1.5. The Earth and its materials

Beyond the questions concerning the fire or heat that the Earth might contain, authors of Antiquity also examined the nature of rocks and metals – in other words, the materials that can be extracted from underground. More than a simple theoretical or speculative issue, the goal was to provide a theoretical framework for practical knowledge derived from mining or quarry exploitation (Craddock 2016). The Earth sciences largely developed on technical and industrial foundations; metallurgy, whose development was considerable throughout Antiquity, provides a set of practical cases and experiences which, without amounting to scientific experimentation, fed theoretical reflection – much of which was later reinvested in its speculative and symbolic branch, alchemy. The same can be said for glassmaking, a field of excellence in Roman industry. To this practical context must be added an ancient imaginary of the Earth's depths, whose assumptions Gaston Bachelard has brought to light (Bachelard 1938): the Earth is perceived both as a matrix – Mother Earth, nourishing and fertile – capable of engendering all kinds of materials useful to human activity, and as a womb – still the image of the Mother – subjecting what it contains to processes of digestion. The Earth is the place where matter is transformed. In both Greek and Latin, the terms describing the formation of rocks or metals underground most often refer to the idea of growth, following the biological metaphor of generation.

Here again, the etiological framework that emerges from Aristotle is that of the four elements and the four primary qualities associated in two pairs of opposites:

61 I, 3, 340b 14–23.

earth–air and water–fire. This theory of the four elements cannot be seen as a rigid or simplistic framework. The physics it implies allows for combinations and makes it possible to conceive of the transformations of matter. For Aristotle, the four elements constituted the principles that allowed one to conceive of the composition of matter in its diversity. To consider that the quaternary system imposes a fixed or static conception of the world would be to disregard the demonstrations of the *Meteorologica* and in particular what Book IV[62] teaches us about the transformations to which matter is constantly subjected; the instability of meteors can only be conceived within such a framework. The elements, according to Aristotle, are neither fixed nor immutable, since they can undergo transformations from one into another; there must exist a substratum that persists through these transformations and constitutes the primary matter. The expression *πρώτη ὕλη* (prime matter) used by the philosopher conveys this idea of matter as a kind of indeterminate and changing substratum, bearer of potentialities[63]. The four elements are the manifestations or most elementary states of this substratum. Concerning the actual existence of bodies, the distinction established between homoeomerous and anhomoeomerous entities accounts for the different states that matter can present, with an increasing degree of compositional complexity:

> I call homoeomerous the bodies such as metals, gold, copper, silver, tin, iron, stone, and other substances of the same kind (Λέγω δ'ὁμοιμερῆ οἷον τά τε μεταλλευόμενα, χρυσό, χαλκόν, ἄργυρον, καττίτερον, σίδηρον, λίθον και τἆλλα τὰ τοιαῦτα), as well as everything that can be extracted from them, and likewise everything found in plants and animals, such as flesh, bones, nerves, skin, viscera, hair, fibers, veins, from which are composed the anhomoeomerous parts (ἐξ ὧν ἤδη συνέστηκε τὰ ἀνομοιομερῆ), like the face, the hand, the foot, and other parts of the same kind, just as in plants, wood, bark, leaf, root, and all parts of the same kind. Since these homoeomerous bodies are constituted by another cause[64], but that by which they are constituted is, as matter, the dry and the moist, that is, water and earth (for each of them possesses this potentiality in the highest degree), and, as agents, the hot and the cold (for these assemble and solidify these bodies), let us examine among the

62 This last section of the *Meteorologica* has often been presented as the first known treatise of "chemistry". The quotation marks are important to avoid any anachronism (see Viano 2015a).

63 Thus, *Metaphysics* Θ, 7, 1049a 25–26, where it is said that wood is not earth but is of earth, the earth being capable of a multiplicity of entities in potential.

64 One must understand the formal cause.

> homoeomerous bodies which belong to earth, which to water, and which to both together[65].

In other words, there exist simple bodies and composite structures combining simple bodies. The example of the human body is indeed illuminating, for one immediately understands that a face combines flesh, bone, nerves, skin, hair, etc. Above all, one understands that every body is made of either earth or water, or a combination of the two elements. This is a first point of importance: there is no necessity that a body to be related to one element alone. In Chapter 4, Aristotle had moreover defined the hard and the soft as qualities of delimited bodies (*τὰ ὡρισμένα*), stating that what is made of the dry and the moist is necessarily hard (*σκληρόν*) or soft (*μαλακόν*[66]). It is therefore possible to conceive of all kinds of bodies presenting an intermediate state between the absolutely hard and the absolutely soft – in this case, the elements earth and water themselves. Later, there is a list of homoeomerous bodies and the mixtures they can produce[67]. In quite a comparable manner, the conceptual framework used by Seneca in the *Natural Questions* will postulate the existence of a finite number of elements, four, and assert the possibility of explaining the entire diversity of phenomena through the variety of combinations and transformations of these. It is evident that a rigid physical system forbidding any consideration of intermediate states between the four elements would not have lasted long in light of experiential data.

One of the most active mythological figures used to account for the subterranean world is that of the god Hephaestus/Vulcan and his forges. The myth can draw its components from the most concrete reality. It is easy to understand that metallurgy, a major activity of ancient industry, provides an essential and immediate explanatory model when one seeks to understand, in particular, what molten volcanic matter, in other words, lava, might be. This craft was indeed the human activity through which the laws of the transformation of matter most strikingly appeared[68]. Anaxagoras had already considered the nature of the sun, that burning star, in terms referring to the

65 IV, 10, 388a 13–26. The proposed definition reprises the causal principle dear to Aristotle, distinguishing formal cause, material cause, and efficient cause. Only the final cause is not considered here; it will apply to anhoméomères, as the author will show later (IV, 12, 389b 23-390b 2).

66 IV, 4, 382a 8–21.

67 IV, 10, 388a 25-389a 23.

68 We borrow this observation and the analyses that follow from Halleux (1974), a fundamental study on the question of metals in Antiquity.

activity of the forges: he compared it to a *μύδρος διάπυρος*[69]. It is difficult to translate the Greek term *μύδρος* otherwise than by molten matter or molten metal. It is the word used in Greek in the context of metallurgy. Plato, in the *Timaeus*, distinguishes two varieties of the watery element: the liquid variety (*ὑγρόν*) and the solid variety, which is fusible (*χυτόν*[70]).

Now, the *Timaeus* is the dialogue that, before Aristotle, helps establish the theory of the four elements. It clearly shows that the philosopher feels the need to think about the conditions of the transformation of matter at the same time as he establishes the quaternary framework destined to endure. It should be noted that he does not conceive of fusion according to the simplistic model of a transformation of the element earth, that is, the one characterized by solidity, into water, but rather as a property of the element water. This same element, says Plato, contains particles of fire that give it its mobility. Metal thus becomes a product of the watery element; the text explicitly mentions gold, "adamant[71]" and copper. Also mentioned in this foundational text are two materials resulting from the action of fire on earth: brick, the result of an earth completely dried by the effect of fire, and what the philosopher describes as a black-colored stone, which is in fact an earth still containing enough moisture to be fusible; Plato specifies that it takes on this black hue once cooled[72]. In the latter, we recognize lava from volcanoes. We must therefore cite here the passage from Book IV of the *Meteorologica* in which Aristotle develops an inventory of earthy bodies containing water, capable of transforming when their moisture disappears under the effect of cold or heat: he listed, among others, iron, which can easily melt and solidify, the refractory or "pyrimachus" stone, pumice and millstone. These latter elements are those that Plato emphasized in the *Timaeus*, which indeed seems to be one of the sources of inspiration for the final section of the *Meteorologica*:

> The refractory ('pyrimachus') stone (ὁ λίθος ὁ πυρίμαχος) also melts to the point of forming drops and flowing; its flow, upon solidifying, becomes hard again. The pumice stones (αἱ μύλαι) also melt to the point of flowing; this flow becomes black upon solidifying and becomes similar to lime (τῇ τιτάνῳ). Mud and earth (ὁ πηλὸς καὶ ἡ γῆ)

69 D.-K. 59 A 1 = Diogenes Laërtius, II, 8; D.-K. 59 A 2 (various citers including Olympiodorus, commentator on Aristotle).

70 58d-59e.

71 The identification of this metal is problematic and various hypotheses have been proposed. It is possible that it does not designate the same reality in all texts.

72 60c-d.

also melt. As for the bodies solidified by dry heat, some are soluble, others insoluble in a liquid. Clay and certain kinds of stones (Κέραμος μὲν οὖν καὶ λίθων ἐνίων γένη), which are formed by the combustion of earth, like the millstones (οἱ μυλίαι), are insoluble, while nitre and salt (νίτρον δὲ καὶ ἅλες) are soluble in a liquid, not in every liquid, but in a cold one[73].

Two rocks of volcanic origin are explicitly named, which some manuscripts and translators may have confused with one another. It is necessary to distinguish *μύλαι* and *μυλίαι*, which designate two different formations with distinct properties, even if they share a common origin. The millstone, which in Latin becomes *lapis molaris*, results, says Aristotle, from the complete combustion of the earthy element. This combustion is produced when the blazing of breath against the rocky walls occurs during an eruption. The author then resumes his discussion of bodies capable of undergoing processes of solidification (*πηκτός*) or fusion (*τηκτός*[74]), under the action of cold for the former and heat for the latter, listing several bodies, some mineral, others metallic. Chapter 2 deals with hard bodies that can soften (*τεγκτός*), particularly under the effect of fire.

It is the moist quality of bodies derived from the earth that determines their capacity for fusion. This quality concerns metals first and foremost, as Aristotle stated[75]. Rocks are the product of the dry exhalation trapped underground; metals are the product of the moist exhalation. This association of metals with the watery element is destined to endure. Thus, in his commentary on the *Meteorologica*, Olympiodorus calls *μέταλλον* the millstone (*μυλίας*), which he considered a fusible stone like metals[76]. To speak of the fusibility of a solid is to speak of its passage to a fluid state, recalling the liquid element. That is why all the sources seem to revolve around this same idea, even if they do not define it in exactly the same way. In his treatise on stones, Theophrastus mentions those that can melt and gives as examples the same two volcanic stones used as fluxes in metallurgy, the pyrimachus stones

---

73 IV, 6, 383b 5–13.

74 IV, 8, 385a 20-385b 5.

75 III, 6, 378a 15–21. Thus, gold can “boil” because it contains moisture (IV, 3, 380b 24–30).

76 Halleux, who emphasized the interest of the Presocratics in metallurgy, often invoked as an analogical argument (Halleux 1974, p. 51). We refer to the reference works on this commentary and the tradition derived from it (Viano 2002, 2006). It should be noted that the Greek term designating metal in Aristotle’s commentator does not occur in the source text. Aristotle used *μεταλλευόμενα* to designate what is extracted from mines.

and the millstones; they can melt because they contain water[77]. The Stoics, for their part, seem to have conceived water as the constitutive element of metals; one cannot say that they repeat exactly the data of the *Timaeus*, but they are inspired by it. A passage from Seneca reflects this definition, which makes metals the juices of the earth:

> But, just as we have not only blood in us, but different kinds of liquids, some necessary for life, others corrupted and somewhat thicker – the brain in the skull, the marrow in the bones, mucus, saliva, tears, as well as another around the joints, thanks to whose lubricating effect they move more easily – so also several kinds of liquids are found under the earth (*sic in terra quoque sunt umoris genera complura*): some harden over time – hence comes the value of metals (*metallorum honos*), from which greed draws gold and silver; and others turn into stone by losing their liquid quality (*quae in lapidem ex liquore uertuntur*). Earth and water decay to produce other substances, such as bitumen (*bitumen*) and other similar materials[78].

1.6. Toward a history of the Earth: floods and variations in sea level

Reading the sources finally helps us understand that the Ancients did not imagine the Earth and the known world as fixed for all time, but that they could conceive their history, even if this history was written outside the framework of human history. Many mythical narratives convey this idea of the ages of the world. The question of sea level – whose periodic variations make lands alternately emerge and submerge – seems to have been a particularly persistent concern. The figure of the cycle is at work here (Le Blay 2015a).

In his *Meteorologica*, Aristotle affirmed the periodicity of phenomena, following a model derived from biology, and applies this principle to the reciprocal transformations experienced by seas and continents[79]. He posited that the history of

77 *De lapidibus*, 9 (see Halleux 1974, pp. 118–120). This kinship between minerals and metals should be related to the kinship between metals and glass, which was so established that, from the Babylonians to European glassmakers of the 19th century, molten glass was called "metal" (Richet 2020). The *De lapidibus* of Theophrastus indeed implements an analogical approach whose model is borrowed from the arts of metal and molten material.

78 *NQ*, III, 15, 2–3.

79 I, 14, 351a 26–30.

the Earth, comparable to that of living beings, could be read as a succession of cycles combining growth and decline. He thus accounted for the cycle of interversion between lands and seas and its effects on the destiny of territories and populations. It is not always the same regions of the Earth that are dry or wet: the areas of the continent or the sea change. These cycles, which affect the appearance of inhabited lands, have major consequences for the history of populations, but, Aristotle observes, the timescale of these meteorological changes does not allow us to perceive all such evolutions. It is wars, diseases, or famines that affect populations massively and rapidly; long-term meteorological change is, for us, most often imperceptible. The history of humankind does not unfold at the same pace as that of the Earth.

The crucial point that the philosopher seeks to refute is the idea of an end of the world[80]. Reading him, it seems that some of his contemporaries were worried about seeing the sea completely disappear, based on local observations that could have suggested such a harmful outcome. The theory of the Great Winter or the Great Year, to which he then refers, was apparently widespread[81]. Aristotle also mentions the gradual drying up of Egypt and of Mycenae, a land once fertile – at the time of the Trojan War – become unfit for cultivation[82]. In Aristotelian cosmology, the sublunary world undergoes the direct influence of the revolution of the solar body, which does not always occur at the same distance from the Earth. This principle is recalled in the first chapters of the treatise[83]. The philosopher insists several times on the perpetual nature of this cycle, in order to dispel any ambiguity about a possible interpretation of the principles of generation and corruption that would tend, by complete analogy with living beings, to posit a foreseeable death of our world[84]. This revolution of the Sun is thus the cause of a perpetual cycle of evaporation and condensation of the moist element that surrounds the Earth.

Aristotle's warnings at least express the awareness the Ancients had of transformations in their environment, whether they witnessed them directly or situated them in the time of their distant ancestors. A passage from Plato's *Critias*

80 I, 14, 352a 15–20.

81 The exact origins of this theory are difficult to identify and trace; it appears in various authors and in multiple contexts. In his *Timaeus* 39c-e, Plato explains it by the return of the seven planets to their starting point. Tacitus (*Dialogue of Orators*, 16) referred to a lost work of Cicero in which he would have estimated the duration of this period at 12,954 years. On this subject, see Van der Waerden (1952) and Neugebauer (1975, pp. 560, 618).

82 I, 14, 351b 29-352a 15.

83 I, 9, 346b 20.

84 Thus, in II, 2, 354b 30 the eternity of the hydrological cycle is reaffirmed.

describes the ancient territory of Attica and the natural upheavals that altered its appearance – an evolution that transformed a land marked by vegetal abundance, covered with pastures and forests, into a harsher soil[85]. Mention is made there of earthquakes and a succession of floods spanning a period of 9,000 years. Indeed, it is from the perspective of catastrophe that these metamorphoses are conceived. In the same Plato (*Critias, Timaeus*), we read the legend relating to the civilization of Atlantis and its destruction by natural disasters, preceding its final submersion. Atlantis is in fact the emblematic example of the often ambiguous link between mythic narrative and the memory of distant events. It is unnecessary to recall how much this myth has given rise to all sorts of interpretations and hypotheses, some of which are above all the product of a fertile imagination[86]. While some still seek to situate geographically and historically the inspiration for Atlantis, not without appealing arguments[87], it is now generally accepted that Plato's narrative is an invention with an allegorical purpose. This does not mean that Plato did not draw from the memory of disasters maintained among his contemporaries. For such a story to seem to contain a portion of truth, the elements composing it must refer to a world and representations that were collectively shared. Among the most credible hypotheses regarding its inspiration, it is certain that the eruption of the supervolcano of Thera (Santorini) is the most convincing. Archaeological excavations have shown that a flourishing civilization had developed there before the destruction of the site by a major volcanic event. The dating of this event remains a matter of debate: radiocarbon (or Carbon-14) analysis indicates the second half of the 17th century BCE, and archaeology favors the second half of the 16th century[88]. Whatever date is retained, it is known that this large-scale eruption was the source of catastrophic episodes across a vast area including the Minoan world. Its various effects (ash cloud, seismic shocks, tsunami) are attested by archaeological findings at different sites in the eastern Mediterranean. It would be

85 110d-112d.

86 On the "archaeological" myths of Atlantis, which enjoyed considerable fortune between the late 19th and early 20th centuries (see Le Quellec 2010). See also the investigation carried out by the geologist Collina-Girard (2009).

87 The most recent hypothesis, to our knowledge, is that of Kieffer (2004), which situates the inspiration of the Platonic narrative in eastern Sicily, on the site of Taormina. He notably recalled Diodorus of Sicily's mention of the cataclysmic eruption of Etna, which would have forced the Sicani to abandon this part of the island.

88 On this point, which is the subject of controversies, see the recent synthesis by Manning (2014) and the very critical review by Bietak in *Bryn Mawr Classical Review* (see: http://bmcr.brynmawr.edu/2016/2016-04-06.html). Also see Jusseret and Sintubin (2017) and O'Connell (2018). See also the summary of the event established by the geologist Caroff (2017).

surprising if a disaster of such magnitude had not left a lasting impression on memory, to the point of giving rise to shared narratives.

The primary function of myths can consist of preserving and maintaining a memory of significant events from the past, a past sometimes too distant to be the subject of a historical account.

In 1973, Dorothy B. Vitaliano published a monograph intended to revive this reading of myths, *Legends of the Earth: Their Geologic Origins* (Vitaliano 1973). Through her reflection and the methodology she implemented, a new field of knowledge was opened, now designated as "geomythology". The original postulate was simple: mythology, like folk traditions based on oral narratives, can refer more or less explicitly to major geological events. They can therefore be studied from the perspective of testimony, even if the witnesses themselves fall outside the scope of History.

Greek sources have transmitted to us a set of mythic narratives referring, in varied but convergent forms, to the universal Flood. The origin of this imaginary is Babylonian (Dumas-Reungoat 2001, 2018). In the Platonic narrative, the world is cyclically subjected to catastrophes that destroy the most developed civilizations and reset the survivors "to zero"; the Greek myths of the Flood, as well as the fiction of Atlantis, are linked to this theory of cyclical time. There thus exists, in Greek thought, an association between cataclysm or catastrophe and the birth of myths. The current world would have emerged from this deluge; it would have been preceded by other worlds. It is generally considered that the rise in sea level accompanying the end of the last major glaciation may have given rise to this deluge imagery; populations who witnessed this phenomenon had to abandon the coastal sites where they were settled and which provided their subsistence through fishing activities. Archaeology preserves visible traces of these forced abandonments, which took place over several generations. The Ancients could find in certain material traces confirmation of this mythic flood, followed by the retreat of the waters rendering the lands habitable again. It is now known that the Black Sea was particularly affected by this rise in water level. It is interesting to note that many ancient sources, Strabo in particular, were already questioning the history of this sea (Dan 2015); some scholars think that the deluge imagery may originate from this part of the world.

In the myths that feature him, the Giant represents the necessary imaginative device by which phenomena of exceptional magnitude, which can only be attributed to superhuman forces, are recounted: mountain formations, volcanoes and earthquakes. For the ancient Greeks, as for their Roman continuators, Giants, far from being supernatural creatures, refer to a bygone age and are therefore inscribed in the history of the world since its origins. In a world that, in its early history, may have included beings of great size, the Giant has its place as an anthropomorphic version of what Man is in the world as we know it. Some sources inform us about the curiosity of ancient peoples regarding the giant bones they could exhume here and there, notably in Crete. These remains, considered as the bones of Giants, were very likely the fossilized bones of dinosaurs[89]. Such fossils were indeed evidence of a prior population of the world and of a change of era from a certain time. Similarly, the pre-Socratic Xenophanes of Colophon refers to marine remains that could be found in the soil inland:

> Xenophanes believes that there is a union of the land with the sea, and that the land is dissolved over time by moisture; he advances as proof that in the middle of the land and on the mountains one finds shells, that in Syracuse, stone quarries yielded the imprint of a fish and a seal, at Paros an imprint of a laurel in the stone, and at Malta shells of all kinds of marine animals. This happened, he declares, when long ago all animals were wallowing in mud and their imprint in the mud dried. All humans perished when the earth, falling into the sea, became mud; then their generation began again, and this change occurs in all worlds[90].

The preserved fragments suggest that this thinker, one of our first "naturalists", understood the emergence of life as resulting from the mixture of earth and water. It is interesting to note here that he envisioned a succession of periods marked by a cycle of deaths and rebirths.

Finally, one should dwell longer on the observations of the Ancients regarding the role of rivers and waterways in shaping lands and landscapes (alluvial plains, islands, coasts, etc.): here again, Strabo is a key witness (Craddock 2016; Irby 2016a). Complementing Herodotus' account[91] of the Nile, he discusses the

89 Libanius referred to it in two of his speeches (*Speech I = Autobiography*, 93; *Speech V, On Artemis*, 41, ed. J. Martin and P. Petit, 2 vols., Collection des Universités de France, Paris, 1979 and 1988). The Roman historian Suetonius recalled that Augustus liked to adorn his villas with collections of natural curiosities. Among the items he prized most were a series of remains of large animals, discovered at Capri, which had received the designation of Giant bones and weapons of heroes (*Aug.* 72).

90 Hippolytus, *Refutation of All Heresies*, I, 14 (D.-K. A 33, trans. J.-P. Dumont, *Les Présocratiques*, Gallimard, coll. "Bibliothèque de la Pléiade", Paris, 1988).

91 II, 4-5 ; 10–12.

formation of deltas[92]. He notably posited that the Black Sea, because of the great rivers flowing into it, such as the Danube, called Ister in Antiquity, could one day become nothing but a surface of land[93]. By the same causes, the delta of the Pyramos, a river of Cilicia, could, according to him, eventually reach Cyprus and transform this island into a coastal promontory[94]. Regardless of the accuracy of these theories, it is clear that the Ancients were attentive observers of their environment and its transformations, and that their conception was not fixist, but genuinely historical.

1.7. Conclusion

We have here only sketched the picture of ancient knowledge and speculation regarding the objects that constitute the current field of geology and Earth sciences. In a presentation aiming at exhaustiveness, one must constantly keep in mind certain methodological precautions, essential for the rigorous writing of any history of science extending to the earliest periods. This approach depends on the preservation of textual sources. Archaeology can complement what the texts tell us, but in this regard, its contribution remains limited. Any perspective on our antiquity, therefore, offers a truncated and incomplete vision. Many witnesses have disappeared that could be decisive in our ability to reconstruct the state of knowledge. Working on Antiquity is more or less equivalent to building hypotheses on a field of ruins.

Thinkers of Greece and Rome also have their own manner since they conceive the intellectual and speculative approach from the angle of doctrinal controversy. This stance certainly has something to do with our definition of science, which cannot do without contradictory debate. Nevertheless, in its own cultural context, it is influenced, or even imposed, by the polemic between philosophical schools. These schools can be the bearers of a worldview involving cosmology and physics.

The allegiance of a given author must therefore be questioned before any reading under the risk of not grasping the doctrinal biases that guide the argumentation. It is rare that, on a given subject, one can conclude unanimity among the sources. Any univocal synthesis would therefore be inappropriate.

92 I, 3, 8.

93 I, 3, 4.

94 XII, 2, 4.

1.8. References

Bachelard, G. (1938). *La formation de l'esprit scientifique. Contribution à une psychanalyse de la connaissance objective*. Vrin, Paris.

Bakker, F.A. (2016). *Epicurean Meteorology: Sources, Method, Scope and Organization*. Brill, Leiden.

Beretta, M., Citti, F., Pasetti, L. (2012). *Seneca e le scienze naturale*. Olschki, Florence.

Bertrand, E. and Compatangelo-Soussignan, R. (eds) (2015). *Cycles de la nature, cycles de l'histoire : de la découverte des météores à la fin de l'âge d'or*. Ausonius, Bordeaux.

Bousquet, B. (2006). Les séismes de l'Antiquité, entre nature et société. In *L'homme face aux calamités naturelles dans l'Antiquité et au Moyen-Âge*, Jouanna, J., Leclant, J., Zink, M. (eds). De Boccard, Paris.

Bousquet, B. and Péchoux, P.Y. (1981). Séismes et espaces séismiques : une incursion de géographes dans le domaine de l'Antiquité classique. *Annales de l'Université Toulouse II-Le Mirail*, 17, 45–57.

Capelle, W. (1912). Μετέωρος-Μετεωρολογία. *Philologus*, 71, 414–448.

Capelle, W. (1913). Zur Geschichte der meteorologische Literatur. *Hermès*, 48, 321–358.

Caroff, M. (2017). *Terres singulières. Quinze lieux remarquables à travers le monde*. Quæ, Versailles.

Collina-Girard, J. (2009). *L'Atlantide retrouvée, enquête scientifique autour d'un mythe*. Belin, Paris.

Compatangelo-Soussignan, R. (2015). La théorie des marées de Poseidonios d'Apamée et les cycles de la nature dans la tradition philosophique des IVe-Ier siècles a.C. In *Cycles de la nature, cycles de l'histoire : de la découverte des météores à la fin de l'âge d'or*. Bertrand, E. and Compatangelo-Soussignan, R. (eds). Ausonius, Bordeaux, 83–96.

Compatangelo-Soussignan, R., Diosono, F., Le Blay, F. (2022). *Living with Seismic Phenomena in the Mediterranean and Beyond between Antiquity and the Middle Ages. Proceedings of Cascia (2019) and Le Mans (2021) Conferences*. Archaeopress Archeology, Oxford.

Craddock, P.T. (2016). Classical geology and the mines of the Greeks and Romans. In *A Companion to Science, Technology and Medicine in Ancient Greece and Rome*, Irby, G.L. (ed.). Wiley Blackwell, London.

Cusset, C. (2003). *La météorologie dans l'Antiquité entre science et croyance. Actes du Colloque interdisciplinaire de Toulouse, 2-3-4 mai 2002*. Publications de l'Université de Saint-Étienne, Saint-Étienne.

Dan, A. (2015). 'Pontos par excellence' : cyclicité et linéarité dans les théories antiques sur l'existence de la Mer Noire. In *Cycles de la nature, cycles de l'histoire : de la découverte des météores à la fin de l'âge d'or*, Bertrand, E. and Compatangelo-Soussignan, R. (eds). Ausonius, Bordeaux.

Dumas-Reungoat, C. (2001). *La fin du monde : enquête sur l'origine du mythe*. Les Belles Lettres, Paris.

Dumas-Reungoat, C. (2018). L'imaginaire de la fin du monde en Grèce ancienne : fin provisoire et fin radicale, deux mythes miroirs. In *A Universal Imagination of the End of the World?*, Le Blay, F. (ed.). Cambridge Scholars Publishing, Newcastle-upon-Tyne.

Ellenberger, F. (1994). *Histoire de la géologie 1 : des Anciens à la première moitié du XVIIe siècle*. Tec & Doc, Paris.

Foulon, E. (ed.) (2004). *Connaissance et représentations des volcans dans l'Antiquité*. Presses universitaires Blaise Pascal, Clermont-Ferrand.

Furley, D. (1983). The mechanics of meteorologica IV: A prolegomenon to biology. In *Zweifelhaftes im Corpus Aristotelicum: Studien zu einigen Dubia. Akten des 9. Symposium Aristotelicum*, Moraux, P. and Wiesner, J. (eds). De Gruyter, Berlin, 73–93.

Gilbert, O. (1907). *Die meteorologischen Theorien des griechischen Altertums*. B.G. Teubner, Leipzig.

Gohau, G. (1990). *Une histoire de la géologie*. Le Seuil, Paris.

Graham, D.W. (2016). Matter. In *A Companion to Science, Technology and Medicine in Ancient Greece and Rome 2*, Irby, G.L. (ed.). Wiley Blackwell, London, 29–42.

Gross, N. (1989). *Senecas Naturales Quaestiones : Komposition, Naturphilosophische Aussagen und ihre Quellen*. Steiner Verlag, Stuttgart.

Gruet, B. (2004). Une pneumatique céleste : volcans et séismes chez Aristote (*Météorologiques*, 365b-369a). In *Connaissance et représentations des volcans dans l'Antiquité*, Foulon, E. (ed.). Presses universitaires Blaise Pascal, Clermont-Ferrand, 205–214.

Guidoboni, E. (1989a). *I terremotti prima del mille in Italia e nell'area mediterránea*. Istituto Nazionale du Geofisica, Bologna.

Guidoboni, E. (1989b). Catalogo delle epigrafi latine riguardanti terremoti. In *I terremotti prima del mille in Italia e nell'area mediterránea*. Istituto Nazionale du Geofisica, Bologna.

Guidoboni, E., Comastri, A., Traina, G. (1994). *Catalogue of Ancient Earthquakes in the Mediterranean Area up to the 10th Century*. Istituto Nazionale di Geofisica, Bologna.

Halleux, R. (1974). *Le problème des métaux dans la science antique*. Les Belles Lettres, Paris.

Hine, H.M. (2002). Seismology and vulcanology in antiquity? In *Science and Mathematics in Ancient Greek Culture*, Tuplin, C.J. and Rihll, T.E. (eds). Oxford University Press, Oxford, 56–75.

Irby, G.L. (ed.) (2016a). Hydrology: Ocean, rivers and other waterways. In *A Companion to Science, Technology and Medicine in Ancient Greece and Rome*. Wiley Blackwell, London, 181–196.

Irby, G.L. (ed.) (2016b). *A Companion to Science, Technology and Medicine in Ancient Greece and Rome 2*. Wiley Blackwell, London.

Jouanna, J., Leclant, J., Zink, M. (2006). *L'homme face aux calamités naturelles dans l'Antiquité et au Moyen-Âge. Actes du Colloque de la Villa "Kérylos", Cahiers de la Villa "Kérylos"*. De Boccard, Paris.

Jusseret, S. and Sintubin, M. (2017). *Minoan Earthquakes: Breaking the Myth through Interdisciplinarity. Studies in Archaeological Sciences*. Leuven University Press, Louvain.

Kieffer, G. (2004). À la recherche des sources de l'Atlantide. In *Connaissance et représentations des volcans dans l'Antiquité*, Foulon, E. (ed.). Presses universitaires Blaise Pascal, Clermont-Ferrand, 85–92.

Le Blay, F. (2005). Microcosm and macrocosm: The dual direction of analogy in hippocratic thought and the meteorological tradition. In *Hippocrates in Context, Papers Read at the XIth International Hippocrates Colloquium, University of Newcastle-upon-Tyne, 27–31 août 2002*, van der Eijk, P. (ed.). Brill, Boston, 251–269.

Le Blay, F. (2006a). Le poème latin sur l'Etna : témoin d'un savoir oublié. In *Musa docta. Recherches sur la poésie scientifique dans l'Antiquité*, Cusset, C. (ed.) Publications de l'Université de Saint-Étienne, Saint-Étienne, 335–361.

Le Blay, F. (ed.). (2006b). Percer les secrets de la nature : la source mystérieuse de Pline le Jeune. In *L'eau, Les eaux, X^es^ Entretiens de la Garenne-Lemot*. Presses Universitaires de Rennes, Rennes, 19–32.

Le Blay, F. (2010). Les pores de la peau : une entité physiologique problématique. In *Body, Disease and Treatment in a Changing World. Latin Texts and Contexts in Ancient and Medieval Medicine*, Langslow, D. and Maire, B. (eds). BHMS, Lausanne, 27–36.

Le Blay, F. (2011). Météorologie et médecine antique : l'alliance de deux savoirs. In *Histoire et nature. Pour une histoire écologique des sociétés méditerranéennes (Antiquité et Moyen Âge)*, Clément, F. (ed.). Presses Universitaires de Rennes, Rennes, 65–86.

Le Blay, F. (2015a). Rythmes du cosmos et rythmes biologiques dans la météorologie antique. In *Cycles de la nature, cycles de l'histoire : de la découverte des météores à la fin de l'âge d'or*, Bertrand, E. and Compatangelo-Soussignan, R. (eds). Ausonius, Bordeaux, 35–45.

Le Blay, F. (ed.). (2015b). L'orgue et le volcan : quand la machine explique la nature. In *Machines et inventions : le mythe et la technique. Actes de la Journée scientifique du XLIVe congrès de l'APLAES. Annales de l'APLAES* [Online]. Available at: http://revues1.aplaes.org:86/index.php/annales/issue/view/2 [Accessed 29 July 2022].

Le Blay, F. (2019). Expliquer les feux des montagnes : la science d'avant les volcans. *Travaux du Comité français d'Histoire de la Géologie (COFRHIGEO)*, 3(XXXI).

Le Blay, F. (2022). Sénèque naturaliste. In *Passeurs de culture : la transmission de la culture grecque dans le monde romain des Ier-IVe siècles après J.-C.*, Favreau-Linder, A.-M., Lalanne, S., Vix, J.-L. (eds). Brepols, Turnhout, 291–303.

Le Blay, F. (2023). *La fascination du volcan. Les mythes et la science*. Vrin, Paris.

Le Quellec, J.-L. (2010). *La Dame Blanche et l'Atlantide : Ophir et le Grand Zimbabwe. Enquête sur un mythe archéologique*. Errance, Paris.

Lloyd, G.E.R. (2006). Diogenes of Apollonia: Master of ducts. In *La costruzione del discorso filosofico nell'età dei presocratici*, Sassi, M.M. (ed.). Edizioni delle Normale, Pise, 237–257.

Manning, S.W. (2014). *A Test of Time and a Test of Time Revisited: The Volcano of Thera and the Chronology and History of the Aegean and East Mediterranean in the Mid-Second Millenium BC*. Oxbow, Oxford.

Martelli, M. (2016). Greco-Egyptian and Byzantine alchemy. In *A Companion to Science, Technology and Medicine in Ancient Greece and Rome*, Irby, G.L. (ed.). Wiley Blackwell, London, 217–231.

Meeusen, M. (2016). *Plutarch's Science of Natural Problems: A Study with Commentary on "Quaestiones Naturales". Plutarchea hypomnemata*. Leuven University Press, Louvain.

Neugebauer, O. (1975). *A History of Ancient Mathematical Astronomy*. Springer Verlag, Berlin.

O'Connell, T. (2018). *Joining the Dots. Plato's Atlantis in the Central Mediterranean*. Tony O'Connell.

Régnier, A. (2022). Living with seismic phenomena in the Mediterranean and beyond between Antiquity and the Middle Ages. In *Proceedings of Cascia (2019) and Le Mans (2021) Conferences*, Compatangelo, R., Diosono, F., Le Blay, F. (eds). Archaeopress Archeology, Oxford, 55–68.

Richet, P. (ed.). (2020). Glass, the wonder maker of science. In *Encyclopedia of Glass Science, Technology, History and Culture*. John Wiley & Sons, New York, 1–26.

Roller, D.W. (2016). Geography. In *A Companion to Science, Technology and Medicine in Ancient Greece and Rome*, Irby, G.L. (ed.). Wiley Blackwell, London, 247–262.

Seneca (1996). In *Naturalium Quaestionum Libros*, Hine, H.M. (ed.). K.G. Saur Verlag, Stuttgart-Leipzig.

Sigurdsson, H. (1999). *Melting the Earth: The History of Ideas on Volcanic Eruptions*. Oxford University Press, Oxford.

Taub, L. (2003). *Ancient Meteorology*. Routledge, London.

Taub, L. (2008). *Aetna and the Moon: Explaining Nature in Ancient Greece and Rome*. Oregon State University Press, Corvallis.

Taub, L. (2016). Meteorology. In *A Companion to Science, Technology and Medicine in Ancient Greece and Rome*, Irby, G.L. (ed.). Wiley Blackwell, London, 232–246.

Thomasset, C., Ducos, J., Chambon, J.-P. (eds) (2010). *Aux origines de la géologie de l'Antiquité au Moyen Âge*. Honoré Champion, Paris.

Van der Waerden, B.L. (1952). Das Große Jahr und die Ewige Wiederkehr. *Hermès la revue*, 80, 129–155.

Viano, C. (2002a). *Aristoteles Chemicus. Il IV libro dei Meteorologica nella tradizione antica e medievale. Colloque de Venise 1-3 décembre 1999*. Academia Verlag, Sanit Augustin.

Viano, C. (ed.) (2002b). Le commentaire d'Olympiodore au livre IV des Météorologiques d'Aristote. In *Aristoteles Chemicus. Il IV libro dei Meteorologica nella tradizione antica e medieval. Colloque de Venise 1-3 décembre 1999*. Academia Verlag, Saint Augustin.

Viano, C. (2003). Corpi e metalli: le "meteore" del Timeo (58c-61c). In *Plato Physicus. Cosmologia e antropologia nel Timeo*, Natali, C. and Maso, S. (eds). Adolf M. Hakkert Editore, Amsterdam, 207–223.

Viano, C. (2006). *La matière des choses. Le livre IV des Météorologiques d'Aristote, et son interprétation par Olympiodore, avec le texte grec révisé et une traduction inédite de son Commentaire au Livre IV*. Vrin, Paris.

Viano, C. (2015a). *Mixis* and *diagnôsis*: Aristotle and the "chemistry" of the sublunary world. *Ambix*, 62(3), 203–214.

Viano, C. (2015b). Les Météorologiques d'Aristote et l'héritage des prédécesseurs. In *Cycles de la nature, cycles de l'histoire : de la découverte des météores à la fin de l'âge d'or*, Bertrand, E. and Compatangelo-Soussignan, R. (eds). Ausonius, Bordeaux, 49–66.

Vitaliano, D.B. (1973). *Legends of the Earth: Their Geologic Origins*. Indiana University Press, Bloomington.

Waiblinger, F.P. (1977). *Senecas Naturales Quaestiones. Griechische Wissenschaft und römische Form*. Beck, Munich.

Williams, G. (2012). *The Cosmic Viewpoint: A Study of Seneca's Natural Questions*. Oxford University Press, Oxford.

Wilson, M. (2013). *Structure and Method in Aristotle's "Meteorologica": A More Disorderly Nature*. Cambridge University Press, Cambridge.

2

Studying and Knowing the Earth in the Middle Ages

Joëlle DUCOS[1,2] and Fleur VIGNERON[3]

[1] *STIH, Sorbonne Université, Paris, France*

[2] *SAPRAT, EPHE, PSL, Paris, France*

[3] *Université Grenoble Alpes, France*

2.1. The Earth and stones: knowledge scattered across multiple fields

The term *geologia* (*geology* in English, *géologie* in French), used in England in the 17th century and present in Diderot's *Encyclopédie*, was not associated with a recognized and distinct discipline before the 19th century. One should not conclude from this that no reflection or knowledge about the Earth and stones existed beforehand, particularly in the Middle Ages, even if they were not the subject of an autonomous science. Pierre Duhem was the first to take an interest in this, with the assumptions that were his own, emphasizing the modernity of medieval scholars' reflections, especially within the University of Paris in the 13th and 14th centuries (Duhem 1913–1955). Other accounts (Hölder 1960; Accordi 1984; Gohau 1987; Ellenberger 1988) continued in this direction based on the same sources, highlighting both the interest of medieval texts and the scarcity of research in this field at the end of the 20th century. Indeed, these studies must be compared with the work of historians of geography (Gautier Dalché 2013), techniques (mining and architecture), medicine and agronomy in order to give an account of the extent of medieval knowledge, something that appears in more recent collective works (Thomasset and James-Raoul 2000, 2010; Thomasset et al. 2010) and in a joint history of geology and medicine (Duffin et al. 2013). Medieval reflection is therefore polymorphous both in its forms of writing, its media and its aims: although

History of Earth Sciences,
coordinated by Pierre SAVATON. © ISTE Ltd 2026.

scholastic thought and writings have been emphasized, along with the model of Aristotelian physics, a diversity of approaches nonetheless, inevitably, arises from the very object of study. Stone and Earth belong to elemental physics, both in their formation and in their constitution, and they are also indispensable material for other sciences (medicine and alchemy) and techniques (architecture, building and agriculture). The importance of lapidaries, those writings describing stones and accounting for their uses, their medical and magical virtues, with symbolic interpretation, is undeniable: they are the earliest attempts to describe and classify stones.

2.1.1. *The three currents of medieval reflection*

It is always difficult to distinguish categories of texts in the Middle Ages, since the boundaries are porous and the genres unclear, and this is precisely the case for lapidaries (Halleux 1985). But three tendencies, nonetheless, appear in the Middle Ages. The first, heir to ancient naturalism, presents the properties of stones. Lapidaries build on Greek and Latin traditions of stone descriptions, particularly Pliny the Elder, or Solinus who gives a reduced image of them, with descriptions transmitted in the Middle Ages by Isidore of Seville. The great encyclopedias of the 13th century, such as those of Thomas of Cantimpré, Vincent of Beauvais, Arnold of Saxony or Bartholomew the Englishman (Draelants 2010), present a synthesis with other works known more recently, classifying gems and minerals often alphabetically and giving their physical, medicinal or other properties. But before these, the *De gemmis* of Marbod of Rennes (late 11th century), describing 60 stones in 734 verses, enjoyed enormous success with more than 250 manuscripts and adaptations in multiple vernacular languages (French, Anglo-Norman, Occitan, Italian, Danish, Hebrew, etc.). It is often the reference work in vernacular literature, especially for descriptions of precious stones.

The second tendency, arising from both the ancient and Alexandrian worlds, places greater emphasis on the magical powers of stones, often linking them to the positions of stars and planets (Weill-Parot 2002). One of the sources is the composite text known as that of Damigeron-Evax, translated in the 11th century, where zodiac signs and planets are associated with stones. In this perspective, stones may serve as amulets or talismans, particularly when engraved. The *Liber sigillorum* of Theel, transmitted in the West in the 12th century, is one of its points of support, as is the *De quindecim stellis, quindecim lapidibus, quindecim herbis* and *quindecim imaginibus*, a Latin translation of a recension of Messahala (8th century). This branch, strongly associated with magic and astrology, was, however, little disseminated outside specialized Latin writings (Weill-Parot 2002, p. 122). There do exist prose treatises on engraved stones (*De sigillis*), especially in Anglo-Norman, but without names of stones, with astral symbolism emphasized rather than the differentiation of stones, the essential point being the relationship established between stars and stones.

The third current, often described as Christian, draws inspiration from the Bible, especially from the two lists of stones found there: the stones of Aaron's Breastplate (Exodus 28, 17–21 and Exodus 39, 8–13) and those of the Heavenly Jerusalem (Apocalypse 21, 19–20). The commentaries on these lists since the Church Fathers provide a symbolic reading of the stones, within the exegetical tradition. The success of these lapidaries can be seen in the adaptations made very early in vernacular languages from the 11th century onward: we may cite the two verse adaptations by Philippe de Thaon in Anglo-Norman, the *Lapidaire chrétien* in verse (Gontero 2010), or the *Lapidaire du roi Philippe* (last quarter of the 13th century), which combines a lapidary of Christian inspiration (17 stones) with mineralogical descriptions (up to 71 stones depending on the versions). The interest in it was considered great enough for it to be translated into Latin. The end of the French Middle Ages continued to produce new lapidaries in the vernacular: the lapidary attributed to Jean de Mandeville, the *Tresorier de philosophie naturelle des pierres précieuses* by Jean d'Outremeuse (between 1382 and 1390), and alphabetical prose collections in the 15th century. Although spiritual interpretation is less present there, magical uses and virtues in relation to the stars are developed, alongside descriptions of colors, shapes and places of origin of stones. This literature, which classifies, names and describes, appears as the beginnings of an emerging mineralogy, with a rationality based on demonstrating cosmological unity, since heavens, stones and living bodies are in relation to one another, and on a search for the forces of nature at work in the sublunary world.

2.1.2. *The uses of stones*

These works show the clear interest in stones, which are listed and described with their colors and their properties, including their effects especially on human beings. The same is true for medicine, whose recipe books attest to knowledge about stones, integrated into a system of humors and qualities. The place of metals and mineral substances is indeed essential in medieval pharmacology (Bénézet 1999; Ausécache 2012), and the School of Salerno (12th century) contributed greatly to their use, by establishing practical and theoretical rules for therapy, according to the precepts of Dioscorides (first century) and Galen (Opsomer 1980). Thus, Gilles de Corbeil cites about 20 mineral substances (salt, salt water, rock salt, sal ammoniac, Armenian bole, sulfur, lead, ceruse, cadmia (zinc oxide), tutty, gold and nitre in particular), which form part of the ingredients in medieval preparations (collyria, powders and ointments, for example). These simples are classified according to their qualities of temperature and humidity and, according to the theory of contraries, must allow a qualitative balance to be restored in the patient. But the texts also draw on lapidaries, integrating another system into medical theory, to the point that one sometimes speaks of pharmacological-scientific lapidaries (Cipriani 2017), where one finds both classifications based on physical principles and astrological-magical developments alongside medicinal properties: the occult virtues of stones also play a role in amulets. This mineral presence from the 12th century increased throughout the Middle Ages. Thus, the *Hortus sanitatis*, a work by Johannes de Cuba, first written in German in 1470, then translated into Latin and other vernacular languages and regularly printed until the mid-16th century, contains a section on stones with 144 chapters, naming them with their medicinal properties. Medical writings do not describe precisely, but rely on knowledge developed in other works, of Arab-Latin origin or written in the Latin West, notably Albert the Great's *De mineralibus*. The porosity between fields is therefore clear, both in the definition and naming of stones and minerals, and in the properties attributed to stones.

These circulations among texts are evident in a work attributed to Albert the Great, the *Liber aggregationis, seu Liber secretorum de virtutibus herbarum* and *lapidum et animalium*. Probably written in the context of Cologne and during the lifetime of Albert the Great, and widely circulated until the 16th century (Draelants 2007), it offers a section on the medicinal, astrological and alchemical properties of stones, drawing on Albert the Great, hermetic texts, Theophrastus and Dioscorides. It is, according to the formulation of I. Draelants, inspired by Thorndike (1958, 1960), experimental literature, meaning descriptions of phenomena based on personal and sensory experience, and not only borrowed from book tradition to indicate effects. This book testifies to an attempt to synthesize all knowledge about stones under the authority of Albert the Great, a major scholar in the scholastic study of minerals and stones. As in medicine, or astrology, or magic, this is not a theoretical explanation, but a writing intended to account for the effective action of stones.

While medicine or this text uses stones and minerals as ingredients, alchemy is also a discipline that combines theoretical and practical knowledge of their qualities and virtues. It is based notably on the mercury-sulfur theory, which is an Arabic reading of Aristotle's *Meteorologica* (Halleux 1974; Calvet 2018). A distinction is made between steel and iron, especially following the translations of Arabic texts (Mandosio 2011). The development of alchemy in the medieval West is primarily due to the transmission of Arabic texts from the 12th and 13th centuries within a significant literature, although strongly contested in the 14th and 15th centuries (Calvet 2018), but with a focus on alchemical operations rather than on minerals and stones.

2.1.3. *Stones as materials*

The importance of writings as witnesses of medieval knowledge can lead to neglecting forms of knowledge that are less easily documented and yet materially obvious: the many monuments inherited from the Middle Ages, cathedrals and churches, castles and fortresses, bear witness to precise knowledge about stones as building materials. It is archival documents that record both local or more general names for stones and the techniques that vary according to the nature of the stone, as well as some texts on quarrying, such as those concerning Paris in 1332 and in 1500 (Benoit et al. 2000). Thus, materials such as *pierre verte* or *pierre marchande* (a form of roofing slate), *pierre de grain* or *gren/grion* (fine-grained sandstone), *groais* (a soft, coarse sandstone used as rubble), *pierre à maçonnage* (general masonry stone), *pierres de cahot* (red schist), *caillou à paver* (paving cobbles) and shell lime (*chaux de coquilles*) obtained from *quarries* in Morbihan are recorded in 14th- and 15th-century archives from Armorica (Le Guay 1985); similarly, dressed building stones for the abbey of Saint-Denis are noted (including *pierre dure de ferrot* (a dense hard limestone) and *pierre de liais* (liais limestone), as well as gypsum plaster (*plâtre*) sourced either from the Ludian marl levels or from dedicated gypsum quarries (*plâtrières*) (Billot 1985). The comparison of archives, particularly construction accounts, with archaeological studies (Viré 1983) makes it possible to assess the knowledge of resources near a building site or farther away and the ability to make use of the properties of stones. However, for stones and soils as for other natural elements, one can only observe the separation between the knowledge of builders (architects, stone merchants, sculptors and stonecutters, and masons) and that transmitted through texts, a situation that continues well after the Middle Ages. The nomenclatures are also divergent between craftsmen and scholars in schools and universities: those found in the archives often correspond to precise uses and local distinctions, in the vernacular or in Latin.

Writings on stones, which are mostly bookish and scholarly in inspiration, show the diversity of medieval knowledge at the turn of the 12th and 13th centuries. From the moment universities developed and knowledge of the Aristotelian corpus spread, scholasticism contributed to the diffusion of the Aristotelian method for explaining *res naturae*: the system of Aristotelian causes, the preference for classifications based on colors or on a qualitative approach (hot, cold, humid, dry), the inquiry into the evolution and change of matter and the reasons for such changes transform the discourse on stones and soils, which fit into a general process, but also stand out by more permanent traits. A naturalistic approach to stones and soils emerges, in which the object of study has value in itself and not for its uses or for some other form of knowledge: the formation of soil and its components, the place of the Earth within a global system which is that of Aristotelian physics and questions about stony or sandy soils, volcanoes, mountains or the presence of fossils – questions that lapidaries focused on describing properties do not address – form a distinct body of inquiry, linked to the diffusion of Aristotle's *Meteorologica* and to attention given to phenomena in the sublunary zone and their explanations. It is in these texts that one finds a rational approach based on Aristotelian etiology, which, as such, belongs to the *philosophia naturalis* of the Middle Ages.

2.2. Sources and authors of the medieval West

2.2.1. *Sources*

The mention of authoritative names reflects the duration and diversity of references, with new elements and original features showing the emergence of domains, methods or questions that did not previously exist. Medieval science, as it appears in medieval scholasticism, is indeed above all founded on commented and discussed texts, and it is striking in the quantity of sources cited, translated and in the boldness with which they may be used. The Middle Ages in fact confront theories and descriptions from different traditions in order to construct knowledge made from an accumulation of repeated, but also transformed, information.

Three types of sources are primarily involved in the Middle Ages: ancient Greco-Latin sources, the texts of the Church Fathers and works of Arabic origin. Certain names recur and mark the limits of fields depending on whether they are used or not. Such is the case with Isidore of Seville, used and cited as a reference authority, and always present in encyclopedic compilations, but gradually abandoned by scholasticism in favor of writings of Aristotelian and Arabic inspiration or medieval treatises. Likewise, Pliny the Elder, Solinus, Damigeron-Evax and Dioscorides are the most commonly cited ancient authors, Dioscorides being more present in medically oriented texts and the Latin lapidary of Damigeron-Evax[1] in works of magic. Beyond descriptions of stones, one must also not forget Seneca, who was read throughout the Middle Ages for more specifically geological aspects, such as the formation of mountains, rivers and seas.

1 It is a composite collection to which are attached the names of the mage Damigeron and of a fictitious king of the Arabs, Evax, author of a letter supposedly addressed to Tiberius; it is the Latin version (5th–6th century) of a Greek original.

If the Church Fathers are cited in lapidaries and encyclopedias, they are much less present in scholarly works from the 13th century onward, where natural philosophy relies rather on philosophical sources, especially Greek and Arabic. The book of a single author is indeed central for the medieval university: Aristotle, especially in the *Meteorologica*, where he presented in many chapters theories on the evolution of the Earth, the interior of the Earth and earthquakes, as well as a brief exposition on the formation of stones and metals at the end of Book III. Knowledge of this treatise, however, comes through Latin versions in the West: Gerard of Cremona translated the first three books in the 12th century from Arabic into Latin, Henry Aristippus, also in the 12th century, provided the translation of the fourth book from Greek into Latin, and William of Moerbeke later wrote a Latin version of the four books based on Greek. This treatise also gave rise to numerous commentaries and collections of questions.

Other works, often later, also circulated under the name of Aristotle, particularly the *De congelatione et conglutinatione lapidum*, a Latin translation around 1200 by Alfred of Sareshel of part of Avicenna's *Kitāb al-Shifā'*, which proposes explanations of the evolution of soils and the formation of rocks. Medieval Western Aristotelianism is therefore not strictly a repetition of the Greek philosopher, but a reinterpretation and set of interpretations based on the fundamental concepts of Aristotelian physics (causality and the matter of the universe, the sublunary world subject to generation and corruption of elements, the action of elemental qualities and motion, etc.).

Besides Avicenna, works of Arabic origin, translated beginning in the late 12th century, have an important place in Western writings. This is the case with the *Lapidary* of King Alfonso X, written in Castilian between 1244 and 1245 by the Jewish scholar Yehuda ben Moshe and the cleric Garci Perez: they compiled and translated texts of Arabic origin, notably those of Muhammad Ibn Mansur (12th century), and present the whole as a synthesis of Aristotle and a Chaldean Abolays. The magical and astrological tendency is strongly inspired by Arab-Latin treatises. But scholastic study of the evolution of soils and natural phenomena on and within the Earth (earthquakes, volcanoes and the sources of rivers, for example) is also based on reading Arab-Latin sources such as Messahala (740–815), Avicenna and Averroes. Medicine also makes broad use of Arabic knowledge concerning the medicinal virtues of stones (Ausécache 2012) and this already in the 12th century with the works of Constantine the African (*Pantegni* and *Ysagogé*), which circulate in Latin in particular a compilation of authors such as Rhazès (865–925). The *Canon* of Avicenna, the treatises of Albucasis (973–1048) and the *Liber agregatus de medicinis* attributed to Serapion further reinforce the importance of this heritage in pharmacopoeia.

With regard to knowledge of soils, the main source for the medieval West is Palladius's *Opus agriculturae* (or *De re rustica*), because it is the agronomic treatise most frequently recopied throughout the Middle Ages in its complete version, whereas the earlier ancient agronomists circulated rather in the form of excerpts. Book I of the *Opus agriculturae* contains in particular considerations on soil quality. Palladius cited his predecessors, which constitutes another means of knowing them for a medieval Western reader. The complete text of Columella only truly becomes accessible again in the 15th century. It is at the end of the Middle Ages, toward the 14th and 15th centuries, that Western authors recover easier access to the other ancient agronomists Cato and Varro. Printing confirms the diffusion of these ancient sources with the *editio princeps* of the *Scriptores Rei Rusticae*, in Venice, in 1472, a work that brings together the treatises of Palladius, Cato, Varro, Columella and Gargilius Martialis (the latter being especially interested in arboriculture and horticulture).

The writings of the medieval West, in which authorities appear as the foundation of thought, nonetheless distinguish themselves, both in their forms and in their content, from earlier or later periods. If one cannot speak of canonical forms, any more than in medicine, certain types of writings are favored and correspond to modes of reflection and discursive aims that differ according to the field.

2.2.2. *Forms and genres of writings on the Earth*

The difficulty of distinguishing a genre of writing in the Middle Ages can be measured with lapidaries, whose common criterion, as the name itself indicates, is the description of stones rather than a specific use, an order of stones, or even the format of the entry. If a formal definition can be made, it would no doubt be around classification, alphabetical or by colors and shapes of stones. One must also certainly take into account the intended audience and the professional milieu in which the writing is situated: thus, adaptations in vernacular languages, often condensed and sometimes in verse, correspond to needs different from those of Latin texts. One may therefore think that it is indeed the milieu in which the writing was produced and circulated that ultimately conditions the form and content of the works.

Thus, university books, which draw in particular on translations of Aristotelian inspiration, correspond mostly to the rhetorical forms of the university (commentaries, questions, expositions), whereas the classified enumerations of lapidaries are of more didactic use for a broader audience, and encyclopedic summae, which integrate a discourse on stone within a work treating the universe or the Earth in a condensed and synthetic form, correspond to the numerous tools developed for students, preaching or other audiences seeking knowledge that is quickly and easily accessible thanks to indexes, chapter headings and other means of reference. One may consider that at the end of the 12th century and especially in the 13th century, a clear distinction takes shape between scholarly literature, in which theories and hypotheses are discussed from authorities, and secondary, above all didactic literature, such as encyclopedic summae or lapidaries, which order knowledge for broader transmission or even practical uses. It is thus that the entry on stones in the encyclopedia of Bartholomew the Englishman is presented by successive evocation of the traits or properties that define them: number and nature of the elementary components (water, fire or earth), appearance and colors, hardness or softness, mode of formation and possible virtues. The whole also reveals, alongside alphabetical order, a hierarchy: common earth, specific earth (clay, orpiment, bitumen, clod of earth, powder, sulfur, terra sigillata, sand, glass, stone, precious stone, and, at the summit, metal; see Corbechon 2021). The intersection between the sources employed and the textual structure of the entry or the book thus reflects the use the author intends for his work.

The books in which the knowledge of soils is discussed are also diverse in form: encyclopedic entries arranged alphabetically, as mentioned above with Bartholomew the Englishman who discusses various types of earth in Book XVI alongside the different stones, or expository works that develop a line of thought according to their own logic, as in Book II of Pietro de' Crescenzi's *Liber ruralium commodorum*, which addresses the question of the nature of soils.

However, university rhetoric, the form of the lapidary or encyclopedic writing do not exhaust the genres of writings on the Earth, and some authors avoid them by producing an original work both in writing and content.

2.2.3. *Notable authors*

Unlike contemporary science, one cannot speak of specialists of the Earth, but rather of authors who offer long and more in-depth developments within the epistemological and intellectual framework that is theirs. Thus, it is not surprising to see cited Hildegarde of Bingen (1089-1179), known for her mystical visions, but also for a scientific œuvre of importance (Hildegarde de Bingen 2003, 2010). The developments on the Earth are essentially present in two very different works: Book II of the *Physica* or *Liber subtilitatum*, where in particular there is a classification of earths according to their colors (white, black, red and greenish), their qualities (cold, hot, dry or moist) and especially their arboricultural and agricultural products; the *Cause et Cure* or *Liber compositae medicinae*, which centers on the human being and on medicine, but which also devotes an important number of chapters to agronomy and agriculture as activities beneficial to humans. One must add finally the fourth part of the *Physica*, the lapidary, where 26 stones are described both for their physical properties and their medicinal virtues. Hildegard of Bingen appears there both as theologian and naturalist (Moulinier-Brogi 2010, p. 230), clearly using various sources, but also giving an important place to local observations that fit within the general framework of Genesis and Christian cosmology.

The work of the Dominican Albert the Great (1193–1280) testifies to the transformation of medieval thought through Arab-Latin writings and an approach to nature founded above all on an etiology that arises both from a syncretism of sources and from observation. It is known that the writing of the *De mineralibus* was accompanied by a visit to mines in Saxony. This treatise is therefore quite different from lapidaries and proposes a genuine reflection on the properties of soils and mineral substances, as well as on their place and mode of formation. The interest that Albert the Great showed there was also found in Book III (treatise 5) of the *De meteoris*, where he treated the generation of minerals and rocks in a more general framework, as well as in three other works, the *De natura loci*, the *De causis proprietatum elementorum* and the *De vegetabilibus*: his study, no doubt the most extensive of all those carried out in the West, even if it was spread over several works, therefore concerned the nature of soils and minerals, the surface and the interior of the Earth, mountains, deserts, sandy or humid places, volcanic and seismic phenomena, as well as the relationships between earth and plants, and also the influence of places on human life. The *De mineralibus* was written shortly before the *De meteoris* (Riddle and Muhlolland 1984, pp. 228–229), and the Dominican provides an overarching framework for reflection in the latter treatise. Indeed, the *De meteoris* deals with broader questions, and the author clearly announced that one should then read his more specific works on stones, plants and animals (Albert the Great 2003, p. 3).

This arrangement shows the breadth of Albert the Great's thought and, a remarkable fact, while the medieval period struggles to define a geology as such, the Dominican uses expressions that demonstrate his perception of a science in its own right: "scientia lapidum et mineralium" and "scientia mineralium" (Albert the Great 2003, pp. 207–208). The existence of a separate work, the *De mineralibus*, in itself signals this distinction of a specific science, and Albert the Great notes moreover at the beginning of his treatise that Aristotle had not devoted a treatise to rocks, minerals and metals (Albert the Great 1890, book I, p. 1). The principles of Aristotelian physics through the lens of those of Arab thinkers constitute the framework on which he constructs a general system based on the elementary qualities, their variations and combinations, a system that is found across the various treatises that respond to one another. Attention to places and to specific variations of temperature and humidity is fundamental and allows him to account for local or seasonal modifications by regarding them as accidents in relation to the general system. The breadth of his reflection makes him a major author for medieval thought on the Earth.

Restoro d'Arezzo, for his part, is known only through one work which is original for the 13th century in being written in the vernacular (the Italian dialect of Arezzo), *La composizione del mondo colle sue cascioni* (1282). It is a cosmography, based on Aristotle and on Arab authors such as Avicenna, Abu-Mashar and Alfaragan, and showing knowledge of the writings of Albert the Great, while also relying on personal observations, for example, of an earthquake in 1279. Nothing is known of his life, but it is supposed that he was also a Dominican. The work is composed of two books, the first describing the general organization of the world, including the Earth. But it is the second that offers the most interesting developments on the Earth, notably on the generation and disappearance of mountains and rivers, the action of waters, the movements of the terrestrial globe and especially the influence of stars and planets on the Earth. Although not widely distributed, this work, by an author who remains mysterious, nevertheless bears witness to the general interest in the Earth, its formation and its place in the universe, and to an attention to geological and seismic phenomena and to stones beginning in the 13th century.

It is therefore not surprising that, notably in commentaries on Aristotle's *Meteorologica*, theories were discussed and hypotheses formed particularly on the movements of the Earth or the formation of mountains, some of which are still to be discovered in the mass of unpublished and little-known scholastic writings. Certain scholars took a particular interest in these subjects, such as Jean Buridan (ca. 1300–after 1358): his *Questions* on the *Meteorologica*, which bear witness to his teaching at the Faculty of Arts of Paris, show this, for they provide the occasion for him to question notably the formation of mountains, the presence of cavities filled with water in the Earth, or the consequences of erosion and the natural displacement of lands on the center of gravity of the Earth. Unlike other authors, he does not mix considerations on Genesis and the Flood with scholastic reflection, and his argumentation rests both on the examination of authorities and on logic supported by precise observations. His book was widely cited, notably by Albert of Saxony (1316–1390) and Nicole Oresme (ca. 1320–1380).

At the turn of the 13th and 14th centuries, Pietro de' Crescenzi, born in Bologna in 1230 or 1233 and died in 1320 or 1321, is notable for his attention to soils in his agronomic treatise, which appears as the major study for the medieval period because of its breadth. The author was a magistrate, and thus had studied law, but also medicine and natural sciences, if we are to believe the presentation of his prologue. Written in Latin between 1304 and 1309, the text is dedicated to Charles II of Anjou, king of Sicily, and to the general of the Dominican Order, Aimeri Giliani. Among the sources of Pietro de' Crescenzi figure the ancient agronomists and especially the medieval authority of Albert the Great, whom he drew on extensively. However, these references are confronted, sometimes to emphasize disagreements, discussed and notably tested through experience, a quite remarkable approach during the medieval period. The author also based himself on his observations during his various travels throughout Italy. Moreover, he owned a farm near his native city, and it is there, at the end of his life, that he undertook his study, which is therefore not only a compilation of authorities, but responds to a need for practical usefulness.

This treatise was read across much of Europe until the end of the Middle Ages and even into the 16th century. We currently have about 100 manuscripts of the Latin text preserved in various European countries. It was also printed several times and benefited from various translations into vernacular languages, including one into French commissioned by King Charles V in 1373. Medieval readers indeed approached this treatise from a general theoretical perspective, considering that it offered general knowledge, their goal being absolutely not, especially for readers outside Italy, to learn about the particularities of soils and agriculture in a specific region: the success of the work shows that it was perceived as delivering knowledge valid throughout Western Europe.

However, one should not think that only a few scholars dealt with the formation of the Earth or its components. Other authors, university-trained or not, took an interest in these questions and left writings that show curiosity, even reflection. Thus, Dante wrote a question on the balance of earth and water in 1320 and sought to demonstrate the emergence of land above the waters due to the influence of the stars, and the *Divine Comedy* incorporates this conception. Petrarch was one of the first to describe Mont Ventoux; Antoine de la Sale, in the *Salade*, described around 1442 the Lipari Islands and the cavernous interior of the Earth. As for Leonardo da Vinci's *Carnets*, they also contain developments on the structure of the Earth, fossils and the movement of water on land. While geology did not exist as a discipline, the Earth was the object of questioning, both regarding its formation and its structure, and regarding the phenomena that occur there (volcanoes, springs, earthquakes, caves, tides, floods and tidal waves, fossils, etc.).

2.3. Medieval debates on the Earth

2.3.1. *The formation and evolution of the Earth*

In medieval physics, earth is first one of the elements, cold and dry by nature, whose place, spherical, is located at the center of the world. But this assertion, repeated regularly, does not correspond to what can be observed and it is from this discrepancy that all medieval questions arise. William of Conches (ca. 1090–after 1154) thus discussed at length in the *Philosophia mundi* (William of Conches 1980) the difference between the element and the elemented, the earth-element being a simple and minimal particle, perceptible by the intellect, but not visible, unlike the elemented earth which, for its part, is visible, but may have by accident other qualities.

This distinction is central in medieval physics and makes it possible to think about the evolution of the Earth and its diversity within the framework of the Aristotelian concepts of matter, form and accidents, as well as of generation and corruption. This analysis, inherited from an ancient philosopher, is also situated within a Christian framework, in particular with two biblical episodes that are central to the conception of the Earth: Genesis and the Flood. It is indeed compatible with Genesis, since the four elements are created from the original chaos, but also with the Flood, which justifies an evolution in the distribution between water and earth. In fact, a syncretism is established between the Christian description of the Flood and Aristotle's assertion in the *Meteorologica* (I, 14, 351a) according to which the moist regions of the Earth can become dry and vice versa. If the earth-element remains identical, the elemented earth evolves and changes, which makes it possible to debate erosion, the influence of the seasons on the Earth, or the geological transformations attested by fossils. However, medieval scholars refused to follow Aristotle when he affirmed the eternity of these cycles, which implied that the universe was eternal, something contrary to the Christian faith and condemned several times. Only Jean Buridan discussed this point, paradoxically contesting the eternity of the world in the name of the Aristotelian principle of generation and corruption: what is born must therefore disappear.

It is therefore not surprising to see the discussion on the Earth and its existence arise in commentaries on the Apocalypse (Boureau 2014), the third biblical episode central to understanding the historicity of the Earth: the question is indeed whether the elements remain after the Apocalypse. It is Thomas Aquinas who, in 1254, lays the foundations for an eschatological physics, not in his commentaries on Aristotle, but in that on the *Sentences*, by affirming the persistence of the elements in their substance and their own qualities after the purgation by the burning of fire. Medieval debates are thus constructed at the boundary of exegesis, Aristotelian commentaries and Arabic contributions, to which are added increasingly frequent observations beginning in the 13th century.

2.3.1.1. *Erosion and the formation of mountains*

The evolution of the Earth is above all measured by its consequences on plants and humans and on three places privileged in studies at the boundary of geography, meteorology and botany: mountains, deserts and humid areas. However, the mountain and its genesis are the object of more in-depth reflection, particularly in two works of Albert the Great, the *De causis proprietatum elementorum* and the *De mineralibus*, where the mountain appears both as a place of stony formation and as the result of a generation. In the *De mineralibus*, the mountain is a favorable place for the generation of rocks, as are the banks of certain rivers that retain vapors that cannot rise. In the *De causis proprietatum elementorum*, it is the very generation of mountains that is the subject of Albert the Great's developments. Everything begins with the *De congelatione et conglutinatione* of Avicenna, where an entire chapter is devoted to orogenesis.

This text gives three explanations for the formation of mountains: one by earthquake that raises the earth (essential cause), another, accidental, by erosion caused by wind or a watercourse that wears away soft earth and leaves hard earth protruding, and the third, also accidental, by a viscous deposit left by watercourses. The Western world favors erosion by water as the origin of mountains, as is shown by the encyclopedias of the 13th century (Ducos 2000, pp. 28–34). Albert the Great, for his part, considered that orogenesis came essentially from earthquakes, and not from erosion by water and wind of a spherical and originally uniform Earth. He also affirmed the importance of accumulation as an accidental cause, the mountain being a favorable place for the generation of rocks. The wind, for its part, can form hills, but only if the earth is sandy, as in the case of dunes. Thomas Aquinas, in his commentary on the Book of Job (1261–1265), a few years after the spectacular collapse of Mount Granier (1248), considered that mountains may move, collapse or form through a process of erosion within the natural framework of generation and corruption ordered by the higher intelligence that is God (Boureau 2014, p. 85). A century later, Jean Buridan took up the discussion again, but considered that the process by accumulation or by erosion could only occur in the case of small mountains, the principal cause being an earthquake of great violence.

These reflections remain within the Aristotelian and scholastic framework. By contrast, Restoro d'Arezzo, apparently less constrained by the scholastic framework, also appealed to astronomy, affirming the influence of the stars on the Earth, in particular on its habitable parts that emerged in the north thanks to the particularly large number of stars and constellations in the northern region of the sky, which caused a greater drying. The most distant stars produce mountains, the element earth being drawn toward them, while the closest stars produce valleys. The intervention of the stars and heavenly bodies on the substance of the Earth is not new: indeed, William of Conches recalled a theory already present in Seneca according to which the proximity or distance of the stars caused variations in humidity on Earth: the farther they were, the more humid the lands were; the closer they were, the more they burned. The novelty of Restoro d'Arezzo is to consider that they also influence the relief. Several accidental causes also transform the relief of the Earth: it is especially water that, when flowing downward, carries away earth and stones and destroys the mountain, hollows the valleys and transports the materials elsewhere for other formations. It is the action of water which, according to him, profoundly transformed the surface of the Earth during the Flood. The author thus combines Aristotelian and Avicennian thought, the contested theory of direct celestial influence on natural objects and the biblical text, but also observations, notably on the soil of mountains or on an earthquake felt in 1265.

2.3.1.2. *Fossils and the generation of rocks*

The presence of fossils in or on the Earth and the importance of rocks and rocks in the mountain lead to reflections on their origin and their causes. While Aristotle, at the end of Book III of the *Meteorologica*, concluded with the generation of rocks, minerals and metals by the same principle of double terrestrial exhalation as meteorological phenomena, and Avicenna specified the processes of formation, it was Albert the Great who formalized the idea that the mountain, rocky, was a place of mineralization and generation of rocks, notably in the *De causis proprietatum* and in the *De mineralibus*: the surface of the mountain is originally a sticky and viscous mud brought by water. The combined heat of the terrestrial exhalation and the sun's rays dries this substance and makes it solid to form large stones. The extreme cold of very high mountains also acts and produces crystal.

The existence of fossils likewise raises many questions, which echo what ancient texts said, notably Pliny the Elder, but also Ovid in Book XV of the *Metamorphoses*, where they appear as the sign of lands once submerged. The Church Fathers, for their part, make them material proof of the Flood. Observations of fossils, notably in Paris by Albert the Great or by Restoro d'Arezzo, on mountains contribute to the interest in these stones containing the image of animals. According to Albert the Great, it is extreme heat or extreme cold, especially on mountains, that has dried and petrified the substance of animals while preserving their forms. According to Restoro d'Arezzo, the presence of these fossils at the top of a high mountain proves the former circulation of the water of the Flood. However, it should be noted that fossils, stones whose origin is thus explained by the laws of natural physics, are also sometimes assimilated to astrological images and associated with the occult virtues of the stars (Weill-Parot 2002).

2.3.1.3. *Earthquakes*

While water and ambient temperature contribute to modifications of the Earth, to its relief, and even to its movements, earthquakes are also the object of reflections in scholastic commentaries on the *Meteorologica*, Aristotle's book proposing an explanation of earthquakes by the principle of double terrestrial exhalation, accumulated inside the Earth and causing earthquakes when it attempts to leave the Earth. Albert the Great discussed this at length in the *Libri Meteororum* and considered that it was above all the dry exhalation, which was also the matter of wind often accompanying earthquakes, that is their origin. He finally proposed a classification of nine earthquakes, undoubtedly following Seneca who distinguished between *succussio* ("jolt"), *inclinatio* ("tilting") and *tremor terrae* ("trembling of the earth"). It is striking for the precision of its analysis, which is based on three criteria: the type of movement, the effect, and the matter. Thus, one goes from the weakest and least destructive movement, with little matter, the *motus agitativus*, to the strongest and most destructive, the *motus depressionis et elevationis*, passing through the *perforativus motus* or the *ruinosus motus* (Ducos 2000, pp. 38–39).

The Aristotelian framework allowed Albert the Great to consider all possible variations that allow the description of earthquakes. However, this nomenclature, however appealing, remains unused outside the commentaries. The places favorable to earthquakes are places where cavities exist and especially where the earth is hard, that is, composed of stone, because soft and muddy earth (*limosa* et *glutinosa*) allows vapor to escape. These are mountains, on the one hand, because they naturally generate stones, according to the repeated assertion in the *De mineralibus* and the *De proprietatum elementorum*, and, on the other hand, because they are hollow.

2.3.1.4. *Volcanoes*

Among the phenomena directly associated with earthquakes, volcanic activity is important, even if it is sometimes poorly distinguished in scholastic commentaries from mountains, following Aristotle who spoke little of them. Albert the Great suggested this association by indicating that the *motus depressionis et elevationis* produces an elevation of the relief with a strong exhalation mixed with dust, fire, sulfur and water. It is, however, in the *De causis proprietatum elementorum* that he devoted a significant development to it, where one sees that the efficient cause is identical to that of earthquakes, the hot vapor enclosed in the Earth. It cannot exit because the pores of the Earth are blocked by seawater, the volcanoes being necessarily contiguous with the sea. This last remark doubtless comes from the abundant literature on Mediterranean volcanoes, notably Etna and Vesuvius. Albert the Great also indicated the material cause, a sulfurous earth mixed with a greasy naphtha. The stones produced by volcanoes are light and porous, because they contain much naphtha before burning. The Aristotelian framework thus makes it possible to go beyond the infernal and catastrophic vision of the volcano to make it an explicable terrestrial phenomenon.

2.3.2. *The interior of the Earth*

The various Latin names for earth in Western medieval texts show different approaches. Thus, the earth is called *humus* because it is moist (Isidore de Séville 2011b, beginning of book XIV), which implies a view of the interior of the Earth and raises the question of the circulation of water. Far from writing, it is also in practice that people in the Middle Ages had knowledge of the depths of the earth due to quarrying and mining. But experience is given little consideration by the learned scholars of the medieval period, who inquire into the interior of the Earth mainly on the basis of written sources.

2.3.2.1. *Mines: the formation of minerals and metals*

In the 13th century, in his *De proprietatibus rerum*, in Book XIV, the encyclopedist Bartholomew the Englishman stated that noble metals are found in mountains (Bartholomew the Englishman 1601, XIV, 2, p. 507). The author then listed mountains nearly all of which appear in the Bible and which he presented as rich in precious metals, particularly in Israel (Bartholomew the Englishman 1601, XIV, 20, p. 605). The biblical source draws on Deuteronomy, where it is written that the Promised Land is a "land whose stones are iron" and that copper is mined from its mountains. The Franciscan surveyed various metals (gold, silver, tin, iron and lead), but did not dwell on their formation, even if he did specify the origin of white lead: the islands of the Atlantic, Gaul and Lusitania (Bartholomew the Englishman 1601, p. 755).

Also in the 13th century, in the *Speculum naturale*, the Dominican Vincent of Beauvais provided details on the silver mines of Spain, recalling the name of Hannibal as the first to have organized the exploitation of this metal in the Iberian Peninsula (Vincent de Beauvais 1624, chap. 17, col. 434–435). He spoke of the search for metals in two other passages: "De metallorum investigatione" (ch. 5) and "De quibusdam modis aurum investigandi" (chs. 9–10). The encyclopedist drew heavily on Book XXXIII of the *Natural History* of Pliny the Elder (Vincent de Beauvais 1624, chap. 10). The ancient view suited the Dominican, who cited an authority. But another ancient source lies at the origin of the explanation of the formation of metals.

At the end of Book III of the *Meteorologica*, Aristotle proposed a theory of the genesis of metals and minerals that medieval scholars took up and made their own. The Stagirite posited a double exhalation inside the Earth: a moist exhalation produces metals, while a dry exhalation produces non-metallic rocks and other mineral substances (in Greek *orukta*). The treatise contains a fourth book that may not be by Aristotle, but which the medievals attributed to him and which deals with the formation of bodily substances, including stones and metals. This ancient thought is continued in Arabic treatises. Avicenna addresses the genesis of metals in the *Kitāb al-Shifā'* (*Book of Healing*), discussing the theory of the Brethren of Purity (*Kitāb Ikhwān al-Ṣafā'*, 10th century), who claimed that metals were born from sulfur and mercury and who envisaged the transmutation of metals by alchemy. His thought on the origin of metals followed in part the text of the Brethren of Purity: for him, metals came from the meeting of mercury and a sulfurous earth and each had particular proportions of mercury and sulfur that were more or less pure; Avicenna also spoke of the transmutation of metals, but to refute this theory, taking a resolutely anti-alchemical position.

This excerpt of Avicenna is known in the West under the name *De congelatione et conglutinatione lapidum*, which in many manuscripts follows Book III of Aristotle's *Meteorologica*. Vincent of Beauvais notably cited passages from this Book III in his *Speculum naturale* (Vincent de Beauvais 1624, VII, 4, col. 427–428). The encyclopedist also alluded to alchemy and the transmutation of metals (Vincent de Beauvais 1624, VI, col. 427–428). Finally, one notes especially that he adopted the Aristotelian distinction between mineral and metal, since the subject is arranged in different sections: Book VI concerns minerals and Book VII is devoted to metals, while stones are reserved for Book VIII (in the final three-part version, called *trifaria*). As a good compiler gathering the knowledge of his time, the Dominican contributed nothing new scientifically; such was not his role.

It was Albert the Great, however, who proposed a more innovative discourse on minerals, metals and stones. From the first chapter of the *De mineralibus*, the scholar highlighted the novelty of his approach compared to alchemy and to medicine, which sometimes uses mineral substances, and he distinguished himself from authorities such as Pliny the Elder who stopped at description (Albert the Great 1890, book I, tr. 1, chap. 1, p. 2). The Dominican adopted an Aristotelian method based on the four causes: material, formal, efficient, final. His approach was also influenced by Avicenna's *De congelatione et conglutinatione lapidum*. This thought inspired by Aristotelian physics emphasizes material change using the notions of generation, corruption and accident. Albert the Great also took from Aristotle the theory of double terrestrial exhalation that generates metals and minerals inside the Earth.

As in the Stagirite's work, the Dominican distinguished minerals and metals: dry exhalation produces minerals, while moist exhalation is the source of metals. For the latter, the exhaled moisture becomes condensed in the dryness of the earth and becomes solid due to its coldness, which creates metals. Then, depending on whether there is more or less moist or dry vapor, more or less dryness or coldness in a given mineral or metal, he established a classification that may be called scientific and not simply an enumeration of different substances (Riddle and Mulholland 1984). Finally, the originality of the Dominican also came from his precise observations, thanks to visits he was able to make to mines. His descriptions were therefore not based solely on written sources; they were also the result of personal knowledge and analysis, the scholar seeking to give scientific coherence to the whole through Aristotelian theories that made it possible to reflect on the causes and formation of everything found within the Earth. Knowing and exploiting the interior of the Earth does not necessarily require digging deep into it; indeed, medieval people were not unaware of caves, which were sometimes exploited, and they wondered about their existence.

2.3.2.2. *Caverns, caves*

In the 13th century, in his *De proprietatibus rerum*, Bartholomew the Englishman devoted two chapters to caverns under the titles "De spelunca" and "De caverna" (Bartholomew the Englishman 1601, pp. 622–623). However, this latter chapter is limited to a list of venomous animals that take shelter in caverns, so the subject is not geological. The encyclopedist noted that metals and precious stones could be found in caves, which implies that humans also dig intentionally in these places. The passage goes on to the difficulty of lighting and moving forward in these sometimes narrow caves and mentions the need for shoring as well as the risks of collapse (Bartholomew the Englishman 1601, p. 622). The emphasis is placed on danger and not on the explanation of the existence of caves.

In the same 13th century, Albert the Great noted that caverns are found mostly beneath mountains (Albert the Great 2003, p. 71). Reasoning still according to Aristotelian principles, the Dominican questioned the efficient cause of these caves; indeed, earth is heavy, it is a compact matter, and it is cold, which also explains the condensation of moisture to form metals, and all of this seems contrary to the formation of cavities within it (Albert the Great 2003, pp. 73–74). The scholar proposed an explanation by reasoning from marshy lands. When these lands dry because of heat, cracks form. The hot vapor inside the Earth is opposed to its moisture; it causes a drying-out that generates cavities, on the condition that this vapor is contained by a solid place such as hills and mountains, otherwise the vapor escapes from the Earth without creating a cavern. Whether for the formation of metals or of caverns, one understands that the question of water is central, for it is always a matter of moisture within the Earth that will condense or dry out, and medieval science questioned the origin of this water.

2.3.2.3. *The circulation of water underground*

Antiquity already sought to understand the presence of water underground. Aristotle addressed this question in the *Meteorologica* and allowed one to consider a dynamic, since some lands dry out and others become moist in a cosmological cycle. In his *Natural Questions*, Seneca thought that rainwater could not penetrate more than 10 feet into soils, based on what he observed in his vineyards, and he questioned the birth of rivers and the existence of underground streams (Book III, pp. 4–16; Book V, p. 15; Book VI, pp. 7–8). For him, the water of rivers was not due only to rain; it also came from the interior of the earth, and the water of wells showed clearly that there was water in the soil at a depth greater than 10 feet. He posited three levels in the earth: the first was porous because of plant roots that created openings which air or water could fill; the second was drier and harder; finally, the third corresponded to underground water reservoirs. The Middle Ages engaged in dialogue with these ideas inherited from Antiquity.

In the 13th century, Albert the Great started from the Aristotelian principle that water could not be generated outside of its natural place: the seas and the ocean. He resumed the discussion in relation to Seneca and opposed this authority by claiming that rainwater could infiltrate to any depth, which also supplied the interior of the earth with water; like Seneca, he called on his observations, for he said he saw in mines drops of water falling to a depth of 130 paces (Albert the Great 2003, pp. 67–69). Furthermore, the caverns located beneath the earth's elevations serve as water reservoirs, and the mountains and hills absorb water within like sponges (Albert the Great 2003, p. 71). This reasoning allowed the Dominican to explain that many springs originated from a hill or a mountain. Albert the Great therefore admitted that rainwater penetrated deeply into the earth, but for him this was not necessarily the only explanation of the presence of underground water (Albert the Great 2003, p. 68). The scholar then came to consider the circulation of water underground through the example of well diggers. Indeed, he explained that when these people dig a well, they encounter many veins of water, which proves that water circulates within the earth through various channels before reaching the underground reservoirs.

In the 14th century, Jean Buridan thought that there existed a water generated inside the Earth itself, which was neither rainwater nor water coming from the seas and the ocean and which did not come from the upper layers of the earth; according to him, it was this water that was the origin of metals in the third layer of the earth (Jean Buridan 1986, Book I, q. 19, p. 269). Like Albert the Great, he thought that there are large reserves of water underground and he took up the division of the subsoil into several levels due to Seneca, expanding it. In the first are the roots of plants, which receive rainwater. The second is located under ponds and marshes and is drier; it is also the level of cellars that are dug so as not to have too much humidity, but it still contains very fine veins of water. The third contains the waters of wells, which do not come from rain, but from this underground generation, from the air that fills the pores of the earth, and it is the level of mines where metals are formed thanks to this water (Jean Buridan 1986, Book I, q. 19). All medieval reflection on the circulation of water in the earth is notably tied to the question of the penetration of rainwater into the soil, meaning that the science of this period does not neglect the study of the surface of the Earth.

2.3.3. *The surface of the Earth*

Among the various names for the earth in the *Etymologies* of Isidore of Seville, several terms bear witness to the importance of the attention paid to the soil: *tellus*, because "we take away" (*tollimus*) its fruits from the earth; *ops*, because it bears the "wealth" (*opem*) of its productions; *arva*, because it is plowed (*arando*) (Isidore de Séville 2011b, beginning of Book XIV). Throughout the Middle Ages, depending on the authors, the names may vary, but overall, these nouns lead to reflections on the different types of soils and their development. Alongside the writings already mentioned since the beginning of this chapter (the encyclopedias, the treatises of Albert the Great, in particular), one may also mention the work of Hildegard of Bingen, who, as early as the 12th century, shows interest in the nature of soils that give rise to plants, in the *Physica* and in the *Cause et cure* (Moulinier 2010, pp. 207–212). Despite the interest and originality of Hildegard of Bingen, it should nevertheless be recalled that she was not widely read and that *Cause et cure* in particular did not enjoy very great diffusion. Moreover, the nature of the earth is one object of study among others for the abbess; it was not a central subject in her writings.

By contrast, agronomy stands out as a new field contributing to the knowledge of the surface of the Earth and considering this point as truly important. From the 11th to the 14th century, Andalusian agronomists who wrote in Arabic were concerned with this question, but their works did not truly achieve dissemination on a European scale. Their treatises are specialized, taking into account the local features of the soil and climate of Spain; some texts were translated into Latin or Castilian, thus passing from the Muslim world to the Christian sphere, but they were truly read only in the Iberian Peninsula, without benefiting from broader transmission. Therefore, we will leave them aside in order to concentrate instead on the *Liber ruralium commodorum* from the very beginning of the 14th century, written by Pietro de' Crescenzi, whom we presented among the significant authors at the beginning of this chapter.

2.3.3.1. *Analysis of soils: between observation and the theory of the four qualities*

Knowledge of soils appears in the *Liber ruralium commodorum*, especially in Book II and in its summary in Book XI, without it being, however, a fully constructed discourse. Indeed, the chapter titles do not highlight this subject and we find remarks that are rather scattered. Nevertheless, the ability to analyze soils is clearly present. Pietro de' Crescenzi used the term *nature*, and this notion of the nature of the earth allowed him to distinguish types of terrain. The author called upon the senses in order to know the earth. Sight then comes into play with names that refer to colors: he spoke of black soil, for example (Pietro de' Crescenzi 1995, vol. 1, Book I, chap. 8), or red soil, black sand and red sand. The agronomists of Antiquity did the same, especially Palladius, who was a major source for Pietro de' Crescenzi. The latter also took up the ancient concept of veins of the earth, which are more or less fine or wide, more or less resistant, and which allow water to pass through the soil in different ways depending on their characteristics. This approach to terrain implies a perception of soil texture. One observes the size of particles, with stones, pebbles and sands, whose components are the largest, in contrast to clay, which has a very fine structure. One also judges the consistency of the earth, which is either hard, medium or soft.

Pietro de' Crescenzi also drew attention to the moisture of the terrain, knowing that, as we saw above, he also thought of two origins of water in the soil: on the one hand, precipitation water, and, on the other hand, water contained in the earth. Thus, some soils are saturated with water and, at the opposite end of the scale, some soils are very dry. The *Liber ruralium commodorum* is quite precise, because one must take into account the distribution of water at two levels in the earth: at the surface and deeper down. Indeed, a soil may be dry at both levels (Pietro de' Crescenzi 1995, vol. 1, Book II, chap. 16), and all variations are conceivable.

The third mode of analyzing the nature of soils is that of soil temperature, which leads to distinguishing terrain as more or less warm. As with water, the author considered two levels in the soil and sometimes spoke of temperature not at the surface but deeper down.

Besides this approach, the adjectives *warm*, *cold, dry* and *moist* were also used to analyze the nature of soils according to a conception inherited from Antiquity, which proposed to understand the world according to these four qualities. This ancient theory also refers to the notion of the four elements, and Pietro de' Crescenzi recalled that, from this point of view, earth is an element and that it is cold and dry (Pietro de' Crescenzi 1995, vol. 1, Book II, chap. 26 and Book XI, chap. 10). But the aim of the *Liber ruralium commodorum* was not to discuss the elements, so the difference is clearly established between earth as an element and the earth that nourishes plants, which is, for the Bolognese agronomist, an object of study. In fact, what remains is reasoning through the four qualities, which the Italian author sometimes applied in his treatise. For example, clay is warm and dry (Pietro de' Crescenzi 1995, vol. 1, Book II, chap. 17). If one performs all conceivable combinations, one may find soils that are cold and dry, cold and moist, warm and dry, or warm and moist.

This scientific theory inherited from Antiquity was therefore combined in a complex way with observational criteria regarding color, water content and temperature. All these analyses of the nature of soils imply naming rock materials at a time when scientific nomenclature was not yet fixed.

2.3.3.2. *The question of names: birth of a scientific lexicon*

Pietro de' Crescenzi wrote in Latin and inherited a vocabulary in use since Antiquity for designating soils and sedimentary rocks. Concerning the hyperonym, in relation to the variety of Latin names listed by Isidore of Seville to which we have alluded, one finds only *terra*, a polysemous noun. In the French translation of 1373, unsurprisingly, it is *terre* that appears. Even if the word is not specialized, one may remark that its exclusive use moves in the direction of a taxonomy, because this situation differentiates the author from everyday language, which allows a proliferation of synonyms.

If one examines the hyponyms, one notes that the treatise of Pietro de' Crescenzi did not yet have a strict terminology, because synonyms may be found to designate or describe the same rock material. For example, the author uses *arena*, *sabulo* and *sabulum*, but it is not at all easy to establish a clear distinction between the terms in each occurrence, notably according to the size of particles, which are more or less fine. The problem extends into the 1373 French translation, with, in addition, variation on the part of the translator: *sablon* may translate *sabulo* or *sabulum*, but also sometimes *arena*. The latter noun may also be translated as *gravier*. This complexity and this lack of coherence throughout the treatise should not, however, obscure the fact that at the level of a sentence or of a restricted passage one may distinguish a more specific meaning of *sablon* meaning “fine sand”, which is different from *arena* in Latin[2] or *gravier* in Middle French, used to designate a sand with slightly larger particles, with other names in Latin and in Middle French taking on the present-day meaning of “gravel”, for still larger components. These tentative distinctions preside over the birth of a scientific lexicon in French; they are all the more valuable because we do not find them in all types of texts. Indeed, an encyclopedia such as that of Bartholomew the Englishman Anglicus proves less precise, because in the French translation by Jean Corbechon, *sablon* and *arène* are rather treated as synonyms. Compared to scientific encyclopedic works and thus general works, the treatise of Pietro de’ Crescenzi shows greater attention to the naming of soils and rock materials.

Since science was first written in Latin, it inherits the division of concepts in that language. While today we clearly distinguish clay and chalk, Latin Antiquity established a connection, probably due to color: chalk is white, like certain kinds of clay. This shared view explains why *creta* and *argilla* may be synonymous in the sense of “clay”, which we find in the treatise of Pietro de’ Crescenzi, where *argilla* always means “clay, soil greedy for water, impermeable and plastic”, while *creta* has two meanings: (1) “chalk, white rock, porous, soft and friable”; (2) “clayey earth”, synonymous with *argilla*. The same synonymy exists in Middle French between *argile* and *craie*[3]. After the Middle Ages, one had to wait for modern science to define clay and chalk in a way that relegated whiteness to the background in order to bring about a conceptual change that no longer allowed synonymy. If synonymic competition prevents access to a strict scientific terminology, it remains the case that the extremely rich and varied pedological vocabulary shows an interest in the quality of soils and aims at a classification of soils, without science being an end in itself, because the approach is above all utilitarian.

2.3.3.3. *Classification of soils: application to questions of fertility*

Knowledge of the nature of soils goes hand in hand with concern for their quality, since farmers need to determine which earth will be favorable to which crop. The notion of the quality of the terrain is understood here as its possibilities for agricultural use. Theoretical knowledge is linked to practice and results in a classification of soils. In the 13th century, in the *De natura loci*, Albert the Great

2 One notes that this is indeed medieval Latin as used by Pietro de’ Crescenzi and not classical Latin, where one would rather use *arena* for “sand” with fine particles, *sabulo* meaning “coarse sand”.

3 *Dictionary of Medieval Scientific French*, *s.v. craie* and *argile*: see http://dfsm.elan-numerique.fr/.

examined the characteristics of soils according to geographical particularities, taking into account especially relief, temperature and the level of humidity. He stated thus that places to the south by the sea were drier and sandy (Albert the Great 1980, tr. 1, chap. 13, pp. 21–23), whereas the situation to the north produced a cold soil and humidity encourages the formation of marshes. In fact, the basis of the classification was binary, dividing soils beneficial for the growth of plants and soils that are hostile to them. This logic is at work in the *De vegetabilibus* of Albert the Great, which places in the harmful category very cold places such as mountain peaks covered with eternal snows as well as very hot, sandy or salty areas such as deserts (Albert the Great 1867, Book IV, tr. 2, chap. 2, p. 235).

The *Liber ruralium commodorum* of Pietro de' Crescenzi stated in Chapter 26 of Book II that each soil may correspond to a use, that one type of earth may suit one sort of crop and another a different one (Pietro de' Crescenzi 1995, Vol. 1). The author indicated that one could know the quality of the soil by observing the plants that grew there naturally. For example, to choose a field where one will cultivate wheat, it is better to choose a land where there grow in particular thick grasses, clover, wild plum trees and other plants that show the rich character of the soil (Pietro de' Crescenzi 1995, Vol. 1, Book XI, Chap. 10).

However, at the end of this same passage, one realizes that the classification of soils according to their possibilities of use is not so neutral. Indeed, we learn that rich soil is the best, even if it requires more work; after it comes thick and rich soil, which requires still more plowing work, the worst being dry, thick and lean soil (Pietro de' Crescenzi 1995, Vol. 1, Book II, Chap. 26). This classification implies in fact a hierarchy of crops, which is not always exactly the same from one treatise to another. In the case of Pietro de' Crescenzi, when one has read Chapter 16 of Book II, one understands what is at stake: rich fields are suitable for vines, rich and thick soils are good for wheat, lean and dry soil is favorable neither for wheat nor for vines, but for woods, whereas nothing can be obtained from dry, thick and lean soil. The order of interest in crops thus places at the top the soil appropriate for the vine, in second place the soil suitable for wheat cultivation and at the end of the list the soil for forests, which has below it only the soil called "sterile", unsuitable for agricultural profit. The classification does not lead to the presentation of a complete grid listing all types of terrain and indicating each time what will grow best there, but one nevertheless perceives the classificatory approach.

This approach is not entirely univocal and the effects of classification sometimes seem to emerge from a hierarchy of crops in order to adopt a more neutral scientific point of view. Thus, in another passage of Book II, in Chapter 15, the author set out four kinds of field, taking into account more criteria than those we have just mentioned, because he combined the nature and the quality of the soil, its agricultural use and the work it required: the field that can be sown every year, which is fertile and warm and moist; the field planted with trees; the meadow for animals; and finally the new field that must be left fallow for a longer or shorter time, after which it may produce and it is then called new, because it has renewed itself. The first and the last require plowing work, which is not the case for the third, the second requiring only work of the earth just around the trees or vines. The notion of "new field" also expresses a non-static approach to the soil, an awareness of possible changes.

2.3.3.4. *Life, evolution and wearing down of soils: a dynamic approach*

As we saw previously, reading Aristotle's *Meteorologica* leads one to be interested in the succession of dry and wet cycles on earth. The notion of "erosion" was known in the Middle Ages, water being presented as hollowing out rocks and the salinity of the sea constituting the proof of this phenomenon.

In the treatise of Pietro de' Crescenzi, changes in the amount of water in the soil were considered according to the seasons: beginning in the fair season, moist earth would be dried out by the heat of the sun (Pietro de' Crescenzi 1995, Vol. 1, Book II, Chap. 17). The author noted that the appearance of the surface of the field changed: moist soil cracks and presents fissures in summer (Pietro de' Crescenzi 1995, Vol. 1, Book II, Chap. 19). The temperature of the soil is not uniform nor always the same. Pietro de' Crescenzi explained that in summer, with the action of the sun, the surface of the earth became warmer (Pietro de' Crescenzi 1995, Vol. 1, Book II, Chap. 17). In Chapter 9 of Book II, we further learn that the temperature of the soil changes with depth according to the seasons: in winter, the interior of the earth is warm, whereas it is colder in summer.

Pietro de' Crescenzi was aware of the nourishing contribution of plants that stagnated and decomposed little by little on the earth in the absence of agricultural activity (Pietro de' Crescenzi 1995, Vol. 1, Book II, Chap. 6). Agriculture impoverishes the soil because there is harvest, thus removal of plant matter that serves to nourish the seeds, which implies the need to ensure fertilization of the soil by other means, to compensate for what the farmer takes from the earth by disturbing natural evolution.

The soil is perceived as a living organism because it is defined as the belly of plants. It is indeed a dynamic vision that understands the earth as the place of a process equivalent to digestion on the human model (Pietro de' Crescenzi 1995, Vol. 1, Book II, Chap. 13). In the soil, the nourishment of plants is digested and is transformed into two substances: a moist matter that the plant absorbs and a dry residue that remains in the earth.

Phenomena of erosion linked to questions of relief are not ignored either. Pietro de' Crescenzi mentioned the problem of sloping fields, for the soil wore away little by little, carried off by the rains (Pietro de' Crescenzi 1995, Vol. 1, Book II, Chap. 18). Given the utilitarian perspective of an agricultural treatise, the author does not stop at the observation but proposes a practical solution: to hold back the soil with wooden stakes. In such cases, water circulates and collects below, and the field then risks becoming too moist. It is then advised to respond by digging ditches to direct the excess water there.

Changes also occur at the surface of the soil. The latter risks hardening due to the trampling of animals, people and also under the action of precipitation. This is moreover one of the reasons given to explain the usefulness of plowing, which allows the soil to remain loose (de' Crescenzi 1995, Vol. 1, Livre II, Chap. 15). Plowing also had the advantage of compensating for small surface irregularities in order to ensure the distribution of water in the soil, without excessive concentration in a given place that would form a hollow.

For the medieval period, the agricultural treatise of Pietro de' Crescenzi was the landmark work that paid close attention to soils, even if this aspect of knowledge at the frontier of geology was also taken into account by other authors. One must also mention Pietro d'Abano who, at roughly the same time as the agronomist from Bologna, commented on Aristotle's *Problems*. One problem raised the question of the reason for the presence of mountains in certain places, and Pietro d'Abano reported two types of explanation with their authorities, while adding a third line of reasoning: the soil could be compact or sandy; in the latter case, it was easily carried away by rainwater, which produced accumulations. The scholar then referred to the cultivation of fields to support his demonstration: the Lombards and the Flemish dug ditches, which channeled rainwater and prevented it from carrying off soil that was too loose, whereas the French and the English did not do so, hence the establishment among them of differences in relief that were nonexistent in the lands of the former (cited by Duhem (1913, Vol. 9, pp. 283–284)). Since Pietro d'Abano was not an agronomist, it is remarkable to note here that the practices of farmers were invoked to try to understand geological phenomena. Thus, agronomy contributed to the earth sciences.

2.4. Conclusion

Can one speak of a medieval system of the Earth and of soils? To attempt to grasp what the people of the Middle Ages knew about this, one must accept this diversity of knowledge, writings and sources, which makes it both difficult and complex: it was in fact less a medieval "thought" than dispersed reflections, which today's historian must assemble, accepting their inconsistencies, silences, repetitive or innovative developments, in order to understand the logics and the rationality that emerge from them, and the fundamental questions that led medieval scholars to seek, beyond the authorities, to understand the Earth, its components, its evolution and its uses.

2.5. References

2.5.1. *Edited sources*

Albert the Great (1867). *De vegetabilibus*. Jessen, Berlin.

Albert the Great (1890). *De mineralibus*. Viviès, Paris.

Albert the Great (1967). *Book of Minerals*. Clarendon Press, Oxford.

Albert the Great (1980). De natura loci. In *Alberti Magni Opera omnia*. Aschendorff, Münster.

Albert the Great (1995). *Le Monde minéral. Les pierres, "De mineralibus", livres I et II*. Éditions du Cerf, Paris.

Albert the Great (2003). *Meteora*. In *Alberti Magni Opera omnia*. Aschendorff, Münster.

Avicenna (1927). *Avicennae de congelatione et conglutinatione lapidum, being sections of the Kitab al Shifa: The Latin and Arabic Texts*. P. Geuthner, Paris.

Isidore of Seville (2011a). *Etimologias. Libro XVI : De las piedras y de los metales*. Les Belles Lettres, Paris.

Isidore of Seville (2011b). *Étymologies. Livre XIV. De Terra*. Les Belles Lettres, Paris.

Jean Buridan (1986). Les "Questiones super tres libros Metheororum Aristotelis". In *Étude suivie de l'édition du livre I*, Bages, S. (ed.). L'École des chartes, Paris.

William of Conches (1980). *Philosophia*. University of South Africa, Pretoria.

William of Conches (2021). *Philosophia et Dragmaticon*. Les Belles Lettres, Paris.

2.5.2. *Encyclopedias*

Bartholomew the Englishman (1601). *Bartholomei Anglici de genuinis rerum coelestium, terrestrium, et inferarum proprietatibus*. Wolfgang Richter, Frankfurt.

Hildegarde of Bingen (1988). *Le Livre des subtilités des créatures divines (Physique)*. J. Millon, Grenoble.

Hildegarde of Bingen (1994). *Holistic Healing*. Liturgical Press, Collegeville.

Hildegarde of Bingen (1997). *Les Causes et les Remèdes*. J. Millon, Grenoble.

Hildegarde of Bingen (1998). *Physica*. Healing Arts Press, Rochester.

Hildegarde of Bingen (2003). *Cause et cure*. Akademie Verlag, Berlin.

Hildegarde of Bingen (2006). *Causes and Cures*. MedievalMS, Charlotte.

Hildegarde of Bingen (2008). *Physica. Edition der Florentiner Handschrift (Cod. Laur. Ashb. 1323, ca. 1300) im Vergleich mit der Textkonstitution der Patrologia Latina (Migne)*. Georg Olms-Weidmann, Hildesheim.

Hildegarde of Bingen (2010). *Physica: Liber subtilitatum diversarum naturarum creaturarum*. De Gruyter, Berlin.

Jean Corbechon (2021). *Le XVIe livre des pierres, des couleurs et des métaux, traduction du livre XVI du De proprietatibus rerum de Bartholomeux Anglicus*. Honoré Champion, Paris.

Konrad von Megenberg (1475). *Buch der Natur*. Johann Bämler, Augsbourg.

Leonardo da Vinci (2019). *Carnets*. Gallimard, Paris.

Pseudo-Albert the Great (2007). *Liber de virtutibus herbarum, lapidum et animalium (liber aggregationis)*. Sismel, Florence.

Restoro d'Arezzo. (1976). *La composizione del mondo colle sue cascioni*. Accademia della Crusca, Florence.

Restoro d'Arezzo. (2007). *La composizione del mondo*. La Finestra, Lavis.

Thomas of Cantimpré (1973). *Liber de natura rerum, editio princeps secundum codices manuscriptos*. De Gruyter, Berlin.

Thomas of Cantimpré (2014). La Place de Thomas de Cantimpré dans l'encyclopédisme médiéval : les sources du Liber de natura rerum. Tommaso di Cantimpré nell'enciclopedismo medievale : le fonti del Liber de natura rerum Cipriani, M. PhD Thesis, Paris, Florence.

Vincent de Beauvais, V. (1624). *Speculum Quadruplex, Naturale, Doctrinale, Morale, Historiale*. Baltazaris Belleri, Douai.

2.5.2.1. *Lapidaries*

Alfonso X (1997). *The Lapidary of King Alfonso X the Learned*. University Press of the South, New Orleans.

Cannella, A.-F. (2006). *Gemmes, verre coloré, fausses pierres précieuses au Moyen Âge. Le quatrième livre du "Trésorier de philosophie naturelle des pierres précieuses" de Jean d'Outremeuse*. Droz, Geneva.

Halleux, R. and Schamp, J. (1985). *Les Lapidaires grecs (Lapidaire orphique, Kérygmes lapidaires d'Orphée, Socrate et Denys, Lapidaire nautique, Damigéron-Évax)*. Les Belles Lettres, Paris.

Marbode of Rennes (1977). *De lapidibus Considered as a Medical Treatise with Text, Commentary and C.W. King's Translation together with Text and Translation of Mabode's Minor Works on Stones.* Sudhoffs, Wiesbaden.

Marbode of Rennes (2005). *Lapidario – Liber lapidum.* Les Belles Lettres, Paris.

Pannier, L. (1973). *Les Lapidaires français du Moyen Âge des XIIe, XIIIe et XIVe siècles.* Slatkine, Geneva.

Studer, P. (1976). *Anglo-Norman Lapidaries.* Slatkine, Geneva.

2.5.2.2. *Agronomics*

Pietro de' Crescenzi (1995). *Ruralia commoda.* Editiones Heidelbergenses, Heidelberg.

Vigneron, F. (2023). *Le Livre des prouffitz champestres et ruraulx de Pierre de Crescens 1. Introduction et texte (livres I-VIII).* Honoré Champion, Paris.

2.5.3. *General histories of geology*

Accordi, B. (1984). *Storia della geologia.* Zanichelli Editore, Bologna.

Adams, F.D. (1954). *The Birth and Development of the Geological Sciences.* Dover, New York.

Boulaine, J.-L.-G. (1984). *Matériaux pour une histoire de la pédologie.* Institut national agronomique, Paris.

Boulaine, J.-L.-G. (1989). *Histoire des pédologues et de la science des sols.* Institut national agronomique, Paris.

Duffin, C.J., Moody, R.T.J., Gardner-Thorpe, C. (eds) (2013). *A History of Geology and Medicine.* The Geological Society, London.

Duhem, P. (1913). *Le système du monde : histoire des doctrines cosmologiques de Platon à Copernic.* Hermann, Paris.

Ellenberger, F. (1988). *Histoire de la géologie.* Tec & Doc, Paris.

Fritscher, B. and Hendresson, F. (eds) (1998). *Toward a History of Mineralogy, Petrology and Geochemistry.* Institut für Geschichte der Naturwissenchaften, Munich.

Gautier Dalché, P. (ed.) (2013). *La Terre : connaissance, représentations, mesure au Moyen Âge.* Brepols, Turnhout.

Gohau, G. (1987). *Histoire de la géologie.* La Découverte, Paris.

Hölder, H. (1960). *Geologie und Paläontolongie in Texten und ihrer Geschichte.* K. Alber, Munich.

Lindbergh, D.C. and Shank, M.H. (eds) (2013). *The Cambridge History of Sciences 2: Medieval Sciences.* Cambridge University Press, Cambridge.

Mather, K.F. (1967). *A Source Book in Geology*. Harvard University Press, Cambridge.

Olroyd, D.R. (1996). *Thinking about the Earth: A History of Ideas in Geology*. Harvard University Press, Cambridge.

Sarjeant, W.A.S. (1996). *Geologists and the History of Geology: An International Bibliography from the Origins to 1978*. Arno Press, New York.

Thorndike, L. (1958 [1923]). *A History of Magic and Experimental Science*. Columbia University Press, New York.

2.5.4. Studies

Alexandre, P. (1995). Les compilations séismologiques et le prétendu cataclysme provençal de 1227. In *Les catastrophes naturelles dans l'Europe médiévale et moderne*, Bennassar, B. (ed.). Presses Universitaires du Mirail, Toulouse.

Anawati, G.C. (1971). Avicenne et l'alchimie. In *Oriente e Occidente nel Medioevo : Filosofia e scienze. Convegno internazionale 9-15 Aprile 1969*. Accademia nazionale dei Lincei, Rome.

Ariew, R. (1990). Fossils in medieval and early modern geology. In *Knowledge and the Sciences in Medieval Philosophy 3: Proceedings of the Eighth International Congress of Medieval Philosophy, Helsinki, 24-29 août 1987*, Finnish Society for Missiology and Ecumenics, Helsinki.

Ausécache, M. (2012). Le minéral au secours du corps dans la pharmacopée médiévale occidentale (XIIe-XVe siècle). In *Substances minérales et corps animés (1100-1500)*, Jacquart, D. and Weill-Parot, N. (eds). Omniscience, Montreuil.

Bailly-Maître, M.-C. (2002). *L'argent. Du minerai au pouvoir dans la France médiévale*. Picard, Paris.

Bénézet, J.-P. (1999). *Pharmacie et médicament en Méditerranée occidentale (XIIIe–XVIe siècles)*. Honoré Champion, Paris.

Benoit, P. (1988). Les techniques minières en France et dans l'empire aux XVe et XVIe siècles. *Journal des Savants*, 1(1), 75–118.

Benoit, P., Blanc, A., Gély, J.-P., Guini-Skliar, A., Obert, D., Viré, M. (2000). La pierre de Paris. Méthode d'étude de la pierre à bâtir depuis son extraction jusqu'à sa mise en œuvre. *Supplément à la Revue archéologique du centre de la France*, 18(1), 121–158.

Berlioz, J. (1998). *Catastrophes naturelles et calamités au Moyen Âge*. Sismel, Florence.

Billot, C. (1985). Les fournisseurs des matériaux de construction de l'abbaye de Saint-Denis, XVe-XVIe siècle. In *Pierre et métal dans le bâtiment au Moyen Âge*, Chapelot, O. and Benoit, P. (eds). Éditions de l'EHESS, Paris.

Blanc, A. (1983). L'utilisation de la pierre dans les monuments de la France médiévale. In *Mines, carrières et métallurgie dans la France médiévale, actes du colloque de Paris 19, 20, 21 juin, 1980*, Benoit, P. and Braunstein, P. (eds). Éditions du CNRS, Paris.

Boureau, A. (2014). Une nature préservée de la transgression humaine des normes : physique des éléments et géologie du monde dans l'exégèse et la théologie du XIIIe siècle. In *La nature comme source de la morale au Moyen Âge*, van der Lugt, M. (ed.). Sismel, Florence.

Braunstein, P. (2003). *Travail et entreprise au Moyen Âge.* De Boeck, Brussels.

Braunstein, P. and Chapelot, O. (1983). Mines et métallurgie en Bourgogne à la fin du Moyen Âge : première esquisse. In *Mines, carrières et métallurgie dans la France médiévale, actes du colloque de Paris 19, 20, 21 juin, 1980*, Benoit, P. and Braunstein, P. (eds). Éditions du CNRS, Paris.

Cailleaux, D. (1985). Les comptes de construction des Célestins de Sens, 1477-1482. In *Pierre et métal dans le bâtiment au Moyen Âge*, Chapelot, O. and Benoit, P. (eds). Éditions de l'EHESS, Paris.

Calvet, A. (2018). *L'alchimie au Moyen Âge (XIIe-XVe siècles).* Vrin, Paris.

Cipriani, M. (2017). Un aspect de l'encyclopédisme de Thomas de Cantimpré. La section *De lapidibus pretiosis* du *Liber de natura rerum. Médiévales*, 72, 155–174.

Connochie-Bourgne, Ch. (2000). Quelques notes sur l'orogenèse chez les encyclopédistes de langue française au XIIIe siècle. In *La montagne dans le texte médiéval : entre mythe et réalité*, Thomasset, C. and James-Raoul, D. (eds). PUPS, Paris.

Deprez-Masson, M.-C. (2004). L'alchimie dans les encyclopédies du XIIIe siècle : Vincent de Beauvais et ses confrères. In *Encyclopédies médiévales. Discours et savoirs*, Baillaud, B., Gramont, J., de Hüe, D. (eds). PUR, Rennes.

Di Gangi, G. (2001). *L'attività mineraria e metallurgica nelle Alpi Occidentali Italiane nel Medioevo : Piemonte e Valle d'Aosta : fonti scritte e materiali.* Hedges, Oxford.

Draelants, I. (1995). Phénomènes célestes et tremblements de terre au Moyen Âge : enquête sur l'historiographie médiévale dans les limites de la Belgique actuelle (600-1200). In *Les catastrophes naturelles dans l'Europe médiévale et moderne*, Bennassar, B. (ed.). Presses Universitaires du Mirail, Toulouse.

Draelants, I. (2007). *Le Liber de virtutibus herbarum, lapidum et animalium (Liber aggregationis). Un texte à succès attribué à Albert le Grand.* Sismel, Florence.

Draelants, I. (2008). Encyclopédies et lapidaires médiévaux : la durable autorité d'Isidore de Séville et de ses Étymologies. In *La réception d'Isidore de Séville durant le Moyen Âge tardif (XIIe-XVe siècles)*, Elfassi, J. and Ribémont, B. (eds). CRM, Gaudonville [Online]. Available at: http://crm.revues.org.

Draelants, I. (2010). La science encyclopédique des pierres au XIIIe siècle : l'apogée d'une veine minéralogique. In *Aux origines de la géologie de l'Antiquité au Moyen Âge. Actes du Colloque international de la Sorbonne (10-12 mars 2005)*, Thomasset, C., Ducos, J., Chambon, J.-P. (eds). Honoré Champion, Paris.

Ducos, J. (2000). Entre terre, air et eau : la formation des montagnes. In *La montagne dans le texte médiéval : entre mythe et réalité*, Thomasset, C. and James-Raoul, D. (eds). PUPS, Paris.

Ducos, J. (2010a). Albert le Grand et la connaissance des sols. In *Aux origines de la géologie de l'Antiquité au Moyen Âge. Actes du Colloque international de la Sorbonne (10-12 mars 2005)*, Thomasset, C., Ducos, J., Chambon, J.-P. (eds). Honoré Champion, Paris.

Ducos, J. (ed.) (2010b). Existe-t-il une géologie médiévale ? In *La Pierre dans le monde médiéval*. PUPS, Paris.

Duffin, C.J. (2005). The western lapidary tradition in early geological literature: Medecinal and magical minerals. *Geology Today*, 21, 60–64.

Effinger, M. and Zimmermann, K. (eds) (2009). *Löwen, Liebstöckel und Lügensteine. Illustrierte Naturbücher seit Konrad von Megenberg*. Universitätsverlag, Heidelberg.

Fery-Hue, F. (2007). La minéralogie selon Jean Corbechon. In *La Traduction vers le moyen français. Actes du 2e Colloque International de l'AIEMF (Association Internationale pour l'Étude du Moyen Français), Poitiers, 27-29 avril 2006*, Galderisi, C. and Pignatelli, C. (eds). Brepols, Turnhout.

Gautier Dalché, P. (2000). La montagne dans la description "géographique" au Moyen Âge. In *La Montagne dans le texte médiéval. Entre mythe et réalité*, Thomasset, C. and James-Raoul, D. (eds). PUPS, Paris.

Gohau, G. (1983). *Idées anciennes sur la formation des montagnes*. Centre de documentation sciences humaines, Paris.

Gontero-Lauze, V. (2010). *Sagesses minérales. Médecine et magie des pierres précieuses au Moyen Âge*. Classiques Garnier, Paris.

Halleux, R. (1974). *Damigéron, Évax et Marbode : l'héritage alexandrin dans les lapidaires médiévaux*. Centro Italiano di Studi sull'alto Medioevo, Spoleto.

Haschmi, M.Y. (1966). Die geologischen und mineralogischen Kenntnisse bei Ibn Sinā. *Zeitschrift der Deutschen Morgenländischen Gesellschaft*, 116, 44–59.

Labbé, Th. (2017). *Les catastrophes naturelles au Moyen Âge*. CNRS, Paris.

Lauzanne, S. (1983). L'apport des cartulaires à l'histoire des mines, des carrières et de la métallurgie dans la France du Nord-Est. In *Mines, carrières et métallurgie dans la France médiévale, actes du colloque de Paris 19, 20, 21 juin, 1980*, Benoit, P. and Braunstein, P. (eds). Éditions du CNRS, Paris.

Lauzanne, S. (1985). Gestion et exploitation des carrières XIIe-XVe siècles : le laconisme des chantiers. In *Pierre et métal dans le bâtiment au Moyen Âge*, Chapelot, O. and Benoit, P. (eds). Éditions de l'EHESS, Paris.

Le Guay, J.-P. (1985). L'approvisionnement des chantiers bretons en matériaux de construction (XIVe-XVe siècles). In *Pierre et métal dans le bâtiment au Moyen Âge*, Chapelot, O. and Benoit, P. (eds). Éditions de l'EHESS, Paris.

Lüthy, C., Murdoch, J.E., Newman, W.R. (eds) (2001). *Late Medieval and Early Modern Corpuscular Matter Theories*. Brill, Leiden.

Mandosio, J.-M. (2011). L'acier dans la minéralogie et l'alchimie médiévales. In *L'acier en Europe avant Bessemer. Actes du colloque international, Conservatoire national des Arts et Métiers, Paris, les 8, 9, 10 décembre 2005*, Dillmann, P., Pérez, L., Verna, C. (eds). CNRS, Université de Toulouse-Le Mirail, Toulouse.

Mieleitner, K. (1922). Geschichte der Mineralogie im Altertum und im Mittelater. *Fortschrifte der Minealogie*, 7, 454–480.

Moulinier, L. (2010). La terre vue par Hildegarde de Bingen (1098-1179). In *Aux origines de la géologie de l'Antiquité au Moyen Âge : actes du colloque international, 10-12 mars 2005, Paris Sorbonne (Paris IV)*, Thomasset, C., Ducos, J., Chambon, J.-P. (eds). Honoré Champion, Paris.

Moureau, S. (2012). Les sources alchimiques de Vincent de Beauvais. *Spicæ, Cahiers de l'Atelier Vincent de Beauvais*, 2, 5–118.

Nobis, H.M. (1998). Der Ursprung der Steine : zur Beziehung zwischen Alchemie und Mineralogie im Mittelalter. In *Toward a History of Mineralogy, Petrology and Geochemistry, Proceedings of the International Symposium on the History of Mineralogy, Petrology and Geochemistry, Munich, 8-9 mars 1996*, Fritscher, B. and Hendreson, F. (eds). Institut für Geschichte der Naturwissenschaften, Munich.

Opsomer, C. (1980). *Livre des simples medecines. Codex Bruxellensis IV 1024*, 2 vols. De Schutter, Anvers.

Ribémont, B. (ed.) (1995). Météorologie et imaginaire : le cas de l'Etna. In *De natura rerum : études sur les encyclopédies médiévales*. Paradigme, Orléans.

Riddle, J.M. (1970). Lithotherapy in the Middle Ages, Lapidaries considered as medical texts. *Pharmacy in History*, 12, 39–50.

Riddle, J.M. and Mulholland, J.A. (1984). Albertus Magnus on stones and minerals. In *Albertus Magnus and the Sciences*, Weisheipl, J.A. (ed.). Pontifical Institute of Mediaeval Studies, Toronto.

Silvi, Ch. (2014). Comment nommer la terre en français ? De quelques signifiants utilisés par les encyclopédistes médiévaux. In *Latinum cedens : le français et le latin, langues de spécialité au Moyen Âge*, Marcotte, S. and Silvi, Ch. (eds). Honoré Champion, Paris.

Thomasset, C. and James-Raoul, D. (eds) (2000). *La montagne dans le texte médiéval. Entre mythe et réalité*. PUPS, Paris.

Thomasset, C. and James-Raoul, D. (eds) (2010). *La Pierre dans le monde médiéval*. PUPS, Paris.

Thomasset, C., Ducos, J., Chambon, J.-P. (eds) (2010). *Aux origines de la géologie de l'Antiquité au Moyen Âge : actes du colloque international, 10-12 mars 2005, Paris Sorbonne (Paris IV)*. Honoré Champion, Paris.

Thorndike, L. (1960). De lapidibus. *Ambix*, 8, 6–23.

Vigneron, F. (2010a). La nature et la qualité des sols dans le *Livre des prouffitz champestres et ruraulx* de Pierre de Crescens. In *Aux origines de la géologie de l'Antiquité au Moyen Âge. Actes du colloque international 10-12 mars 2005, Paris Sorbonne (Paris IV)*, Thomasset, C., Ducos, J., Chambon, J.-P. (eds). Honoré Champion, Paris.

Vigneron, F. (2010b). Le roi René et la pierre en Anjou : de la réalité du mécène à l'expression littéraire. In *La Pierre dans le monde médiéval*, James-Raoul, D. and Thomasset, C. (eds). PUPS, Paris.

Viré, M. (1983). Les carrières de pierre de Paris au Moyen Âge. In *Mines, carrières et métallurgie dans la France médiévale, actes du colloque de Paris 19, 20, 21 juin 1980*, Benoit, P. and Braunstein, P. (eds). Éditions du CNRS, Paris.

Viré, M. (2000). Un terrain assez bien connu et peu de textes (Moyen Âge et Renaissance). In *La pierre de Paris. Méthode d'étude de la pierre à bâtir depuis son extraction jusqu'à sa mise en œuvre*, Benoit, P., Blanc, A., Gély, J.-P., Guini-Skliar, A., Obert, D., Viré, M. (eds). Fédération pour l'édition de la Revue archéologique du Centre de la France, Paris.

Weill-Parot, N. (2002). *Les "images astrologiques" au Moyen Âge et à la Renaissance. Spéculations intellectuelles et pratiques magiques (XIIe-XVe siècles).* Honoré Champion, Paris.

3

Geology from the Renaissance to the Enlightenment

Gaston GODARD
IPGP, Université Paris-Cité, France

In 1603, Ulisse Aldrovandi created in Bologna the word *Giologia* (Vai 2003). Yet this "science of the Earth" was only in its infancy. Moreover, the term coined by Aldrovandi was not used until the end of the 18th century, and this new science did not truly flourish until the 1830s. The creation of the word geology nevertheless reflected a growing interest in this discipline, especially in the origin of fossils and the genesis of volcanoes.

Here, we address the developments of geology from the 16th to the 18th century, which led to the emergence of the discipline at the dawn of the 19th century. However, this presentation can only be partial. While books have been devoted to the history of geology in this period, we can provide only an overview of it in a single chapter. It therefore seems preferable to approach the subject through selected examples that fueled debate at the time, rather than attempting in vain to reconstruct an encyclopedic vision of this history.

In a first part (see section 3.1), we address ideas about the origin of fossils, a word that comes from the Latin *fossilis* ("taken from the earth") and had a broader meaning than its modern sense, since it referred to any object extracted from the ground: crystals were also fossils. The examination of these fossils, alongside observations of volcanic eruptions (notably that of Vesuvius in December 1631) and earthquakes (such as the Lisbon earthquake of 1755), fed more general conceptions

History of Earth Sciences,
coordinated by Pierre SAVATON. © ISTE Ltd 2026.

about the structure and evolution of the Earth, known as “Theories of the Earth”, which we discuss in section 3.2.

3.1. The origin of fossils and stones

Among fossils in the 16th-century sense, the “figured stones” that imitate living beings by their shape – thus our fossils – have been the subject of endless debates since Antiquity. Many explanations were proposed to account for their origin: spontaneous generation within rocks (the freak of nature or *lusus naturae*), failed attempts of the Creator God, the action of the stars, imprints or remains of dead organisms, etc. (Rudwick 1985; Ellenberger 1988; Gohau 1991). The origin of the most enigmatic fossils, such as “giants” bones and “fossil wood”, was debated. Finally, empirical knowledge of crystals and rocks, supported by mining activity, fueled speculations about the formation of “stones”.

3.1.1. *Figured stones*

From the 16th to the 18th century, the origin of “figured stones” was debated. Conrad Gesner devoted a fine book to them, titled *De rerum fossilium* (Gesner 1565). Three main theories were put forward about their origin: they were either the result of *lusus naturae* (the freak of nature), the remains of marine organisms deposited by the Flood, or the remains of marine organisms deposited in the sea and then brought to the surface by the emergence of the land. The thesis of an organic origin still stumbled on the presence of fossils of marine origin up to the tops of mountains, a fact then scarcely understandable.

We cannot retrace here, nor even list, all of these studies, which were already numerous at the time. We have chosen to present significant examples that reveal the evolution of ideas on the formation of fossils.

3.1.1.1. *The shells of the Montagne de Reims*

On the slopes of the Côte de l’Île-de-France and the Montagne de Reims, in the middle of the Champagne vineyards, a layer of fossil-bearing sands containing giant *Cerithium* shells from the Lutetian crops out, which has intrigued naturalists and fed their thinking for more than four centuries. Jean-Claude Plaziat showed that the shells of “Venteul en Valois”, described by Bernard Palissy in his *Discours admirables* (Palissy 1580) and which adorn some of his “*rustiques figulines*”, originate from this layer (Plaziat 2011). Palissy attributed an organic origin to them, but contrary to what has long been claimed, he imagined them living in freshwater basins rather than in a sea – he thought that the biblical account implied the constancy of lands and seas (Ellenberger 1988; Plaziat 2011).

Around 1630, the Jesuit Jacques Vignier drafted a handwritten report titled *Discours sur les coquilles de mer qu'on trouve en terre ferme, particulièrement en Champagne* (Discourse on the seashells found on dry land, particularly in Champagne), which dealt with the discovery of fossil gastropods and bivalves at the Cosson near Nogent-Sermiers, on the northern slope of the Montagne de Reims (Godard 2017). He proposed three hypotheses for their origin:

– These shells were generated by terrestrial animals.

– They were the result of "the plays and fancies of an industrious earth, and of a nature that relaxes". This *lusus naturae* theory denied the organic origin of fossils, attributing them to spontaneous generation in situ within the mass of rocks.

– They were marine shells, brought by the biblical Flood. Vignier adopted this third thesis and supported it by citing numerous ancient texts, the same, more or less, that were used in every argument about the origin of fossils in the 16th and 17th centuries (e.g. Rudwick 1985; Ellenberger 1988, pp. 164–195).

A copy of the report was sent to the Provençal scholar Nicolas-Claude Fabri de Peiresc, who replied by letter on April 26, 1635 (Godard 2017); Peiresc did not doubt the organic origin of the shells and held that "before the Flood, [...] the waters had originally covered the whole surface of the earth". This response may have influenced Jacques Vignier, who, in a second version of the manuscript drafted about twenty years later, did not abandon the thesis of the shells having been brought by the Flood, but equated it with an "inundation" of the sea. He added that "the sea has contributed here not only by its underground channels, but also by its overflowing, and that the tide left here these traces of the inundation".

In the 18th century, the *Cerithium*-bearing layer of the Montagne de Reims was highly appreciated by connoisseurs, who came to visit the fossil collection assembled by Marie-Catherine Le Franc at her Courtagnon estate (see Figure 3.1). Among them, Dom Sébastien Dieudonné, of the abbey of Hautvillers, described in 1751 and 1752 several fossil deposits in the region for his superior, the theologian Dom Augustin Calmet. Dieudonné considered them traces of the Flood, while Calmet argued for an in situ generation of the fossils in the ground. Conversely, F. Musard wrote in the *Mercure de France* of May 1753 that the shells of the Montagne de Reims were "in the places where the sea deposited them"; he even recognized fish eggs in them – possibly foraminifera. Jean-Étienne Guettard, of the Paris Academy of Sciences, also studied the shells collected by Marie-Catherine Le

Franc at Courtagnon; he identified certain anatomical features, such as the scar of the adductor muscle, and concluded that they were of marine origin (Guettard 1759, 1765).

Figure 3.1. *Marie-Catherine Le Franc in her cabinet at Courtagnon (Champagne). Engraved vignette by N.-C. Varin in 1763, after Dieudonné, showing the fossils (left) and the quarries (right) from which they were extracted (© Bibliothèque nationale de France (BnF))*

Guettard described in detail the different layers of the Montagne de Reims and recognized their superposition above the Cretaceous terrains of Champagne (Guettard 1759). A century earlier, Vignier had not perceived the interlayering of the fossiliferous sand layer within the Montagne de Reims; he instead imagined a thin film of sand deposited by the Flood or the sea in the hollow of a gulf, formed here by the Côte de l'Île-de-France. It was above all Nicolas-Antoine Boulanger who, around 1753, clearly explained in his *Anecdotes physiques de l'histoire de la nature* the shaping by erosion of the Côte de l'Île-de-France and its erosional outliers (Hampton 1955; Ellenberger 1994; Boulanger and Boutin 2006):

> The Montagne de Rheims, which [...] dominates this region, is built of stones of a nature entirely different from that of chalk, and it appears that it must once have extended much farther than it does now toward Rheims and toward Châlons, and consequently have formerly covered the surface that is presently exposed. (Boulanger and Boutin 2006, p. 260)

The renown of the fossiliferous Lutetian deposits of the region culminated in the 19th century, a period during which they were studied within the framework of modern paleontology and stratigraphy (Plaziat and Guérin 2011).

3.1.1.2. *The oysters and shells of Guyenne*

On October 2, 1620, the young King Louis XIII and his entourage were astonished before the troglodyte chapel carved into the cliff of fossil oysters at Loubens (see Figure 3.2), in Sainte-Croix-du-Mont in Guyenne – a famous deposit of Burdigalian faluns (Lower Miocene), located on the right bank of the Garonne 40 km southeast of Bordeaux. These gentlemen wondered "whether these Oysters or shells had formerly been carried here by the Flood, or whether it was a simple frolic of Nature, which, wanting to form hard stone, had stopped short, and had formed only shells". Thus, Pierre de L'Ancre recounted the episode in his demonology work titled *L'incredulité et mescreance du sortilege* (de L'Ancre 1622, pp. 17–42 of the *Advertissemens*; Godard 2018a, 2018b). He questioned the origin of the shells and sediments, hesitating between an ancient stay of the sea followed by consolidation "through long centuries", a transport by the Flood from the ocean, and finally an in situ formation by "frolic of nature", a pleasant phrasing of the *lusus naturae* or "freak of nature". De L'Ancre also described the succession of sedimentary layers, which he listed from bottom to top as a modern geologist would, and called them "stages", which suggests he had intuitively understood the principle of superposition of layers that Steno had not yet stated (Steensen 1669).

Figure 3.2. *Troglodytic chapel of Loubens carved into the oyster bed of Sainte-Croix-du-Mont, visited by Louis XIII and his entourage on October 2, 1620 (postcard, late 19th century)*

Written a few years later, an anonymous manuscript titled *Brief reflections on some stones of the land of Sales* addressed the origin of the faluns of Salles (Godard 2018b, 2019), a neighboring locality south of Bordeaux that was once the stratotype of the "Sallomacien" (Serravallian, Middle Miocene). The author favored the thesis of the "freak of nature" rather than that of a marine organic origin. He remained faithful to the archaic "theory of signatures", which established an analogy between the appearance of objects used in pharmacopoeia and the illnesses they were supposed to cure – according to him, the "serpent tongues" (glossopetrae) were used to treat tongue ailments – although these beliefs, inherited from the Middle Ages and developed by Paracelsus in the previous century, were no longer really current in the 17th century.

In 1718, Abbé Jules Bellet and Isaac Sarrau de Boynet, of the Royal Academy of Bordeaux, took up the debate on the origin of the faluns of Sainte-Croix-du-Mont. Abbé Bellet was convinced that the shells were those of marine oysters that had been left behind as the sea receded. He cited ancient and modern authors who had mentioned the presence of such shells inland. He believed that the presence of the sea at Sainte-Croix-du-Mont dated back to the biblical Flood, which he thought had occurred 4,115 years earlier. Sarrau de Boynet also thought the shells had an organic origin, as evidenced by the imprint left by the adductor muscle and their effervescence in vinegar, but he refused to admit that the sea could have reached that place and adopted the extravagant thesis of an anthropic pile of oysters, which Montesquieu did not endorse.

The Academy of Bordeaux sent the Regent Philippe d'Orléans a map and a report on the subject, from which an account appeared in the *Journal des Sçavans* in March 1719. Benoît de Maillet referred to it in his famous manuscript *Telliamed*, where he advanced bold hypotheses concerning the origin of the Earth and the formation of fossils. Here is what he wrote about the shells of Sainte-Croix-du-Mont:

> There are some even in France at six leagues from Bordeaux, in the parish of Sainte Croix du Mont [...]. There, on the crest of a rather high hill, [...] one sees between two beds of stone, the upper of which may be five or six feet thick, a bed of oysters that is twenty or twenty-four feet thick [...]. A chapel fifteen feet deep has been carved into it, where Mass is celebrated. Most of these oysters are closed; and in these one finds a kind of clayey earth in small quantity. It is doubtless the substance of the oyster that has melted. [...] The scholars working in Bordeaux on the History of the Earth will doubtless explain to you how this bed was formed, if the prejudices in which they were raised

> do not prevent them from recognizing the reason for this phenomenon. As for me, I think it is impossible not to remain convinced, at the sight of these oyster beds […] that they were all oyster banks when the sea completely covered them, similar to those that it contains today in countless places, from which we obtain the oysters we eat (Maillet 1748, pp. 102–103).

During the 18th century, the idea of the organic origin of fossils became established. Starting in 1745, Jacques-François Borda d'Oro undertook the study of the terrains and fossils (Eocene) of the region of Dax. The observation of traces of lithophages on these fossils, incompatible with transport by the Flood, strengthened his conviction that they came from the sea. Nicolas Desmarest, who had observed fossils during two journeys undertaken in 1761 in Guyenne, particularly in the surroundings of Saint-Émilion, then in Périgord and in Gascony, did not doubt their marine origin. In 1743 and 1745, the Academy of Bordeaux organized dissertation competitions on the origin of "figured stones". The candidates supported an organic origin but ignored the region's faluns and confined themselves to generalities. One of them, Pierre Barrère, surprises us by applying the principle of uniformitarianism as early as 1745. These writings, among the earliest devoted to the geology of the Aquitaine Basin, demonstrate that certain modern concepts of modern paleontology and stratigraphy were already perceived in the 17th and 18th centuries.

3.1.1.3. *The fossils of Provence and Italy*

In 1603, the Roman Federico Cesi founded the Accademia dei Lincei, one of the first scientific academies in Europe. In the first half of the 17th century, its members included such prestigious figures such as Galileo Galilei, Fabio Colonna, the Provençal Nicolas-Claude Fabri de Peiresc and Cardinal Francesco Barberini, nephew of Pope Urban VIII. Among many other subjects, they discussed the Earth sciences, in particular the origin of figured stones (Godard 2005). Peiresc exchanged important correspondence on this topic with Claude Menestrier, a man from Franche-Comté who worked in Rome as librarian to Cardinal Barberini. In January 1629, Menestrier used the "*lunettes de Drebels*", that is, the magnifying lenses of Cornelis Drebbel, to observe the "marine shells" of Monte Mario, near Rome – a deposit of Plio-Pleistocene sands and marls. Despite the distance from the sea, Peiresc and Menestrier were convinced of the marine origin of these shells.

Peiresc described repeatedly the fossil-bearing layers around his residence of "Boisgency" (Belgentier), north of Toulon. He mentioned in particular "limassons" (coiled fossil shells) that he compared to the *Nautilus* of Basile Besler (1622) and in which one can recognize ammonites with various "estages" (chambers) separated with "feuillages" (suture lines). Although these species were not present in the

Mediterranean, Peiresc thought they might have survived in distant seas, an idea that several authors, such as Bernard Palissy (1580), had put forward before him (Ellenberger 1988, p. 143). By hiring a naval surveyor, he was able to estimate the elevation of these shells relative to the Mediterranean and concluded that "the waters had originally covered the entire surface of the earth", before emergence occurred.

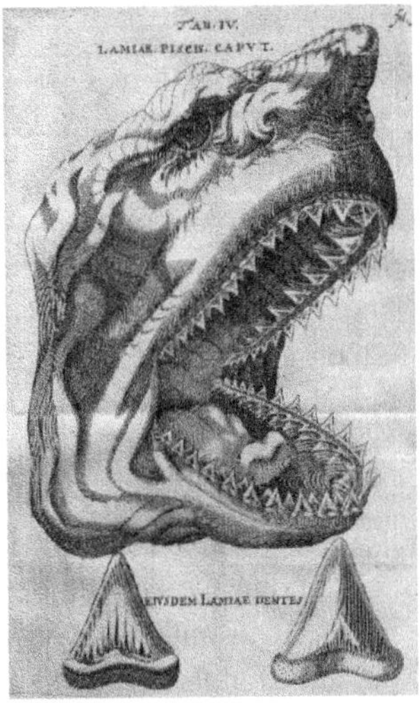

Figure 3.3. *Plate IV of Elementorum myologiae specimen…, cui accedunt canis carchariae dissectum caput… (Steensen 1667), where Steno identified the glossopetrae as shark teeth (© Bibliothèque nationale de France (BnF))*

The Linceans also wondered about the origin of the glossopetrae, whose name derived from the Greek *Γλώσσα* (tongue) and the Latin *petrus* (stone). Known in Antiquity and often called "petrified serpent tongues" – in reality, they were fossil shark teeth – they had long been used as remedies against snakebites and tongue ailments, because of the powers attributed to them by the theory of signatures already mentioned. Their true origin was not understood until the 17th century.

In 1616, Fabio Colonna (or *Columna* in Latin), a member of the Accademia dei Lincei, published a memoir titled *De Glossopetris Dissertatio*, in which he refuted the hypothesis of spontaneous generation of glossopetrae in the soil and asserted that they had an organic origin, assigning them to the teeth of squales (Colonna 1616). In 1667, at the end of his memoir on the *Dissection of the Head of the Dogfish*, Nicolas Steno wrote a famous text on the glossopetrae that he identified as shark teeth (see Figure 3.3). This origin was very quickly accepted (Rivière 1766 [written in 1708]), despite attempts at refutation such as those by Johann Jacob Reiske (1687) and Elias

Camerarius (1712) (see Gaudant and Bouillet 2004; Gaudant 2012). According to Ellenberger (1988), historians have sometimes neglected Colonna's contribution in favor of Steno's contribution, which came later.

3.1.2. *The quarrel of the "giants"*

The Renaissance inherited a centuries-old myth, that of a race of giant men who, according to the Bible and certain ancient texts, religious or secular, had once populated the Earth. Thus, whenever chance brought to light bones or teeth of gigantic size, they were invariably attributed to a "giant" (Céard 1978; Schnapper 1986; Cohen 1994; Godard 2009; Gaudant 2010).

The most remarkable "giant" discovery of this period was that of the supposed Theutobochus, a case very well recounted by Georges Cuvier and later by Léonard Ginsburg (Cuvier 1812, t. 2, pp. 14–17; Ginsburg 1984, 1986). On January 11, 1613, in a sand pit near the Château du Langon in Bas-Dauphiné[1], bones of gigantic size and a few coins were discovered at a depth of 18 feet beneath a tombstone bearing the inscription THEVTOBOCHVS REX. Let us remember that around 102 BCE, the Roman consul Marius defeated the Teuton army in southern Gaul. A legend reported that the king of the Teutons was a giant named Theutoboch. The bones, handed to the Marquis Nicolas de Langon, were examined by scholars from the University of Montpellier, who concluded that they were human. Some of the bones were entrusted to Pierre Mazurier, a barber-surgeon from the region who headed toward Paris. He exhibited them under a tent and sold a booklet titled *True History of the Giant Theutobocus, King of the Teutons, Cimbri, and Ambrosians, defeated by Marius, Roman consul, one hundred and five years before the coming of our Savior*... Displayed for a time at the Château de Fontainebleau, the bones were presented to the young King Louis XIII, then Mazurier left Paris to continue his tour in England and in Flanders. It was then that a heated controversy broke out between Nicolas Habicot, a Parisian surgeon and Jean Riolan, a professor of anatomy and botany at the Royal College of Medicine. Between 1613 and 1618, these gentlemen exchanged invectives, insults and slanders in no fewer than 12 works and pamphlets. Habicot supported the existence of giants, in particular the attribution of the bones exhibited by Mazurier to the giant Theutobochus, whereas Riolan thought that these bones might belong to an elephant and accused Mazurier of having invented the epitaph.

In 1634, more than 20 years after the discovery at Langon, a man named Nivolet sent Peiresc information about the circumstances of that discovery together with one

1 Today in the commune of Valherbasse (Drôme, France).

of the coins found in the pit (Godard 2009). In his reply, Peiresc disputed that the bones could have belonged to Theutobochus. He was surprised that an epitaph had been written in Latin when "Theutobochus did not speak Latin any more than the people of his nation", and he stated that the coin came from a treasure found in Dauphiné even before the discovery of the supposed tomb. "I am not yet altogether well convinced that it is truly from a Giant", Peiresc concluded with understatement. According to Pierre Gassend (Lassalle 1992, pp. 127–129), the examination of a few bones convinced Peiresc that it was the burial of an elephant. We know today that the bones hawked by Mazurier and rediscovered around 1833 in Bordeaux actually belonged to a *Deinotherium giganteum*, a fossil proboscidean of the Upper Miocene (Ginsburg 1984, 1986).

Peiresc took part in another "giant" affair of great importance, because it challenged the testimony of Saint Augustine regarding the existence of giants (Godard 2009). In his book *De Civitate Dei* (XV, 9), written around AD 412, Augustine mentioned the discovery of a giant molar on the shore of Utica (modern-day Tunisia), and drew from it an argument in favor of the past existence of giants. In the Christian world, the tooth of Utica had not been completely forgotten, when in June 1630 Thomas d'Arcos, who lived near Tunis, wrote to Honoré Aycard of Toulon to inform him that a "Tomb of a Giant of immeasurable size" had been discovered near Utica, "in the same place where St. Augustine says in the book *City of God*, book 15, chap. 9 [...] that he saw another human tooth that would have made a hundred of ours". D'Arcos specified that "his body, that is to say the bones only, was 40 cubits long [5.9 m...], his head larger than a wine cask of 12 *millerolles*[2]. I have seen and weighed one of his teeth, and it weighed 2 and a half pounds, which is 40 ounces [1.2 kg], the bones of this body being partly rotten and partly whole". Peiresc became interested in this discovery and obtained from d'Arcos in May 1631 a tooth of the supposed giant (see Figure 3.4). Peiresc then had the opportunity to examine an elephant that was being exhibited in Provence, and he reported his observations to Pierre Dupuy:

> I let myself be carried to such a point of curiosity (or rather of folly), as to put my hand in his mouth and handle and grasp one of his maxillary teeth, to better recognize its shape [...]. This was to verify, as I did, that they were entirely similar in form, though of smaller size, to the tooth of the supposed giant from the coast of Tunis or Utica. [This observation] made me call into doubt that several elephant bones found buried in various places do not often pass for giant bones.

2 A millerolle was an ancient unit of measurement equivalent to about 60 L.

(Letter from Peiresc to Dupuy on December 26, 1631[3]; Tamizey de Larroque 1890, pp. 287–295)

However, Peiresc did not imagine that the bones of Utica came from a fossil elephant, but rather from the burial of an elephant: "The great closeness of ancient Carthage, he thought, should make it seem less strange that one had buried the elephant in question, and even built for him a sort of tomb", he continued. Whatever the case, Peiresc had just shown that Saint Augustine, Father of the Church and Doctor of the Faith, had been mistaken about the Utica find and the question of giants. As Galileo's trial was beginning in Rome, he kept silent about this matter, perhaps practicing a kind of self-censorship. He spoke of it only to the *Lincean* Cassiano dal Pozzo three years later, on August 2, 1635[4]: "It was a real elephant that was thought to be a giant, practically in the same place, or not far from where Saint Augustine said he had seen some relic". Cautiously, he added: "I would not wish to deny, in any way, the truth of the tradition about giants; nevertheless, I would easily doubt that all those discovered in various places are bones of giants". The drawing of the tooth sent to Peiresc (see Figure 3.4) made it possible to identify the upper left molar of an elephant, likely *Loxondonta* (or *Mammutus*) *africana*[5], which must be attributed to the Plio-Quaternary fossil proboscidean deposits of Tunisia.

At the same time, the members of the Accademia dei Lincei were trying to understand the origin of various "giants" found in Italy, particularly in Sicily (Godard 2009). The past existence of giants on the island, affirmed by Boccaccio[6], seemed to have a solid basis, since giant carcasses exhumed near Castelvetrano were exhibited at fairs. Vicenzo Mirabella sent to his colleagues of the Accademia dei Lincei some of these bones, including "a fragment of a jaw with a tooth as large as a fist". These remains, wrote Menestrier in 1634, came from a mountain full of "elephant teeth and jaws that the tricksters sell throughout Italy as unicorn", although it was "fossil *ebur*" (ivory taken from the ground). Some interpreted these remains as proof of the island's past occupation by Homer's Cyclopes (Fazello 1558; Mongitore 1704); today it is accepted that the nasal cavity of fossil elephants, mistaken for a single eye socket, is the origin of this myth.

3 BnF, Dupuy Collection, ms 717, f. 121.

4 Translated from Italian; letter from Peiresc to Cassiano dal Pozzo, August 2, 1635; Montpellier School of Medicine, ms H271 vol. 2, f. 164r–168r.

5 Pascal Tassy and Léonard Ginsburg in Godard (2009).

6 Boccaccio, G. (toward 1360). *Genealogia deorum gentilium*... Venice edition, 1472, lib. IV, cap. 67–68.

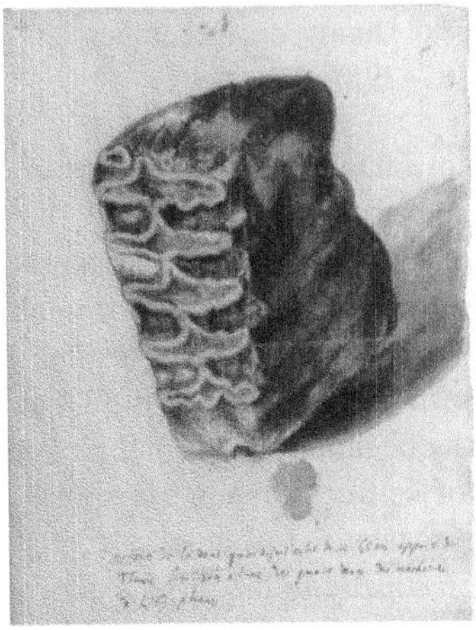

Figure 3.4. "*Drawing of the tooth said to be of this giant brought from Tunis, similar to one of the four teeth of the elephant's jaws" (© Bibliothèque nationale de France (BnF), Dupuy Collection, ms 488, f. 171r). This tooth studied by Peiresc in 1631 was found at Utica, "in the same place where St. Augustine says [...] that he saw another human tooth that would have made a hundred of ours"*

About 50 years later, Giovanni Giustino Ciampini described bones of extraordinary size, found at Vitorchiano near Viterbo (Latium, Italy), likely in sediments or cinerites of the Lower or Middle Pleistocene, which are abundant in this region (Langenmantel 1689). Ciampini concluded that these remains were those of an elephant, after comparing them with the drawing of an elephant skeleton kept in Florence in the collections of the Grand Duke of Tuscany, Cosimo III de' Medici (Gaudant 2010).

In February 1696, the college of physicians of the Thuringian town of Gotha (Germany) published, at the request of Duke Frederick II of Saxe-Gotha, the description of a skeleton that had just been discovered in a sand pit in Tonna, north of Gotha. After having considered that the skeleton might be that of a giant, then of an elephant, they finally attributed it to a "unicorn" (Gaudant 2010). Wilhelm Ernst Tentzel, historiographer to the duke, examined these remains and concluded that they were the true skeleton of a large elephant, basing his opinion on the description of an elephant skeleton published by an Irish physician a few years earlier. Since the remains had been found within "layers piled one upon another, in order", he had to admit that this elephant had been carried and deposited by the Flood. His demonstration did not convince the physicians of Gotha, who published a second brochure to defend their interpretation, asserting that the animal was a "unicorn" or "fossil unicorn", whose resemblance to an elephant could be explained by the ancient theory of the *vis plastica* or freak of nature. The spokesman for the physicians noted anatomical differences between these remains and the African elephant – it was in fact an *Elephas* (= *Mammuthus*) *meridionalis* – and emphasized the implausibility of transport by the Flood from Africa. Tentzel nonetheless received support from Gottfried Wilhelm Leibniz, Giovanni Battista Scaramucci, physician in Urbino (Italy), and the Royal Society of London (Gaudant 2010).

These few examples show that the 17th century marked the questioning of the myth of giants, already criticized in the previous century by Girolamo Maggi (Maggi 1564, 1603) and Jan van Gorp (Goropius 1569). The giants, however, persisted – if one may say so – and the myth endured until the 18th century (Murray 1904; Céard 1978; Schnapper 1986, 1988; Cohen 1994; Godard 2009). Religious writers in particular applied themselves to defending the Bible and Saint Augustine. In chapter IV of book VIII of his *Mundus subterraneus*, Athanasius Kircher worked to reduce the importance (and the size) of the giants, without denying their reality. Others were not convinced that these remains should be attributed to elephants, such as Robert Plot and Abbé Antoine Gachet d'Artigny, who even persisted in believing in Theutobochus (Kircher 1665; Plot 1677; Gachet d'Artigny 1749); Benoît de Maillet did not doubt, in his famous *Telliamed*, the past existence of giant men (Maillet 1748). In addition to religious beliefs, the absence of living proboscideans in Europe contributed to the persistence of the myth, which was not definitively dispelled until the emergence of modern paleontology at the beginning of the 19th century.

3.1.3. *Fossil woods*

Within the Accademia dei Lincei, they debated the origin of fossil woods found in the region of Acquasparta in Umbria (Italy) – in reality, Pliocene lignites (Alessandrini 1978, pp. 111–142; Scott and Freedberg 2000; Scott 2001; Cerqualia 2005; Godard 2005; Ottaviani 2021). Francesco Stelluti, one of the founders of the academy, undertook a detailed study of them, which he published in 1637. In his book (Stelluti 1637) (see Figure 3.5), Stelluti described fragments of wood; identified knots, barks and branches; confirmed that the wood was combustible; and suggested that it had undergone spontaneous underground combustion – an idea borrowed, it seems, from Agricola. He reached the conclusion that in this place, the earth had transformed itself into wood.

Intrigued, several travelers undertook the journey to Acquasparta. The Lincean Cassiano dal Pozzo went there in October 1635 with Jacques de La Ferrière, physician to Alphonse de Richelieu, cardinal of Lyon and brother of the principal minister of Louis XIII. In a manuscript account addressed to Peiresc[7], La Ferrière was critical of the hypothesis of spontaneous generation of wood in the soil, estimating "that it is easier for nature to convert wood into baked earth or into stone than to make wood out of earth". Alessandrini (1978, pp. 111–142, 214–217) suggested that Stelluti might have deliberately ignored a copy of this manuscript[8], dated 1635 (and therefore earlier than his treatise), and coming from the archives of the Lincei. In November 1635, Gabriel Naudé went there in turn and sent Peiresc a very evasive report.

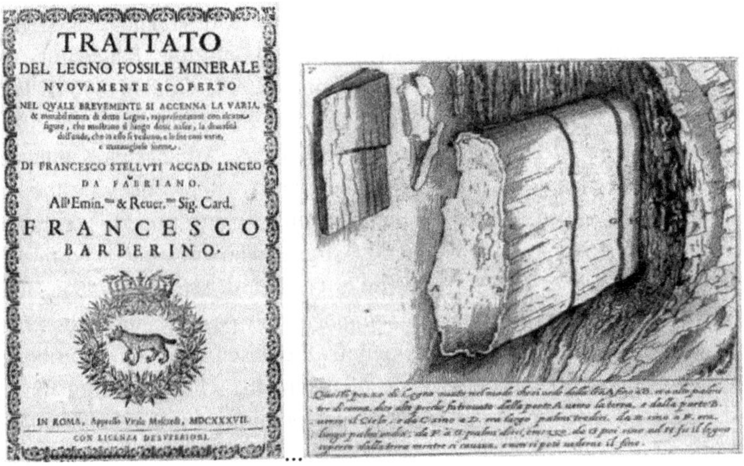

a) b)

Figure 3.5. Treatise on Fossil Wood *by Francesco Stelluti (source: Stelluti 1637, © Bibliothèque nationale de France (BnF))*

COMMENTS ON FIGURE 3.5.– *The title page (a) bore the Lynx, emblem of the academy and the dedication to Cardinal Francesco Barberini, nephew of Pope Urban VIII. Plate 7 (b) showed a trunk and fragments of wood taken by some for sawn boards. One recognizes in this treatise part of the 133 drawings of fossil wood from the collections of Cassiano dal Pozzo preserved at Windsor (United Kingdom) and published by Scott and Freedberg (2000).*

A few years later, while he was in the service of Cardinal Mazarin in Paris, Naudé published a curious and amusing account of his observations:

7 Bibliothèque Mazarine, Paris, ms 4377 (1761A), 14 f. 188.

8 Bibliothèque of the Montpellier School of Medicine, ms H170, f. 11r–12r.

> But although this whole Academy [of the Lincei], and a certain Claude Menestrier of Besançon, who was said to have been a great Naturalist, or rather a great Fabulist, since everything suited him, as also the Physician to the Cardinal of Lyon [La Ferrière], had judged that this wood was fossil [formed in the ground], I found, after having observed it well and having dug while excavating on the site, sticks, planks, staves, beams, tree trunks, and many kinds of wood, knotty, polished, forked, straight, twisted, furnished with bark, or stripped of it, lying in one place and standing in another; I recognized, I say, that all this wood came from some forest crushed, with all the timber yards and stores that were in it, under the fall and overturning of higher and nearer lands, in the furious shocks and agitations of an earthquake, as in the year 1618 the one that crushed under the collapse of a mountain the town of Pioury in the Grisons. (Naudé n.d., pp. 667–668)

Pierre Gassend, known as Gassendi, proposed an explanation very similar to that of Naudé in his work *De Lapidibus, ac metallis* (Gassend 1658), to the point that it seems that one had inspired the other. In his biography of Peiresc, he stated that Peiresc had aligned himself with Stelluti's opinion:

> He learned exactly what he had desired to know: that there existed only, in fragments, kinds of stumps, but no traces of branches, knots, or roots, which could be proof that woods of this nature had been born and had grown as such, but had not been piled up and petrified with, as their origin, ordinary wood. (Lassalle 1992, p. 258)

The hypothesis of a spontaneous, but unfinished, generation of trees in the soil, proposed by Stelluti and perhaps shared by Peiresc, fell within the tradition of the "freaks of nature" and of the *vis plastica*. The degradation of the lignites likely influenced this opinion, because the woody fragments resembled the original trees only from a distance. When, in 1630, Peiresc had observed leaf impressions from the Aix Basin, on the contrary clearly recognizable, he had not hesitated to name the plants that had produced them (see Figure 3.6).

Believing they had seen sawn boards among the woody fragments, Naudé and Gassend argued in favor of an origin that was both organic and anthropic, already imagined by Agricola in the previous century for another lignite deposit. This idea reflected the tendency of men of the 16th and 17th centuries to reason within the framework of a short chronology, combining geological phenomena with human history.

Figure 3.6. *Fossil flower from the Aix Basin*

COMMENTS ON FIGURE 3.6.– *Impression and counterimpression of a flower from the Aix Basin: "Flos lapidescens, found on the estate of Mr. Porciolx in the diocese of Aix", excerpt from: "Antiquities of the Cabinet of Peiresc, drawn by Poussin, Rubens and others..." (© Bibliothèque nationale de France (BnF), Estampes, ms Aa54res). The fossil belongs to the flora of the lacustrine basin of Aix and might be that of a calyx of* Heterocalyx ungeri *Saporta (an Oligocene Anacardiaceae). Peiresc did not doubt its plant origin.*

3.1.4. *Stones, crystals and petrification*

3.1.4.1. *The formation of "stones"*

Starting in the Renaissance, "stones" aroused the interest of scholars, the enthusiasm of collectors and an entire trade among jewelers and even apothecaries. The proto-mineralogy of the time, tinged with esotericism, stood at the crossroads of various concerns, such as jewelry and mining operations, but also medicine, since certain stones were considered to have therapeutic virtues (Duffin et al. 2013).

Every prince or eminent scholar possessed a cabinet of curiosities (Schnapper 1988) whose *naturalia* included, besides "figured stones", stones used in the pharmacopoeia of the time, such as "bezoars" (stony concretions of animal origin), "variolite" (variolitic metabasalt), "aetite" (or eagle stone) and also "jade" imported into Europe by the Spanish under the name *piedra de ijada*, that is, stone of colic (nephritic), which it was supposed to cure.

Among these were also collected certain "stones" with singular characteristics. Father Marin Mersenne received samples of the "floating stone of Poitou", which was lighter than water (perhaps a nectic flint). Jacques Hallé held a "bending stone" (a mica?). Girolamo Aleandro il Giovane, Cassiano dal Pozzo, Fortunio Liceti and Nicolas Fabri de Peiresc exchanged fragments of the *pietra di Bologna*, discovered in 1602 at Monte Paderno near Bologna (a barite luminescent when heated) (Marsigli 1698; Galvani 1780; Bombicci 1873–1874, t. IV, pp. 121–124; Godard 2005). Among the "extravagant stones" he held in Rome, Claude Menestrier mentioned a "*pierre aymant*" (magnetite)… In the 18th century, Dezallier d'Argenville (1742) described and attempted to classify these stones in a rational manner.

Finally, cabinets of curiosities contained gemstones exchanged among collectors, jewelers and lapidaries. Peiresc's papers were rich in information in this regard, both on his correspondence with various jewelers and lapidaries and on the gemstones in his cabinet (Schnapper 1988, pp. 237–240). Peiresc exchanged information on the minerals of Provence with Johannes de Laet of Antwerp, received a "recipe to whiten amethyst" by heat, and also copied a *Memoire des diamans* dictated by Fernand Nuñez, the brother of a lapidary from Amsterdam. He exchanged many letters with the Parisian jeweler Alvares and, above all, tried to correspond with Augustin Herryard, a Bordeaux jeweler who was staying in the East Indies, where he had entered the service of the Great Mughal in Lahore (Sneyders de Vogel 1955). Herryard's fabulous tales about the sapphire, diamond and ruby mines of the "great mougoul" prompted his colleagues Philippe Nuñez and Manuel da Costa Casserets to leave for the East Indies in July 1630.

Mineralogists and lapidaries at the time wondered how "stones" formed, whether extravagant, precious or ordinary. Understanding processes of petrification was also an obsession at a time when people suffered greatly from "stone disease" (kidney/bladder stones). It was then commonly accepted that the "coagulation" of stones was caused by a "seed", a "lapidifying juice" or a "germ" (Oldroyd 1974; Halleux 1979, 1982; Hirai 2005).

This notion was presented by Agricola in his *De re metallica* (Agricola 1556) concerning mineralized veins, then developed by Étienne de Clave (Clave 1635) in his *Paradoxes ou Traittez philosophiques des pierres et pierreries*. Peiresc also explained in this way the formation of stalactites and stalagmites in caves, where water loaded with a "stone germ […] curdles such that it forms candles or gutters at the place from which it drips above, and below in its fall" (Bib. Carpentras, ms 1821, f. 78r–81v). He extended this conclusion to kidney stones, gemstones and then to all stones, including stratified sedimentary rocks whose induration he attributed to a "seed". Pierre Gassend presented the same idea in his biography of Peiresc (Lassalle 1992, pp. 196–200) and in his work *De Lapidibus ac metallis* (Ellenberger 1988, pp. 224–232).

3.1.4.2. *Crystals*

The scholars of the time largely confused the modern concepts of crystal, mineral species (to which a crystal belongs) and rock (an aggregate of crystals), which they referred to indiscriminately as "stones". However, certain scholars took interest in crystals displaying regular shapes. In 1611, Johannes Kepler examined the arrangement and hexagonal form of snowflakes in *Strena seu de nive sexangula* (see Chapter 4, section 4.1.1). Kircher also devoted several pages to salts and gems in Book VIII (*De Lapidibus*) of his famous *Mundus subterraneus*. In 1635, Gassend carried out experiments on the crystallization of salts (Rochot 1944, pp. 66–69; Bloch 1971, pp. 244–269):

> These large solids, whether cubic [salt], octahedral [alum], or others, are all composed of smaller ones of the same figure, and these of still smaller ones down to a resolution into parts so minute that they are almost imperceptible and always of the same figure, from which I concluded that these resolve further down to the atoms, which by some sort of necessity must be of the same figure. (Letter from Gassend to Peiresc dated July 6, 1635; (Rochot 1944, p. 66-67))

Pierre Gassend's conception of matter, which supported the notion of the atom in its etymological sense (that which cannot be divided), lay behind this argument. In 1688, Domenico Guglielmini offered a reasoning entirely similar in his *Riflessioni* on salt crystals:

> If all salt crystals each have a form specific to their species, and these can be and are composed of other smaller and similar crystals, it follows that the indivisible parts of matter possess a form they acquired at their formation… . (Guglielmini 1688, p. 32)

With his atoms that "must be of the same figure", Gassend came close to the modern concept of the unit cell of crystals, which is attributed to René-Just Haüy (1801). Haüy's "integrant molecule" was both a chemical entity (for us, a grouping of atoms) and a unit cell whose shape and packing determined the faces and cleavages of crystals (Haüy 1801; Maitte 2001). The texts of Gassend and Guglielmini showed that they understood what this unit cell was, though they were obviously unable to assign it a chemical composition. Pierre Gassend extended his observations on salts to certain minerals (Lassalle 1992, p. 98): "Rhomboidal stones [calcite] unearthed on the territory of Rians [are] admirable in that, if they are broken into small parts, each particle retains this rhomboidal figure". It is said that by accidentally breaking a calcite crystal, René-Just Haüy had the revelation of his "integrant molecule". It is undeniable that Pierre Gassend had a similar intuition a century and a half earlier.

Another important step in the history of crystallography was taken by Steno in his *Prodromus* (Steensen 1669). His analysis of the geometry of quartz, hematite and pyrite crystals showed that he understood the concept of truncation and had a clear intuition of the law of constancy of angles between crystal faces (Ellenberger 1988, pp. 276–289), like Guglielmini (1688) after him. This law was formally stated by de Romé de L'Isle in his *Essai de cristallographie*: "Whatever the relative dimensions of two specific faces of the same crystal, they always present between them the same dihedral angle" (de Romé de L'Isle 1772).

3.1.4.3. *Mining operations*

In the 16th century, Saxony and Bohemia had well-established mining practices and traditions, as shown by the *Bergbüchlein* of Ulrich Rülein von Calw (around 1500) and especially the famous *De re metallica* of the Saxon George Bauer known as "Agricola" (Agricola 1556). The gold and silver mines of South America, such as those of Potosi, contributed to the wealth of Spain. Mining operations were, on the contrary, little developed in England, France and Italy.

Mining science was then rather empirical and often embellished with esoteric or astrological considerations. In France, it was marked by a pair of adventurers, Jean du Chastelet and Martine de Bertereau, baron and baroness of Beausoleil, who had the reputation of having handled the divining rod more than the compass or the hammer.

In 1640, the baroness published *La restitution de Pluton*, a book that mixed astrology, alchemy, mineralogy and politics (Bertereau 1640). Cardinal Richelieu had these unfortunate people arrested for charlatanism and suspicion of heresy. The baron died in the Bastille in 1645 and the baroness suffered the same fate at Vincennes (Vallet 1976; Halleux 1982, pp. 121–124; Routhier 1987). The tragic story of this family inspired the novel of Thibaud d'Anthonay, *Le baron de Beausoleil* (d'Anthonay 2014).

As in many other fields, the 18th century marked an effort to rationalize mining research and operations. In Germany, this resulted in the creation of the renowned Freiberg School of Mines (*Kurfürstlich-Sächsische Bergakademie zu Freiberg*) in 1765. This institution saw the flourishing of the Wernerian geognostic school, which had a considerable impact across Europe. In France, The Royal School of Mines was founded in Paris in 1783.

3.2. From an immutable Earth to a dynamic Earth

In this section, we discuss how the ideas presented in section 3.1 on the origin of "fossils" and "stones", as well as observations of volcanoes and earthquakes, shaped more general conceptions of the structure, dynamics and evolution of the Earth, which, in the 17th and 18th centuries, fed numerous "theories of the Earth".

3.2.1. *Freaks of nature, flood inundation or invasion by the sea?*

In the 16th and 17th centuries, two main theses opposed each other regarding the origin of "figured stones". The thesis of "freak of nature" or *lusus naturae* sought to explain the origin of fossils by a form of spontaneous in-place generation within the rocks. It opposed the organic origin, which was hindered by the distance from the sea and – supposedly – required shells to be transported from the sea (Rudwick 1985; Ellenberger 1988). As we noted earlier (see section 3.1.1.2), Pierre de L'Ancre hesitated in 1622 between the two theses. Ulisse Aldrovandi in the 16th century and then Gaffarel (1629) and Lister (1678) in the following century maintained the idea of in-place generation within rocks, but they gradually became a minority. Fracastoro, Alessandri, Palissy and Cesalpino in the 16th century, then Colonna, Peiresc, Vignier, Menestrier, Gassend, Steno, Hooke, Woodward and some others in the 17th century did not doubt the organic origin of fossils. Nevertheless, these authors did not always adopt the bold hypothesis of a stay of the sea in the interior of the land: Bernard Palissy made the shells live in freshwater "receptacles", and Pierre Gassend in underground cavities filled with water (Palissy 1580; Gassend 1658).

Most supporters of an organic origin argued that the shells had been transported from the sea during the Flood. Some were fervent diluvianists, while others were described as "hesitant" diluvianists by Ellenberger (1988), because, being convinced of the marine origin of the figured stones, they invoked the Flood as a convenient device accepted by all, thus making a concession to the dominant opinion of the time.

Some advocates of an organic origin, however, had the audacity to invoke fluctuations in sea level. This was the case of Claude Menestrier around 1635 to explain the fossils of Monte Mario, near Rome. Peiresc thought that the fossils of Provence had been exhumed by vertical movements of the soil. In 1622, Pierre de L'Ancre also considered a mobility of the seas operating over "long centuries" rather than transport by the Flood, suggesting a dynamic evolution of the Earth: "We see the mountains and rocks grow each day, and the depressions fill themselves" (de L'Ancre 1622).

With the contributions of Steno (Steensen 1667), Ray (1673), Woodward (1695), Hooke (1705) and Scheuchzer (1708), the thesis of an organic origin became established around the turn of the 18th century. At the same time, the so-called "giant bones" and remains of "unicorns" were most often attributed to elephants. However, most authors still believed in transport and deposition by the Flood (Gaudant 2008; Godard 2018b), including for elephants!

During the 18th century, the organic origin prevailed and the idea of the Flood was finally set aside. Here are some examples already mentioned for France: Benoît de Maillet affirmed in his *Telliamed* that the oysters of Sainte-Croix-du-Mont had a marine origin. He could draw on studies of the Bordeaux Academy, where Isaac Sarrau de Boynet had demonstrated in 1718 the organic origin of oysters thanks to the scar left by the adductor muscle on the inner surface of the shells. A few decades later, Jean-Étienne Guettard (1765) used this same argument for the (Eocene) fossils of Champagne. In 1767, Jacques François Borda d'Oro firmly asserted that the fossiliferous (Eocene) sediments of the Dax region in Gascony had been deposited in the sea, because he observed traces of lithophages incompatible with the Flood. In 1761, Nicolas Desmarest was convinced of the marine origin of Aquitaine fossils. Later, in his *Encyclopédie méthodique*, he dismissed any geological implication of the biblical Flood and adopted a uniformitarian approach: "The course [of phenomena] has always been such as we observe it today, simple, regular, and subject to ordinary laws" (Desmarest et al., an III-1828, Vol. 3, Article "Déluge", p. 612).

Despite great confusion of minds and the confrontation of contradictory theories, some of them innovative and others obsolete, one perceives a significant evolution of ideas from the 16th to the 18th century: the Earth, first considered fixed and immutable, to the point of imposing the denial of the marine origin of fossils, gradually became a dynamic Earth, where the seas are mobile. During the transition between these two conceptions, the biblical Flood was used by some as an indisputable event or by others as a convenient process. It would be definitively replaced by the concept of marine transgression with the birth of modern geology at the beginning of the 19th century.

3.2.2. *Volcanoes*

Volcanoes have aroused interest and great fear since time immemorial. Etna, Vesuvius and the volcanoes of the Aeolian Islands, in particular, have always fed stories, legends and myths. At the beginning of the 17th century, one event marked people's minds: on December 16, 1631, Vesuvius erupted, in one of the most catastrophic eruptions in its history, causing the death of several thousand people.

Bibliographies on Vesuvius mention more than 100 accounts of this eruption written by contemporaries: 113 books or pamphlets, 14 manuscripts, 30 poems, five songs and a dozen engravings (Ricciardi 2009; Godard et al. 2017). It is therefore no exaggeration to say that the event marked its time.

One of the episodes that surprised witnesses was the temporary withdrawal of the sea from the port of Naples on the third day of the eruption, confirmed by several testimonies, including that of Gabriel Naudé.

> The sea withdrew from the port for the space of a Credo only, to the length of six picques [about 10 m…], so that all the small boats and ships that were at the edge of the little Mole were left dry without a drop of water. […] After which, the sea returned promptly, and exceeded its ordinary bounds by far more than it was accustomed. (Naudé 1632, p. 22)

As simultaneously with this *maremoto*, it began to rain mud and ashes, and it was imagined that "the said water was carried under the ground and poured like a deluge upon the said mountain", generating torrents of mud. These were called "*lava*" by the Neapolitans, and the misinterpretation of this word is the origin of the modern word "lava".

A few months later, the Reverends Fathers Gilles de Loches and Césarée de Rosgo, returning from the Orient, recounted to Cardinal Francesco Barberini in Rome, and then to Peiresc, the eruption of a volcano in Abyssinia. The apparent simultaneity of this eruption with that of Vesuvius led them to suggest that the two volcanoes might be connected in depth and that "the source of this fire came […] from very near the center of the earth". This is a clear allusion to the "central fire" found in the writings of many authors, such as Kircher, who, in his *Mundus subterraneus* (Kircher 1665), calls it "*ignis centralis*" (see Figure 3.8). From the 16th century to the first half of the 18th, the word "fire" was taken in a literal sense, which was no longer acceptable at the end of the 18th century, when Lavoisier's chemistry prevailed (Ellenberger 1994, pp. 19–21). Buffon, de Romé de L'Isle and then Hutton argued that, lacking air, combustion is impossible at depth. The "central fire" was then replaced by "underground heat".

Scholars of the 16th and 17th centuries, probably influenced by the texts of Antiquity (see Chapter 1, sections 1.2 and 1.3), imagined cavities and underground conduits from which "liquid bitumen" (what we call "lava") poured out and where seawater could be drawn in, which could explain variations in sea level. Peiresc also circulated winds there, which he believed were responsible for earthquakes. Others thought that such underground circulations of water could compensate for the enigmatic sea current that, through the Strait of Gibraltar, flows constantly from the Atlantic into the Mediterranean. This conception of an Earth full of cavities and conduits then allowed many phenomena to be explained. Kircher popularized it with his *Mundus subterraneus*, published in 1664–1665.

Figure 3.7. *Vesuvius before and during the December 1631 eruption, according to Giuliani (source: Giuliani 1632) (© Bibliothèque nationale de France (BnF))*

The 18th century was marked by the discovery of extinct volcanoes in Europe, particularly those of the Chaîne des Puys, the Velay and the Vivarais, as well as in Tuscany, Germany and Bohemia. Antonio Micheli recognized in 1733 the volcanic origin of Monte Amiata (Tuscany), but his observations were published only four decades later by Giovanni Targioni-Tozzetti in his *Relazioni d'alcuni viaggi fatti in diverse parti della Toscana*. Jean-Étienne Guettard (Guettard 1756) recognized the stone of Volvic as lava and, although the discovery's priority is disputed, it led to recognition of the volcanoes of the Chaîne des Puys. It was especially Nicolas Desmarest who, from 1763 onward, conducted numerous observations on the volcanoes of Auvergne, for which he produced a map with François Pasumot; he recognized the volcanic origin of the prismatic basaltic flows, which he derived from the melting of the underlying granite (what we call trachytes resulted, according to him, from the fusion of granite), and distinguished three "epochs" of volcanism. Genssane, Faujas de Saint-Fond, Mortessagnes and Giraud Soulavie in turn discovered extinct volcanoes in the Velay, the Vivarais and southern France.

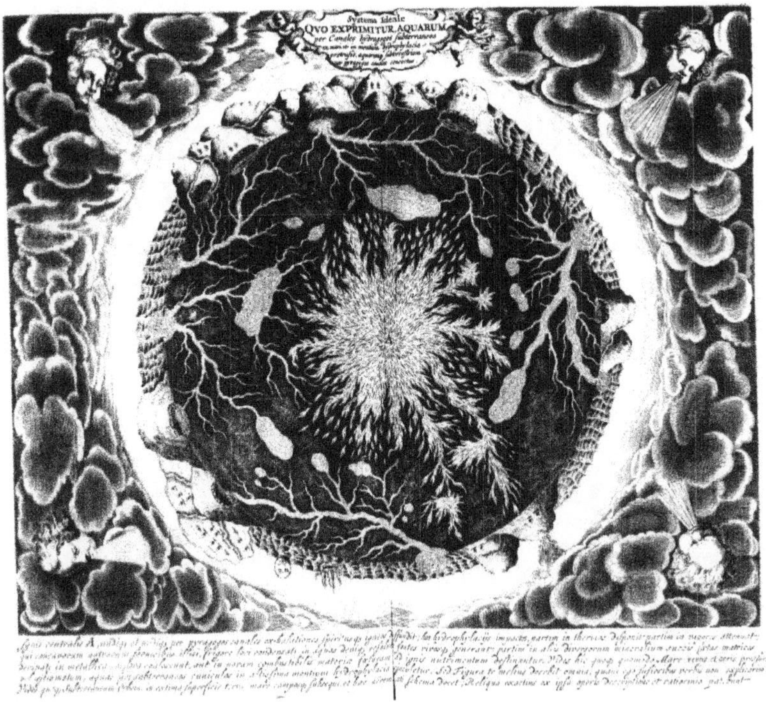

Figure 3.8. *Section of the Earth according to Kircher's* Mundus subterraneus *(vol. I, p. 175 of the 1668 edition). The Earth is represented with cavities and conduits filled with water and "fire", around the cavity of the "central fire" (A) (© Bibliothèque nationale de France (BnF))*

More than volcanoes, earthquakes, such as that of Noto (Sicily) on January 11, 1693, and especially that of Lisbon on November 1, 1755, impressed minds through their unprecedented destruction and innumerable victims. In the 17th century, circulations of "fire", water and wind in underground cavities were invoked to explain these quakes.

This conception seems to evolve in the 18th century, where Desmarest, for example, mentioned an underground explosion in the Atlantic whose shock propagated through solid earth to explain the Lisbon earthquake. The strongest questions raised by this earthquake were, however, philosophical, since European Enlightenment thinkers, particularly Voltaire in *Candide*, questioned the role of divine will and providence in these phenomena.

3.2.3. *Theories of the Earth*

An important innovation in Earth sciences in the 17th and 18th centuries was the publication of theoretical systems that aimed to provide a coherent and comprehensive model of the formation and structure of the entire globe. These "theories of the Earth" sought to explain the uplift of mountains, the existence of fossils, the formation of terrains, the origin of earthquakes and the cause of volcanic eruptions (Roger 1973; Ellenberger 1994). They were often speculative, to the point of being sometimes criticized, even by contemporaries. Nicolas Desmarest, for example, considered that "the theory of the Earth is to physical geography what fable is to history". In his *Éloge de Guettard*, Condorcet praises him for having "advanced the true theory of the Earth more than the philosophers who produced brilliant hypotheses". Jacques Roger explains that Werner coined the word "geognosy" precisely to distinguish concrete science from systematic speculation, yet he could not help himself from also producing a theory of the Earth (Roger 1973).

Let us briefly trace some of these theories of the Earth, in chronological order, to perceive an evolution of ideas. In the 1630s, Peiresc conceived such a global theory, which emerges in the hundreds of letters he exchanged with scholars across Europe and the Orient, though he never wrote it formally (Godard 2005). He sometimes evoked uplifts and sometimes collapses to explain here the emergence of marine shells and there the immersion of Baltic amber, in which one observes, he wrote, "gnats". At the time of Galileo's trial, he appeared as a Copernican *in pectore*, imagining that the Earth's rotation on its axis caused ocean currents and winds oriented "from the Levant to the Ponant", a direction which was also, according to him, that of mountain chains. He did not realize that such a conception implied the solid Earth was decoupled from its fluid envelopes, which was incompatible with Copernican heliocentrism.

The first theory of the Earth recognized as such is the one, so often cited, published by René Descartes in his *Principia Philosophiæ* (Descartes 1644). For Descartes, the Earth was a cooled star, of which volcanoes were a remnant. Vertical movements were important since he imagined an external solidified crust which, collapsing into an underlying watery layer, would have been submerged in places and remained emergent elsewhere, producing seas and mountains. Rather than a continuous aqueous layer at depth, most of Descartes' contemporaries, such as Peiresc and especially Kircher, envisioned a spongy Earth, formed of a network of cavities where "fires", waters and winds circulated, connected to the "central fire", the sea bottoms and the emerged lands (see Figure 3.8). They admitted vertical movements, like Steno, who in his *Prodromus* (Steensen 1669) attributes the tilting of horizontally deposited sedimentary layers to collapses.

According to Roger (1973), it is possible that the expression "theory of the Earth" was used for the first time in the title of *Telluris Theoria sacra* by Thomas Burnet (1681). This author revisits the ideas of Descartes' *Principia Philosophiæ*, in particular the concept of a solid envelope collapsing into a watery layer, and attempts to reconcile them with the biblical text by adopting a short chronology compatible with it (Gohau 2000). John Woodward (*An Essay towards a Natural History of the Earth*) and William Whiston (*A New Theory of the Earth*) were also respectful of the biblical account (Woodward 1695; Whiston 1696). These three English naturalists piously referenced Genesis and the Flood, which they did not intend to depart from much.

Such concern is absent from Benoît de Maillet. A toned-down version of his *Telliamed* was printed by Abbé Le Mascrier in 1748, but the manuscript had circulated clandestinely since the 1720s (Maillet 1748; Maillet and Cohen 2023). In the manner of Galileo's *Dialogo*, the author made a French missionary converse with an Indian philosopher named Telliamed (an anagram of De Maillet), who advanced audacious hypotheses on the origin of the Earth and species, without concern for the Bible, which he effectively treated as a collection of myths. De Maillet believed that lands, once entirely covered by water, gradually emerged through the evaporation of this water into space, which explains the presence of marine shells on the highest mountains and a kind of mutation of marine beings into terrestrial ones. He did not hesitate to adopt a long chronology of several billion years.

In his *Protogaea*, written in 1693 but published in 1749, Gottfried Wilhelm Leibniz drew on Descartes (the proto-Earth is a cooled star) and Steno (sedimentary strata tilted during collapses), believed that waters had covered much of the proto-Earth before being partially engulfed in underground cavities, admitted the organic origin of fossils and avoided giving great importance to the universal Flood. In a word, he approached De Maillet's *Telliamed* without adopting its most radical and extravagant ideas.

In 1749, Buffon published *Histoire et théorie de la Terre*, followed by 19 "Proofs", in the first volume of his *Histoire naturelle* (Buffon 1749). This work, completed by *Les Époques de la nature* in 1778, was intended as a compilation of the earlier writings and offered a synthesis of the avant-garde ideas of the time: the organic origin of fossils, the past presence of the sea on lands now emergent, the horizontal deposition of strata that are now inclined, the slow and continuous cooling of the Earth, a long chronology (Earth's age estimated at 75,000 years) and the actualistic conception of Earth's evolution ("The effects that occur every day, the movements that succeed and renew themselves without interruption, the constant operations always repeated, these are our causes and our reasons…").

Was Buffon a precursor of uniformitarian thought, which would dominate in the 19th century, especially through Lyell? Uniformitarianism involved two distinct ideas (Rudwick 1985): (1) geological agents operating in the past were the same as today (actualism); (2) they operated slowly over very long durations (uniformitarianism sensu stricto). Some 17th- and 18th-century pioneers, such as Steno, Barrère, Buffon, Desmarest and Hutton, more readily accepted the first idea than the second, as many were inclined to think that geological agents could have acted in the past with high intensities rather than over long periods.

The memoir and subsequent book *Theory of the Earth* by James Hutton (1788, 1795), as well as the work of his disciple John Playfair, who published *Illustrations of the Huttonian Theory of the Earth* in 1802, marked a turning point toward modernity (Dean 1992). These authors did not content themselves with pure speculation, as many earlier theories of the Earth did, but brought new observations from which they deduced concepts still valid in modern geology. These include unconformities, which are evidence of what are known today as orogenic cycles, and the intrusive origin of granites. Having observed the granite veins of Glen Tilt in the Scottish Highlands, Hutton argued for their injection in a molten state. This proposal of an igneous origin of granite gave rise to plutonism, a theory that, at the beginning of the 19th century, would render neptunism obsolete. Adopted by Werner and his disciples, the latter theory, which still prevailed at the end of the 18th century, implied that crystalline rocks, including basalts, had precipitated in a primordial ocean.

3.3. Conclusion

We have just surveyed three centuries of the history of Earth sciences in a succinct and therefore incomplete manner. Great names have not been cited (Varenius, Arduino, Fontenelle, Réaumur, de Luc, Saussure, Dolomieu, etc.); others have barely been mentioned. More complete or specialized works, listed in the bibliography, may be consulted to learn more. We have also omitted discussion of advances in stratigraphy, which emerged at this time, particularly with Nicolaus Steno, since this topic is treated in a specific chapter (see Chapter 6).

Despite these omissions, this chapter presents an overview of the evolution of Earth sciences from the Renaissance to the Age of Enlightenment. During this period, awareness grew of the numerous upheavals the Earth had undergone in the past: the definitive rejection of the freak of nature theory in favor of the organic origin of fossils implies a mobility of the seas; mountains have risen, since marine fossils are observed there and the strata are tilted and folded; fossil species have disappeared from present-day seas; elephants once lived in Europe; volcanoes now extinct were active; granites intruded older terrains; etc.

In short, the idea of a dynamic and changing Earth emerges. This conception requires adopting a chronology longer than that of the biblical account. Sacred texts, as well as rediscovered ancient secular texts from the Renaissance, reflect humanity's unconscious tendency to adopt an age of the world acceptable to the mind, that is, neither too short (it must at least encompass history) nor too long (at most a few millennia), in order to spare it the vertigo of an immensely long time. This short chronology was challenged, especially in the 18th century (Richet 1999): the upheavals of sedimentary layers, the salinity of seawater (Edmond Halley) and the carving of valleys (Giraud Soulavie) required immense time. The human mind did attempt to resist the vertigo of this new chronology by adopting certain artifices. The transport of fossils by the Flood, for example, was a convenient way to explain their presence in emergent terrains, but it was also a means of compressing time. The uplift of mountains during cataclysms and earthquakes was another. Thus, catastrophism arose, persisting into the 19th century, while a few pioneers laid the groundwork for uniformitarianism, which, with long chronology, would ultimately prevail.

It therefore appears that the perception of Earth's history evolved greatly since the Middle Ages. Despite a chaotic path marked by many back-and-forths, the human mind passed through several stages of a complex game of goose that, by the early 19th century, brought it to the threshold of a decisive stage in the history of geology, which François Ellenberger very aptly called "the great emergence" (Ellenberger 1994). It was emphasized in the introduction of this book that, after a long prehistory, geology crossed a threshold to become a modern science in which the contemporary geologist recognizes the beginnings of the discipline. This was in fact the case for many scientific disciplines, such as alchemy being replaced by Lavoisier's chemistry. For geology, this threshold was crossed at the end of the 18th century and certainly by the beginning of the 19th century, as evidenced by the definitive acceptance of the organic origin of fossils, the perception of terrestrial dynamics (this chapter), the beginnings of geological mapping (see Chapter 5), the emergence of stratigraphy (see Chapter 6) and emergence of crystallography (see Chapter 4).

3.4. References

Agricola, G. (1556). *De re metallica*. Mining, London.

Alessandrini, A. (1978). *Cimeli Lincei a Montpellier*. Accademia nazionale dei Lincei, Rome.

d'Anthonay, T. (2014). *Le baron de Beausoleil*. Albin Michel, Paris.

de Bertereau, M. (1640). *La restitution de Pluton à Mgr l'éminentissime cardinal duc de Richelieu des mines et minières de France*. Hervé du Mesnil, Paris.

Besler, B. (1622). *Continuatio rariorum et aspectu dignorum varii generis quae collegit et suis impensis aeri ad vivum incidi curavit atque evulgavit Basilius Besler*. Nuremberg, Berlin.

Bloch, O. (1971). *La philosophie de Gassendi. Nominalisme, Matérialisme et méta-physique*. M. Nijhoff, La Haye.

Bombicci, L. (1873–1874). Descrizione della mineralogia generale della provincia di Bologna. *Memorie dell'Accademia delle Scienze di Bologna*, IV, 57–139, V, 105–222.

Boulanger, N.-A. and Boutin, P. (2006). *Anecdotes physiques de l'histoire de la nature, avec la Nouvelle Mappemonde, et le Mémoire sur une nouvelle mappemonde. Édition critique, textes établis et commentés par Pierre Boutin*. Honoré Champion, Paris.

Buffon, G.-L. (1749). Histoire et théorie de la Terre. In *Histoire naturelle générale et particulière*, Buffon, G.-L. (ed.). Plassan, Paris.

Burnet, T. (1681). *Telluris theoria sacra, orbis nostri originem et mutationes generales, quas aut jam subiit aut olim subiturus est, complectens. Libri duo priores de diluvio et paradiso*. G. Kettilby, London.

Camerarius, E. (1712). *Dissertationes taurinenses Epistolicae physico-medicae*. Impensis G. Cottae, Tübingen.

Céard, J. (1978). La querelle des géants et la jeunesse du monde. *Journal of Medieval and Renaissance Studies*, 8, 37–76.

Cerquaglia, Z. (2005). Francesco Stelluti ed i metallofiti nell'Umbria meridionale. In *Federico Cesi e i primi lincei in Umbria : atti del Convegno di studi nel IV centenario della fondazione dell'Accademia dei Lincei, Terni, 24-25 ottobre 2003*, Pirro, V. (ed.). Thyrus, Arrone.

de Clave, É (1635). *Paradoxes ou traittez philosophiques des pierres et pierreries, contre l'opinion vulgaire*. Vve P. Chevallier, Paris.

Cohen, C. (1994). *Le destin du mammouth.* Le Seuil, Paris.

Colonna, F. (1616). De Glossopetris Dissertatio. In *Fabii Columnae. Purpura, hoc est de purpura ab animali testaceo fusa, de hoc ipso animal*, Colonna, F. (ed.). Mascardum, Rome.

Cuvier, G. (1812). *Recherches sur les ossemens fossiles de quadrupèdes, où l'on rétablit les caractères de plusieurs espèces d'animaux que les révolutions du globe paroissent avoir détruites.* Deterville, Paris.

Dean, D.R. (1992). *James Hutton and the History of Geology.* Cornell University Press, Ithaca.

Descartes, R. (1644). *Principia philosophiae.* Elvezirium, Amsterdam.

Desmarest, N., Doin, G.-T., Ferry, C.-J., Huot, J.-J.-N., Bory de Saint-Vincent, J.-B.-G.-M. (1828). *Encyclopédie méthodique. Géographie-physique.* H. Agasse, Paris.

Dezallier d'Argenville, A.-J. (1742). *L'histoire naturelle éclaircie dans deux de ses parties principales, la lithologie et la conchyliologie, dont l'une traite des pierres et l'autre des coquillages.* De Bure l'aîné, Paris.

Duffin, C.J., Moody, R.T.J., Gardner-Thorpe, C. (2013). *A History of Geology and Medicine.* Geological Society, London.

Ellenberger, F. (1988). *Histoire de la géologie 1 : des Anciens à la première moitié du XVIIe siècle.* Tec & Doc, Paris.

Ellenberger, F. (1994). *Histoire de la géologie 2 : la grande éclosion et ses prémices 1660-1810.* Tec & Doc, Paris.

Fazello, T. (1558). *De Rebus siculis de cades duae, nunc primum in lucem editae.* J.M. Maidam et F. Carraram, Panormi.

Gachet d'Artigny, A. (1749). *Nouveaux mémoires d'histoire, de critique et de littérature 1.* Debure, Paris.

Gaffarel, J. (1629). *Curiositez inouyes sur la sculpture talismanique des Persans, horoscope des patriarches et lecture des estoilles.* H. Du Mesnil, Paris.

Galvani, C. (1780). *Della pietra fosforica Bolognese.* Longhi, Bologna.

Gassend, P. (ed.) (1658). De Lapidibus, ac mettalis (Phisicae Sectio III, Liber III). In *Opera Omnia*. Anisson & Devenet, Lyon.

Gaudant, J. (2008). *Les fossiles témoins du déluge*. Mines ParisTech, Paris.

Gaudant, J. (2010). Wilhelm Ernst Tentzel (1659-1707) et le Collège des médecins de Gotha face à l'éléphant fossile de Tonna (1696-1697). *Travaux du Comité fr. d'Histoire de la Géologie*, 3(XXIV), 117–130.

Gaudant, J. (2012). Une nouvelle contestation de la nature organique des fossiles : les Dissertationes taurinenses Epistolicae physico-medicae d'Elias Camerarius (1712). *Travaux du Comité fr. d'Histoire de la Géologie*, 3(XXVI), 235–240.

Gaudant, J. and Bouillet, G. (2004). Johannes Reiskius (1687). Une tentative de réfutation aristotélicienne des idées de Fabio Colonna et Nicolas Sténon sur la nature organique des glossopètres. *Travaux du Comité fr. d'Histoire de la Géologie*, 3(XVIII), 93–102.

Gesner, C. (1565). *Conradi Gesneri de Rerum fossilium, lapidum et gemmarum maxime, figuris et similitudinibus liber*. Gesner, Tiguri.

Ginsburg, L. (1984). Nouvelles lumières sur les ossements autrefois attribués au géant Theutobochus. *Annales de Paléontologie*, 70(3), 181–219.

Ginsburg, L. (1986). La fabuleuse histoire du géant Theutobochus. *Univers du vivant*, 15, 44–51.

Giuliani, G.B. (1632). *Trattato del monte Vesuvio e dé suoi incendi…* Egidio Longo, Naples.

Godard, G. (2005). Peiresc et la proto-géologie du début du XVIIe siècle. *Sciences et techniques en perspective*, 9(1), 63–110.

Godard, G. (2009). The fossil proboscideans of Utica (Tunisia), a key to the "giant" controversy, from Saint Augustine (424) to Peiresc (1632). *Geological Society, London, Special Publications*, 310, 67–76.

Godard, G. (2017). The discourse on the seashells […, found] in Champagne, written by Vignier to Peiresc (1635): An early text supporting the organic origin of fossils. *Comptes Rendus Palevol*, 16, 122–131.

Godard, G. (2018a). Premières descriptions des faluns miocènes de Sainte-Croix-du-Mont en Guyenne (1622, 1718) : Pierre de L'Ancre a-t-il perçu le principe de superposition 45 ans avant Sténon ? *Bulletin de la Société Linnéenne de Bordeaux*, 153(46), 297–310.

Godard, G. (2018b). Early texts on the Cenozoic fossils of Aquitaine (1622-1767) and pioneering debates on the organic origin of fossils, the superpositioning of strata and the mobility of the seas. *BSGF-Earth Sciences Bulletin*, 189(8). doi: 10.1051/bsgf/2018007.

Godard, G. (2019). Un manuscrit sur les faluns miocènes de Salles en Gascogne, écrit vers 1660 : "Reflections sommaires sur quelques pierres de la Terre de Sales". *Bulletin de la Société Linnéenne de Bordeaux*, 154(47), 249–260.

Godard, G., Guzzetta, G., Frati, G. (2017). Manuscript on the Vesuvius eruption of 1631 by the Carthusian Dom Severo Tarfaglione. *Analecta Cartusiana*, 291, 16–66.

Gohau, G. (1990). *Les sciences de la terre aux XVIIe et XVIIIe siècles : naissance de la géologie.* Albin Michel, Paris.

Gohau, G. (1991). *A History of Geology.* Rutgers University Press, New Brunswick, New Jersey.

Gohau, G. (2000). Thomas Burnet, la première Théorie de la Terre : entre science et religion. *Travaux du Comité fr. d'Histoire de la Géologie*, 3(XIV), 49–59.

Goropius, J. (ed.) (1569). Gigantomachia. In *Origines Antwerpianae II.* Brill, Leiden.

Guettard, J.-E. (1756). Mémoire sur quelques montagnes de la France qui ont été des volcans. *Mémoires de l'Académie Royale des Sciences*, 27–59, 4–5.

Guettard, J.-E. (1759). Mémoire où l'on examine en général le terrein, les pierres & les différens fossiles de la Champagne & de quelques endroits des Provinces qui l'avoisinent. *Mémoires de l'Académie Royale des Sciences*, 435–494, 17–18.

Guettard, J.-E. (1765). Mémoire sur les accidents des coquilles fossiles, comparés à ceux qui arrivent aux coquilles qu'on trouve maintenant dans la mer. *Mémoires de l'Académie Royale des Sciences*, 189–226, i-iv; 329–357, i-iii; 399–419, i-ii.

Guglielmini, D. (1688). *Riflessioni filosofiche dedotte dalle figure de' Sali.* A. Pisarri, Bologna.

Halleux, R. (1979). L'œuvre minéralogique d'Anselme Boèce de Boodt (1550-1632). *Histoire et Nature*, 14, 63–78.

Halleux, R. (1982). La littérature géologique française de 1500 à 1650 dans son contexte européen. *Revue d'histoire des sciences*, XXXV, 111–130.

Hampton, J. (1955). *Nicolas-Antoine Boulanger et la science de son temps.* Droz, Geneva.

Haüy, R.-J. (1801). *Traité de minéralogie.* Louis, Paris.

Hirai, H. (2005). *Le concept de semence dans les théories de la matière à la Renaissance de Marsile Ficin à Pierre Gassendi.* Brepols, Turnhout.

Hooke, R. (1705). *The Posthumous Works of Robert Hooke. To These Discourses is Prefixt the Author's Life. Publish'd by Richard Waller.* S. Smith and B. Walford, London.

Hutton, J. (1788). Theory of the Earth or an investigation of the laws observable in the composition, dissolution, and restoration of land upon the globe. *Trans. Roy. Soc. Edinburgh*, 1, 209–304.

Hutton, J. (1795). *Theory of the Earth with Proofs and Illustrations.* W. Creech, Edinburgh.

Kircher, A. (1665). *Mundus subterraneus, in XII libros digestus.* Janssonium & Weyestraten, Amsterdam.

de L'Ancre, P. (1622). *L'incredulité et mescreance du sortilege plainement conuaincue, ou il est amplement et curieusement traicté de la verité ou illusion du sortilege et d'vne infinité d'autres rares & noueaux subjects.* N. Buon, Paris.

Langenmantel, H.A. (1689). De ossibus elephantum. *Miscellanea curiosa medico-physica Academiae Naturae Curiosorum sive Ephemeridum medico-physicarum Germanicarum curiosarum*, 2(VII), 446–447.

Lassalle, R. (1992). *Peiresc : 1580-1637. Vie de l'illustre Nicolas-Claude Fabri de Peiresc, conseiller au parlement d'Aix par Pierre Gassendi.* Belin, Paris.

Leibniz, G.W. (1749). *Summi polyhistoris Godefridi Guilielmi Leibnitii Protogaea, sive de Prima facie telluris et antiquissimae historiae vestigiis in ipsis naturae monumentis dissertatio, ex schedis manuscriptis in lucem edita a Christiano Ludovico Scheidio.* Sumptibus J.G. Schmidii, Goettingen.

Lister, M. (1678). *Historiae animalium Angliae tres tractatus : unus, de araneis, alter de cochleis tertius de cochleis marinis.* J. Martyn, London.

Maggi, G. (1564). *Hieronymi Magii Variarum lectionum seu Miscellaneaorum libri IV.* Zileti, Venice.

Maggi, G. (1603). *Hieronymi Magii de Gigantibus; accedit Eilardae Almae Bellum giganteum.* Commelinianos, Heidelberg.

de Maillet, B. (1748). *Telliamed ou Entretiens d'un philosophe indien avec un missionnaire françois sur la diminution de la mer, la formation de la terre, l'origine de l'homme, etc./ Mis en ordre sur les mémoires de feu M. de Maillet. Par J. A. Guer.* L'Honoré & fils, Amsterdam.

de Maillet, B. and Cohen, C. (2023). *Telliamed. Entretiens d'un philosophe indien avec un missionnaire français sur la diminution de la mer.* Jérôme Millon, Grenoble.

Maitte, B. (2001). René-Just Haüy (1743-1822) et la naissance de la cristallographie. *Travaux du Comité fr. d'Histoire de la Géologie*, 3(XV), 145–179.

Marsigli, L.F. (1698). *Dissertazione epistolare del fosforo minerale o sia della pietra illuminabile bolognese.* Leipzig.

Mongitore, A. (1704). *La Sicilia inventrice, o vero, le inventioni lodevoli nate in Sicilia.* Felice Marino, Palermo.

Murray, D. (1904). *Museums: Their History and Their Use.* J. MacLehose & Sons, Glasgow.

Naudé, G. (1632). Discours sur les divers incendies du mont Vesuve et particulièrement sur le dernier qui commença le 16 décembre 1631, Paris.

Naudé, G. (n.d.). *Jugement de tout ce qui a esté imprimé contre le cardinal Mazarin : depuis le sixieme januier, jusques à la declaration du premier avril mil six cens quarante neuf.* Mascurat.

Oldroyd, D.R. (1974). Some neo-Platonic and Stoic influences on mineralogy in the sixteenth and seventeenth centuries. *Ambix*, 21, 128–156.

Ottaviani, A. (2021). The opposite poles of a debate. Lapides figurati and the Accademia dei Lincei. *Substantia*, 5(1), 19–28.

Palissy, B. (1580). *Discours admirables de la nature des eaux et fontaines, tant naturelles qu'artificielles, des métaux, des sels et des salines, des pierres, des terres, du feu et des émaux.* Martin le Jeune, Paris.

Plaziat, J.-C. (2011). Bernard Palissy (1510-1590) and the French geologists: A critical reappraisal concerning the founding naturalist and his rustic ceramics. *Bulletin de la Société géologique de France*, 182(3), 255–267.

Plaziat, J.-C. and Guérin, H. (2011). Le Lutétien de la Montagne de Reims, du XVIIIe siècle (Courtagnon) à nos jours (Fleury-la Rivière) : un apport essentiel à la connaissance des fossiles de l'Éocène et un espoir de renouveau. *Bulletin d'Information des Géologues du Bassin de Paris*, 48(4), 3–15.

Plot, R. (1677). *The Natural History of Oxfordshire, Being an Essay toward the Natural History of England.* Theater, Oxford.

Ray, J. (1673). *Observations Topographical, Moral and Physiological, Made in a Journey Through Part of the Low-Countries, Germany, Italy and France.* J. Martyn, London.

Reiske, J. (1687). *Commentatio physica æque ac historica de Glossopetris Luneburgensibus.* Ex Officina Johannes Ziegeri, Nuremberg.

Ricciardi, G.P. (2009). *Diario del Monte Vesuvio : venti secoli di immagini e cronache di un vulcano nella città.* Edizioni scientifiche e artistiche, Naples.

Richet, P. (1999). *L'âge du monde : à la découverte de l'immensité du temps.* Le Seuil, Paris.

Rivière, G. (1766). Mémoire sur les dents pétrifiées de divers poissons, comparées avec les dents des mêmes poissons nouvellement pêchés. *Histoire de la Société royale des sciences, établie à Montpellier*, 1, 75–84.

Rochot, B. (1944). *Les travaux de Gassendi sur Épicure et sur l'atomisme. 1619-1650.* Vrin, Paris.

Roger, J. (1973). La théorie de la Terre au XVIIe siècle. *Revue d'histoire des sciences*, XXVI, 23–48.

de Romé de L'Isle, J.-B.L. (1772). *Essai de cristallographie, ou Description des figures géométriques, propres à différens corps du règne minéral, connus sous le nom de cristaux.* Didot, Paris.

Routhier, P. (1987). Deux "mineurs" spoliés et emprisonnés ou "la Restitution de Pluton" (1640) par Madame la Baronne de Beausoleil. *Travaux du Comité français d'Histoire de la Géologie*, 3(1), 1–8.

Rudwick, M.J.S. (1985). *The Meaning of Fossils. Episodes in the History of Paleontology.* University of Chicago Press, Chicago.

Rülein Von Calw, U. (1500). *Ein nützlich Bergbüchlein.* Martin Landsberg, Leiden.

Scheuchzer, J.J. (1708). *Piscium querelae et vindiciae expositae a Johanne Jacobo Scheuchzero.* Sumtibus authoris, Tiguri.

Schnapper, A. (1986). La persistance des géants. *Annales : Economie sociétés civilisations*, 40(1), 177–200.

Schnapper, A. (1988). *Le géant, la licorne et la tulipe. Collections et collectionneurs dans la France du* XVII*e siècle. Histoire et histoire naturelle.* Flammarion, Paris.

Scott, A.C. (2001). Federico Cesi and his field studies on the origin of fossils between 1610 and 1630. *Endeavour*, 25(3), 93–103.

Scott, A.C. and Freedberg, D. (2000). *Fossil Woods and Other Geological Specimens, Part Three (The Paper Museum of Cassiano dal Pozzo. Series B: Natural History).* Harvey Miller, London.

Sneyders de Vogel, K. (1955). Une lettre de Herryard, joaillier du Grand Mogol. *Neo-philologus*, XXXIX, 1–8, 95–103.

Steensen, N. (1667). *Nicolai Stenonis Elementorum myologiae specimen, seu Musculi descriptio geometrica, cui accedunt Canis Carchariae dissectum caput et dissectus piscis ex canum genere.* Ex typ. sub signo Stellae, Florence.

Steensen, N. (1669). *Nicolai Stenonis de Solido intra solidum naturaliter contento dissertationis prodromus.* Ex typ. sub signo Stellae, Florence.

Stelluti, F. (1637). *Trattato del legno fossile minerale nuovamente scoperto.* Mascardi, Rome.

Tamizey de Larroque, P. (1890). *Lettres de Peiresc aux frères Dupuy [II: janvier 1629-décembre 1633].* Imprimerie nationale, Paris.

Vai, G.B. (2003). Aldrovandi's Will: Introducing the term "Geology" in 1603/Il testamento di Ulisse Aldrovandi e l'introduzione della parola "Geologia" nel 1603. In *Four Centuries of the Word Geology; Ulisse Aldrovandi 1603 in Bologna*, Vai, G.B. and Cavazza, W. (eds). Minerva Edizioni, Bologna.

Vallet, M. (1976). *L'aventure magique de Martine de Bertereau, suivi de la réédition de ses œuvres.* Repro 2000, Nice.

Whiston, W. (1696). *A New Theory of the Earth from Its Original to the Consummation of All Things.* B. Tooke, London.

Woodward, J. (1695). *An Essay Toward a Natural History of the Earth and Terrestrial Bodies, Especially Minerals; As Also of the Sea, Rivers and Springs, with an Account of the Universal Deluge.* R. Wilkin, London.

4

Crystallography: 17th–19th Centuries

Bernard Maitte
Université de Lille, Villeneuve d'Ascq, France

4.1. The first studies

4.1.1. *Kepler in a lineage*

> To the illustrious Aulic Councillor of His Majesty the Emperor […] Maecenas of men of letters and philosophers, my beneficent protector. I am well aware how much you love Nothing, not, to be sure, because of its slight worth, but on account of the play […]. Thus it is easy for me to presume that a gift will be all the more agreeable and pleasant to you the closer it comes to Nothing. (Kepler 1609)

Johann Kepler thus addressed Johannes M. W. von Wackhenfels in the dedication of the work he offered him for his New Year's gift of 1610, the *Strena*, a scholarly entertainment whose theme was a snowflake, so beautiful, so fleeting, that, if one takes it in one's hand, it melts, disappears and nothing remains. Kepler explained this "nothing". He first described the flakes: they "take on the shape of an asterisk with six angles", then made various conjectures that might explain this shape, as well as those of other crystals, which all present themselves in hexagonal or parallelepipedal forms. The first is a material necessity of the mechanical arrangement of small spheres: spherical, non-deformable globules enclosed in a

History of Earth Sciences,
coordinated by Pierre Savaton.

round vessel assemble under compression with a triangular, square or hexagonal geometry, depending on the disposition of the layers of spheres in relation to one another. Second, in the Aristotelian lineage, the assertion that snow, like all crystals, including quartz and gems, is water frozen by the action of centuries. Third, in a very Neoplatonic manner – Kepler's writings are all tinged with mysticism – a "formative faculty" existing in bodies, the "soul of the world", "a reflection of the (geometric) perfection of God".

These conjectures recall that in the 17th century different traditions coexist: they go back to Antiquity and the Middle Ages. Those of Pythagoras, Plato, Aristotle, and the atomists, which one recognizes in Kepler's sentences. That of Pliny, whose *Natural History* sees in gemstones the realization of the highest degree of perfection of the Beauty of nature. Those of Marbode (1035–1123) and his poem in praise of precious stones, of Hildegard of Bingen (1098–1179), who describes with lyricism in her *Book of the Subtleties of Divine Creatures* and *Book of Stones* the multiple facets of creation, including crystals, minerals, their forms, uses in pharmacopoeia, roles for salvation or damnation. That of Albert the Great, who benefits from the contributions of the sciences of the Islamic world, from knowledge gathered from miners, ironworkers, or alchemists, and from his own many observations as well, to compose his magisterial *De mineralibus et Rebus Metallicis*. Finally, those of the *Studioli*, *Wunderkammern*, or "cabinets of curiosities" with which princes, merchants or bankers enrich their residences in the Renaissance.

4.1.2. *From Galileo to Steno*

At the time of Kepler, Galileo gave a decisive impulse to modern science, which he founded on mechanics in his *Discorsi* (1638). Gassendi was very likely the first to look under the microscope at snowflakes, small crystals and crystalline germs appearing in salt solutions, and to describe them in a letter to Mersenne in 1629 – which the latter made known to Descartes – then in his inaugural lecture at the Collège Royal (1645).

In the *Météores* (1647), Descartes opposed Gassendi's atomism, but explained the formation of snowflakes and hail by small balls furnished with hairs that assemble.

Robert Hooke's *Micrographia* took on an entirely different dimension: he promoted the microscope, observed the crystallization of salts and the exterior form of these crystals, and accounted for it through demonstrations "by the eye", showing that assemblies of small identical spheres coming into contact with one another formed figures with determined angles (see Figure 4.1).

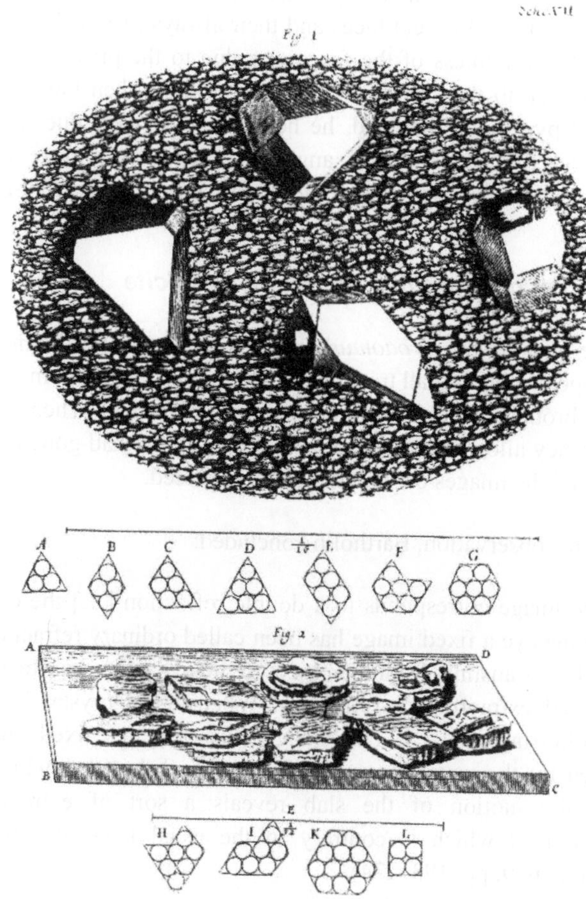

Figure 4.1. *Observation of a sulfate and explanation of its forms ((Hooke 1665, p. 82); source: gallica.bnf.fr/Bibliothèque nationale de France)*

Niels Stensen (Steno) (1638–1686) went further: in his *Prodomus* (1669), he studied the forms, always hexagonal, of quartz from various deposits, wanted to explain them, rejected all earlier explanations and held that all individual observations had to be connectable and lead to their causes. He thus defined his angle of research: "Given a body having a certain shape and formed by Nature, to find in this body itself the sure signs of the place of its formation and of the manner in which it was formed" (Sténon 1669). To answer, he made the following conjectures: quartz had formed within a nourishing fluid, their growth took place by the juxtaposition of small particles coming from the fluid and applying themselves to an initially formed germ already possessing geometric forms, the successive increments explained the different faces and their always identical angles (which he did not measure), the flatness of the faces was due to the pressure of the external fluid and their polish to the slowness of growth. Taking then the examples of the angular forms of pyrite and diamond, he held that they were due to the action of several fluids and that the study of angular stones would make it possible to determine the varieties of movements of the fluids causing the crystallizations.

4.1.3. *Bartholin and the double refraction of calcite*

In the same year as the *Prodomus*, in 1669, Erasmus Bartholin (1625–1698) published in Copenhagen a small treatise that would make history: in it, he described calcite crystals brought back from an expedition to Iceland. Their large size and perfect transparency allowed him to notice a property that had gone unnoticed until then: they doubled the images on which they were placed.

Specifying this observation, Bartholin concluded:

> A double image corresponds to a double refraction […] the one that sends to the eye a fixed image has been called ordinary refraction, the other, which transmits from the object to the eye a moving image, has been called extraordinary refraction. Whence this crystal has been called birefringent […] The image that appears fixed may be explained by the usual laws of refraction […], but that which […] follows the motion of the slab reveals a sort of extraordinary refraction […] which is contrary to the usual laws of refraction. (Bartholin 1669, pp. 193–224)

Bartholin also observed that calcite crystals always broke by cleavages while keeping the same rhombohedral shape, these rhombohedra always showed birefringence and the separation of the two images was always proportional to the thickness traversed. He believed he could affirm that the extraordinary rays were always parallel to the inclined edges of the rhombohedron and conjecture that the particles of light were refracted in channels oriented in two different ways in the crystal. These surprising and new observations attracted the attention of physicists and, notably, of Christiaan Huygens.

4.2. 17th century: the intervention of physicists; Huygens contradicted by Newton

4.2.1. *Double refraction and the structure of crystals: Huygens*

When he read Bartholin, Huygens reflected on the nature of light. Like Descartes and Hooke, he supposed space to be filled with ether; like Hooke and against Ibn al-Haytham, the atomists, Newton and Bartholin, he supported a wave theory of light: for him, the corpuscular theory – which supposed an empty space and could explain only that two light rays propagating in the same direction but in opposite senses did not weaken – is implausible. In his *Traité de la lumière* (Huygens 1690), he succeeded in explaining the existence of light rays, the rectilinear propagation, the reflection and the refraction of light by the propagation of waves, which he conjectured to be perpendicular to the rays and to propagate at a speed inversely proportional to the refractive index of media (the corpuscular theory predicted proportionality).

Seizing upon Bartholin's results, he made precise measurements on "Iceland spar" that, on the one hand, invalidated those results (the extraordinary rays were not parallel to the sides of the rhombohedra) and, on the other hand, brought many entirely new experimental results, which he interpreted. He thus demonstrated that ordinary rays corresponded to refracted spherical waves, extraordinary rays were associated with ellipsoidal waves of revolution (the rays were then no longer perpendicular to the waves), these two categories of waves could be represented by surfaces with two sheets tangent at two opposite points, in this direction there was only one refractive index and that an incident ray along this direction therefore should not split, which he verified. He measured that calcite maximally doubled the image of a straight line on which it was placed when that line was the acute bisector of the faces of calcite, that there was no doubling when the line was the bisector of the obtuse angles, and that the direction of isotropy, which had never been noticed before him, lay exactly along an axis equally inclined with respect to the rhombohedral faces of calcite, which he verified. The representation of the wave surface allowed him to predict what the propagation indices of the ordinary and extraordinary rays would be for arbitrary sections of calcite: his measurements verified his predictions. There was thus a relation between the geometric forms of the crystal and its optical properties.

Huygens then conjectured that calcite was formed by the stacking of small masses invisible with the best microscopes and bathing in ether: the geometry of each of them was compatible with an ellipsoid (Huygens 1690) (see Figure 4.2); crystallization consisted of the orderly addition of small masses to the edifice; and crystalline forms, cleavages and optical properties were determined by this stacking. If this was indeed so, other crystals had to show the phenomenon of birefringence: Huygens looked for it and brought it to light in quartz and micas. These birefringences were much weaker than that of calcite, which explained why they had until then passed unnoticed. He conjectured that ordinary rays propagated in ether, and extraordinary rays in ether and in the small masses arranged according to the precise order in which they were placed.

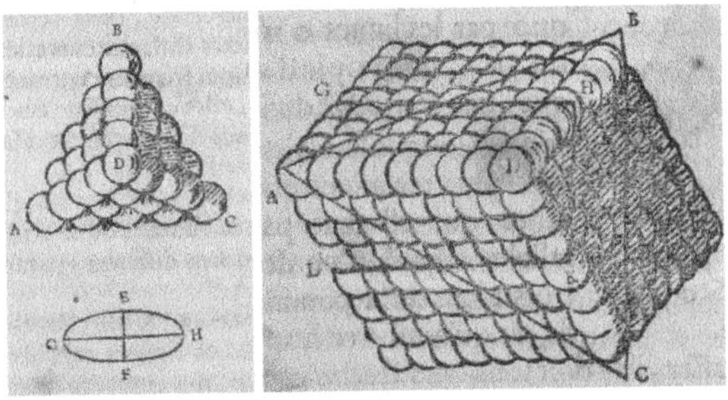

Figure 4.2. *Structure of calcite (Huygens 1690, pp. 92–93) (source: gallica.bnf.fr/Bibliothèque nationale de France)*

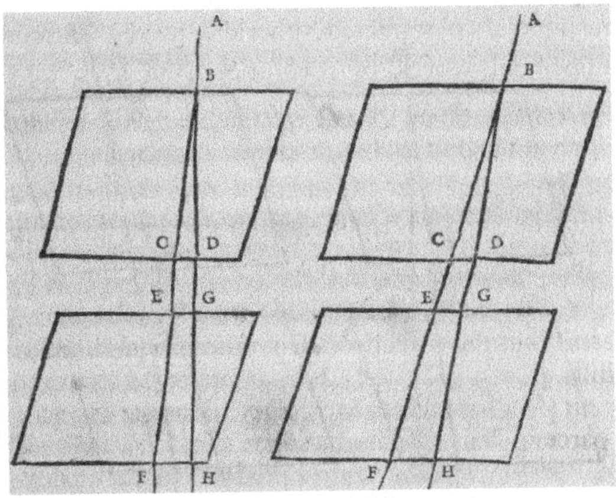

Figure 4.3. *Figure: Refraction of light in two superposed calcite crystals ((Huygens 1690, p. 89); source: gallica.bnf.fr/Bibliothèque nationale de France)*

The studies of Huygens left one phenomenon, which he discovered, unexplained: if two calcite crystals were superposed, the rays emerging from the first crystal, corresponding to ordinary or extraordinary rays, arrived at the second, were doubled there by refraction, but no longer had the same intensity. When he rotated the second crystal on itself, Huygens noticed that there existed two positions, orthogonal to each other, in which the rays were no longer doubled, although they should have been (see Figure 4.3). Huygens described the phenomenon and wrote:

> It seems that one is obliged to conclude that the waves of light, after having passed through the first crystal, acquire a certain form or disposition in virtue of which, when meeting the texture of the second crystal, in certain positions, they can move the two different kinds of matter which serve for the two species of refraction; and when meeting the second crystal in another position are able to move only one of these kinds of matter. But to tell how this occurs, I have hitherto found nothing which satisfies me. (Huygens 1690, p. 91)

4.2.2. *Newton's stinging refutation*

In his *Opticks* (1704), which he published after the deaths of Huygens (in 1695) and Hooke (in 1703), Newton, a supporter of a corpuscular theory of light, very quickly addressed crystals. Like Huygens, he supposed them formed by small masses arranged regularly, but surrounded by void. Their uniform arrangement would explain shapes, cleavages and physical properties. Newton, a marvelous experimenter, said above all that he had redone all of Huygens's experiments and had found results that did not agree, sometimes by several degrees, with those obtained by his predecessor: it was therefore wrongly that the latter asserted that his wave theory was predictive! And Newton insisted on the lack of explanation of the possible non-splitting of rays by the second calcite and justified his rejection of the wave theory by repeating, the only quotation in the 500 pages of the Opticks, Huygens's admission of incomprehension: "But to tell how this is done, I have hitherto found nothing that satisfies me" (Newton 1704).

Newton's rejection of the wave theory of light and of Huygens's results was explicable: his physics was incompatible with that of Descartes-Huygens. The latter supposed a full world in which all physical actions were explained mechanically by contacts and transmissions: God had created a perfect world analogous to an immense clock and rested on the seventh day, no longer intervening in physics. By contrast, Newton postulated an empty world in which all physical actions were explained mechanically by forces acting on masses. This allowed him, in the *Principia*, to postulate a structure of solid matter: it was composed of small masses that attracted each other by gravity.

But, since corpuscular light traversed transparent bodies in straight lines and emerged with roughly the same intensity, it was not hindered in its path by the masses: these left in such bodies "63 times more empty than full"; they attracted each other at a distance, but there were also repulsive forces that became dominant at very small distances, the masses then repelling one another and remaining separated, not collapsing upon one another to give infinitely dense solids. What were these repulsive forces? Newton could not answer and looked for leads in alchemy. What was universal gravitation? It acted in the void and was therefore not mechanical in nature: for Newton, it represented the constant action of God upon the world. The Newtonian system of the world triumphed in England; Voltaire, who became acquainted with it during an exile, was charmed by this approach that affirmed the role of God; he called it "British philosophy" and helped introduce it among the intellectuals of the continent. On the latter, it was the philosophy of Descartes that was generally accepted: How could the Cartesians accept the introduction of "occult forces" into physics? The two systems did not lead to the same predictions: their opposition was going to concentrate on the crucial problems of the movements of the stars and of the shape of the Earth. This orientation brought about the physicists' lack of interest in crystals and, after the success of Newtonian mechanics, the neglect of a phenomenon (the anisotropy of crystals) whose explanation had been labeled an error by Newton. As for the universities of Christian countries, they taught exclusively Aristotelianism.

4.3. 18th century: from the crystals of the natural sciences to mathematics

4.3.1. *Sensualists*

Against the claim of the "great systems" of Descartes and Newton, and also of the atomists, to want to explain the world in terms of mechanics, there stood at the beginning of the 18th century not only Aristotelianism but also the philosophies of John Locke (1632–1704) and Nicolas Malebranche (1638-1715). In his *Essay Concerning Human Understanding* (1690), Locke showed himself to be an empiricist in that, if determining a system of the world remained the objective to be attained, this ideal, in order to be conquered, required patient analyses. For him, all our ideas came from two sources: sensations, which made ideas enter our soul, and reflection on these ideas, which produced others: truth was only the provisional concordance between what we apprehended; to give our knowledge this characteristic was to guarantee tolerance. The "Age of Enlightenment" was passionate about Locke, practiced him, admired him. Malebranche undertook, for his part, to reform Descartes's physics by developing geometrization in it: we had clear ideas only of numbers and their relations, not of the soul. After them, philosophers and scholars, deists and atheists studied with passion everything that was neglected by the "great systems", notably the natural sciences. Collections of plants, animals, rocks, minerals and crystals were assembled: they were housed in cabinets of curiosities that multiplied. In salons, the richest and wildest ideas were elaborated. Journals, reviews and scientific works were published and met with great success. A new stage in the study of crystals was born from this orientation.

How to present the pieces in cabinets of curiosities? Leibniz (1646–1716) showed himself to be an advocate of expressing the continuity of nature, going from the least organized matter to Man, each degree differing hardly from those preceding or following it. Charles Bonnet (1720–1793) shared this position and proposed a "Scale of Natural Beings" going, in order of increasing complexity, from the "elements" (fire, air, water and earth) to crystallizations, then to fossils, stones, corals, mushrooms and lichens, before passing to plants, animals and then humans. Buffon, Diderot, Kant and Rousseau also showed themselves to be partisans of this continuity; Voltaire, on the contrary, mocked it by comparing it to the Catholic hierarchy, from the Pope to the Capuchins: he defended the necessity of introducing discontinuities, characterizing and then classifying. But how to classify? For minerals, this could be done, depending on the cabinets, by provenance, by colors and other sensible properties, by chemical compositions, according to uses, etc. All set aside a large number of properties in order to emphasize the one that seemed essential to them. Observations under the microscope multiplied. Axel-Frederik de Cronstedt (1722–1765) brought mineralogy and chemistry closer, and following Anton von Swab (1702–1768), perfected new ways of testing minerals to determine them: pulverizing a small part, projecting it into the flame of a blowpipe and observing the flame color to characterize certain chemical elements present, or identifying them by fusion, making beads of them in borax (Cronstedt 1771). Tongue tests, odors, hardnesses, etc., were also evaluated. To rationalize the evaluation of these, Friedrich Mohs (1773–1839) would establish in 1822 a scale based on 10 minerals, from talc to diamond.

Catalogs and dictionaries appeared, but a general work was lacking: D'Holbach was troubled by this and translated (1753) J.G. Wallerius's (1709–1785) *Treatise on Mineralogy*, which gave a philosophical overview of this science, precisely described the morphology of minerals, classified them according to the chemical bodies they contained, and achieved well-deserved success throughout Europe. Beginning in 1767, Jean-Étienne Guettard undertook field surveys throughout France: they resulted in the publication of a *Mineralogical Atlas of France* in 1777. In botany, Carl von Linné (1707–1778), in his *Systema naturae* (1735), wanted to unify the nomenclature of thousands of plants and construct a classification as an exact science: after having discovered the sexuality of plants, he posited that a species was the set of individuals that reproduced among themselves, therefore privileged the study of the sexual organs (stamens and pistils), characterized them by their number and figure, and established a classification based on these arithmetic and geometric data, expressible in univocal and rigorous language. This allowed a unification of nomenclature and enabled any amateur to identify a plant even if seeing it for the first time. He extended his classification with less success to animals and to crystals, but decisively influenced his mineralogist successors, Jean-Baptiste Romé de l'Isle (1736–1790) and René-Just Haüy (1743–1822).

4.3.2. *Romé de l'Isle and the law of constancy of angles*

Romé de l'Isle began his career as the author of catalogs of mineral collections intended for sale: they became references for merchants and collectors. In 1772, he published the *Essai de cristallographie*, the fruit of his reflections on crystals: it was adorned with 10 plates representing 175 specimens (as compared with about 40 in Linné). Through this work, he wanted "to push back the limits of a science whose first foundations were laid by the celebrated Linné". What struck him most were "the regular and constant forms that certain bodies which we call CRYSTALS naturally take on": he systematically studied the number of their faces, their arrangements, imagined that their growth took place by juxtaposition in the nourishing medium of "integrant molecules" possessing a geometric form "according to their own nature", and held that the ultimate parts of matter, on which chemical properties depended, "would always remain unknown", since they were invisible with the best microscopes (Romé de l'Isle 1772). He grouped the different forms in a "Crystallographic Table" and accompanied it with plates representing idealizations with harmoniously developed faces of the crystals described: they made with one another the exact angles observed (the engraver had used for representing them clay models made by Romé with cut playing cards to transfer these angles). The *Essai* immediately gave its author international renown, but also caused division among taxonomists, systematists and chemists: Romé, in fiercely defending his approach, made enemies who barred him from the Academy. Yet he wanted to clarify and extend his study and continued to examine and describe more and more minerals.

From 175, the number of specimens studied reached 440. He wanted to publish his study and to accompany it, once again, with plates by providing the engraver with clay models: faced with the magnitude of the task, he took on the services of one of his students, Arnould Carangeot (1742–1806). Exhausted by the difficulty of the work, the latter had the clever idea of inventing an applied goniometer (see Figure 4.4), whose arms were brought into coincidence with the faces, enabling the direct reading of their angle.

Through this use, Carangeot and Romé realized that the angles between the homologous faces of a given crystalline species were constant, whatever the developments of these faces and the habits. Romé published in 1783 this "law of constancy of angles" in the "Dedicatory Epistle" of his *Crystallography, or Description of the Forms Proper to All the Bodies of the Mineral Kingdom* (Romé de l'Isle 1783). To be in agreement with the Solomon of the Bible, he characterized all crystalline species by means of three parameters, density, hardness and polyhedral form: this was what he thought would make it possible to guide chemistry, which was too complex.

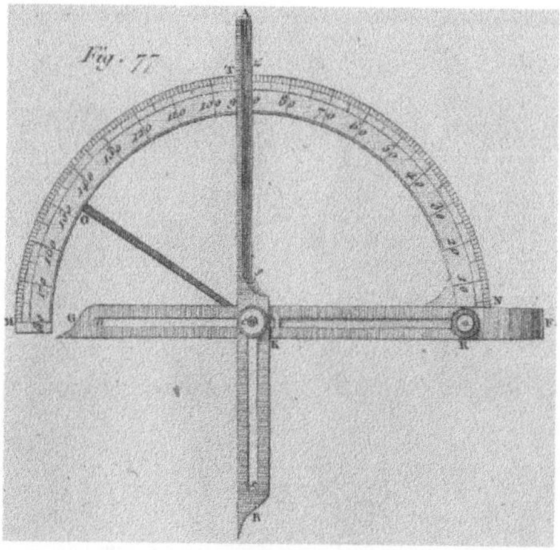

Figure 4.4. *Carangeot's goniometer ((Haüy 1801, t. 5, pl. VIII); source: gallica.bnf.fr/Bibliothèque nationale de France)*

Romé de l'Isle, aware of founding a "New Science", read the predecessors he knew (the describers), criticized them, pursued his research and derived from it a coherent reflection: crystals were formed of small corpuscles (the chemical parts) invisible with the best microscopes. During crystallization, they approached one another, remained at a distance; their union gave a mass that was the smallest part the crystal could present and possessed all its properties: the "integrant molecule", which might one day become accessible through observation.

These integrant molecules of geometries, variable according to species, stacked regularly to give crystals. Romé de l'Isle enumerated them: they could be of six categories, which were the six convex polyhedra whose assemblages allowed space to be filled (tetrahedron, cube, octahedron, rhomboidal parallelepiped, rhomboidal octahedron, dodecahedron with triangular faces). Romé de l'Isle gave a geometric proof of his conjectures: one could recover all the idealized forms of natural crystals by intersecting all the vertices and/or all the faces of the six polyhedra with mathematical planes (which he called "truncations") (see Figure 4.5). He made these six categories the basis of a classification of minerals. To make himself clearly understood and add to the interest of his work, he made and had others make terra-cotta models, which he offered to all subscribers of his *Crystallography*, reserved brass models for his own use (to handle them without risk), had them cut in wood and, like those in terra cotta, marketed them throughout Europe. Note that 448 different models came out of his workshop.

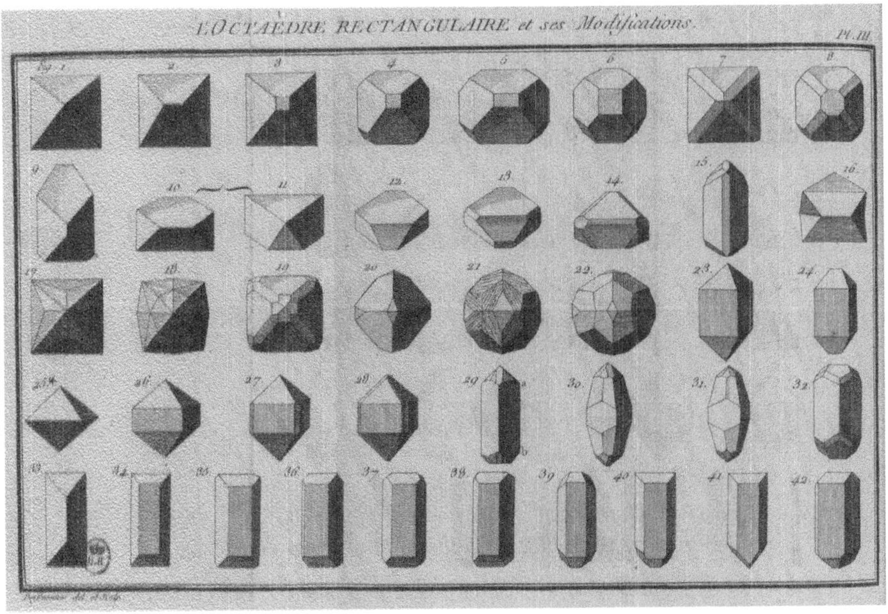

Figure 4.5. *The idealized forms of Romé de l'Isle and the forms explained by truncations, on an octahedron (first row) and on other forms ((Romé de l'Isle 1783, vol. IV, pl. III); source: gallica.bnf.fr/Bibliothèque nationale de France)*

When Romé de l'Isle described the formation of crystals by small corpuscles that approached one another and remained at a distance to form the integrant molecules, the influence of Newtonian physics was felt. By the late 18th century, the successes of this branch of mechanics had accumulated: the determination of the Earth's shape, and above all, the verified prediction of Halley's Comet's return. Pierre-Simon Laplace (1749–1827) expanded upon Newtonian predictions, developing a physics governed by reversible laws, identified within a continuous space and time. His *Exposition of the System of the World* made gravity an inherent property of matter, which dispensed with divine intervention. By carrying out numerous weighings on bodies that combined, Antoine de Lavoisier (1743–1794) contributed to founding a new chemistry and integrated it into this system. Even more than Romé, Haüy took his place in this movement.

4.3.3. *The first works of René-Just Haüy*

Haüy had Mathurin-Jacques Brisson as a professor, who found him gifted for physics experiments, carried out in the tradition of Nollet. He was a friend of Lhomond, who interested him in botany. He was admitted to the Academy of Sciences as an assistant botanist, discovered the works of Linné, and was captivated by a course by Daubenton that made crystals the assemblage of molecules of inorganic matter approaching one another in a fluid, joining when they had the necessary time, rest and space, and classed according to von Linné's reasoning on flowers. He became passionate about this subject, new to him and, without ever having read anything on the theme, carried out observations on the most common of minerals, calcite. He discovered its very numerous forms, its cleavages, extracted from these very small rhombohedra, always of the same geometry.

Observing several minerals that could be easily divided, he always extracted from them, by successive mechanical divisions, "nuclei that had other forms". He then undertook the reading of Tobern Bergman (1735–1784), made, as the latter did, calcite the assemblage of parallelepipedal lamellae.

Haüy noted Bergman's observational errors and distinguished various forms of lamellae constituting different types of crystal "nuclei", whose stackings "as if in tiers" decreasingly caused the crystalline forms, which he classified according to the geometry of their nuclei. Always plane geometry and precise measurements, with no mention of face angles in his publications prior to 1784.

4.3.4. *Crystal classifications in 1784*

In 1784, the French translation of the *Manual of the Mineralogist or Sciagraphy of the Mineral Kingdom, Distributed According to the Chemical Analysis* is provided (Bergman 1784). This translation was the book of Mongez the Younger, who preceded it with an important introduction in which he listed, in a manner that did not claim to be exhaustive, the classifications of "the dead productions of the earth, those buried in its bosom" (they mixed rocks, minerals, crystals): these were "the most famous systems of Mineralogy" established since 1730 "that he had been able to know and study", which he presented to us.

This book deserved mention in that it gave the reflection of the intense activity in which European mineralogists were engaged in 18th century (they described more and more minerals) and their hesitations concerning how to classify them, which we evoked above. The book was quickly exhausted and had to be reissued, but Jean-André Mongez (1750–1788) embarked in 1785 with Lapérouse on *La Boussole*, as chaplain and physicist. For a long time, no news of the expedition was heard, and so Jean-Claude Delamétherie (1743–1817) was asked to carry out this reissue: wishing for the return of Mongez, he postponed this task, to which he agreed in 1792, when it was "almost no longer permissible… to hope". He then published a work considerably augmented by new discoveries, which he inserted while noting and commenting on Bergman, all while respecting the first edition, notably the introduction of Mongez and his enumeration of classifications (Delamétherie 1792). He nevertheless added at the end of the work a reflection on these classifications and proposed his own, which he left open: even after the death of Romé de l'Isle, the problem of classifying remained intact.

Mongez listed 22 different classifications, which he cited, for they bore witness to the "increase in knowledge": they ranged from that of Bromel (1730) to those of Daubenton (1784) and Le Sage (1784). He divided them into two classes. The first grouped the 16 based on external and sensible characters, including the systems of Henckel (1747), Lehman (1759), von Linné (1770) and Werner (1774) "totally founded on the characters apparent to the five senses; but it is so complicated that it can be of no use", of Wallerius (1778) "the best known in Germany and the North", Romé de l'Isle (1783) and Daubenton (1784). The second enumerated the six founded on "constituent principles", hence chemistry: they followed the foundational work of Cronstedt (1771) and included, notably, the classifications of Fourcroy (1780), Bergman (1784) and Le Sage (1784).

Among the systems based on external characters, Mongez (Delamétherie 1792) noted concerning that of Daubenton that "there is none in this genre that is as perfect": this system distinguished:

1) stones, subdivided into:

– 1-1 those that spark by the strike of a flint (including Quartz and sandstones),

– 1-2 those that do not spark and do not effervesce with acids (including clays and Asbestos),

– 1-3 those that effervesce (including chalk and calcareous spar),

– 1-4 those that are mixed (including breccias);

2) fossil salts (including nitre and gypsum);

3) combustible substances (including diamond, sulfur and coal);

4) metallic substances:

– 4-1 semi-metals (arsenic, cobalt, etc.),

– 4-2 mercury,

– 4-3 metals,

– 4-4 mixed metallic substances (several metals present).

Then, there are no differences between rocks, minerals and crystals. As Mongez noted in the following quotation:

> However, the systems founded on chemical analysis teach one more thing, the intimate composition of the substance; that is why two classifications are needed, one for Naturalists, who consider minerals in their natural state, the other for Mineralogists, which indicates the different integrant parts. (Delamétherie 1792)

He adopted Bergman's chemical classification while enriching it, for the description of "varieties", thanks to Daubenton's system. This showed to what degree of confusion all the systems by which naturalists classified still stood in 1784.

Eight years later, Delamétherie ended the second "considerably augmented" edition of the work by publishing a new classification in it. He introduced this work with these considerations:

> The object of the Mineralogist is to know minerals; it is a matter of seeking the surest method […] Do the external characters of minerals suffice to recognize them? or must one resort to chemical analysis? It is well demonstrated that analysis alone can reveal the nature of any mineral substance. But that substance, once analyzed, will always be easily recognized by the external characters, as daily experience proves […]. (Delamétherie in Bergman (1784, vol. 2, p. 329))

These external characters "which seem to merit confidence are the following: specific gravity […], hardness […], figure, refraction […], electricity, magnetism, phosphorescence", and Delamétherie detailed each of them. Concerning figure, he noted that "Romé de l'Isle must be regarded as its true creator", that Haüy had made new applications of Bergman's ideas, notably by replacing the latter's lamellae with "small solids", but added: "I think on the contrary that crystals are composed of lamellae [...] which seem to me reducible to three principal forms [...] triangular, rectangular, rhomboidal" that stack differently by "advance or retreat of the lamellae on one another [...]". Figure being only one of the external characters, Delamétherie could conclude by confirming the primacy of chemistry:

> Mineralogists have been obliged to do as zoologists and botanists did, to establish methodical divisions in order to know mineral substances. Every good mineralogical method must be founded on analysis, as Cronstedt proved; we shall have a true mineralogical system only when the analysis of all minerals has been well done... It is from the known analyses that I establish the method I propose here. If some substance has been badly analyzed, it will be restored to its natural order when it is better known. (Bergman 1784)

Then, an open chemical classification is used and one that aimed to be general. The author distinguished nine classes in it: airs, waters, sulfur and phosphorus, metallic substances, acids, alkalis, earths, neutral salts (metallic, alkaline, and earthy) and fossils. He added: "One should perhaps include in this method light [...], the electric fluid [...], the magnetic fluid [...], the inflammable principle [...], the matter of heat". The airs ranged from "pure air" to "atmospheric air", the waters from "rainwater" to "water of hot springs". Delamétherie justified his classification: "the airs [...] are very abundant in the mineral kingdom [...] the waters follow them [...] Finally, the class of fossils well ends the mineral kingdom, since peat contains many plants; thus I have reason to believe that my method comes close to the natural method, and indicates the transition from the mineral kingdom to the vegetable kingdom". Always this idea of continuity of nature. This classification appeared when Haüy was in the process of publishing articles and works concerning the geometry of crystalline stackings and was going to make him, according to Georges Cuvier (1769–1832), the "legislator of mineralogy" (Cuvier 1823).

4.4. René-Just Haüy's study

4.4.1. *From the* Essai *(1784) to the* Exposition abrégée *(1792): rejection of the possibility of a crystallographic classification*

In 1784, Haüy published his *Essai d'une théorie sur la structure des cristaux* (Haüy 1784). Clearly, he had then read Romé de l'Isle, whom he did not cite: he used the law of constancy of angles, replaced his plates and lamellae with volumes described and generalized by geometry, and answered Romé point by point. He did not want to deal with what crystallization is, but to concern himself with the structure of crystals. By observing the cleavages, inclusions and growth striations of a crystal, he wanted "to determine the precise shape of its constituent molecules, their respective arrangement, and the laws that govern (their) variations". He used Carangeot's goniometer without citing him, explained all observed crystal forms by the three-dimensional stacking of "primitive forms", which could present decreases by "stair-step degrees", invisible under the microscope and constituting the faces, appearing flat, of "secondary forms". The measurement of angles and geometric deduction were necessary to define the "pitch" of the decreases. All of Haüy's book consisted of wanting to determine, for each mineral species studied and the crystals it presented, what "the precise shape of its constituent molecules is, their respective arrangement, and the laws that govern the variations of the plates of which it is composed". But he knew that different crystals could have similar constituent molecules and different chemical compositions, and that they could also have the same chemical composition but different constituent molecules (calcite and aragonite): in a field of study that was very vast and too difficult to treat, Haüy carried out a reduction of the properties of reality, took no account of color, taste, hardness, density, chemical affinities, etc., in order to consider only the constancy of angles and explained it by a geometrization that accounted for it. In doing so, he accepted the fact that "crystallography will never be the basis of a classification".

The success of the *Essai* was great among geometers. Never having heard of the describer Romé – who worked at the Palais-Royal, whereas Haüy worked at the Montagne-Sainte-Geneviève and at the Academy – they grew enthusiastic when they saw that their science penetrated those of nature, which had seemed to them far from lending themselves to mathematization: Fourcroy, Guyton de Morveau, Lagrange and Laplace met in Haüy's small room and wanted to help extend this new science "very capable of being perfected". Haüy made and had made idealized wooden models on which to reason, sold them and competed with those of Romé, who, upset by plagiarism and commercial competition, called him a "cristalloclaste". He especially pointed out that Haüy was too general when he dared to base himself on only those crystals that cleave in order to extend his reasoning to all crystals.

In 1792, Haüy generalized the deductions made in the *Essai*. He conjectured that the division of a crystal could not be carried out to infinity: there existed a stage at which the molecules obtained could no longer be divided without destroying the substance of the crystal. If one went beyond it, one "analyzed" the crystal: one separated its various chemical constituents, the "principle molecules"; in the solid, these remained separated from each other, since light (corpuscular) passed without attenuation through transparent crystals. The stage was thus constituted by a molecule still possessing all the properties of the crystal: Haüy called it an "integrant molecule e", taking up the term coined by Romé, who had died in 1790 and was still not cited in this regard. Crystals were constituted by the three-dimensional assembly of these integrant molecules that filled space: Haüy noted that it would be tedious to reconstruct all possible assemblies of all crystals in order to explain their faces. It was better to proceed geometrically: to abstract a "primitive form" and recover from it the ideal forms observed by means of a single "law of decrease" (Haüy 1792).

To illustrate his approach, Haüy showed that the pentagonododecahedra of pyrite resulted from the three-dimensional assembly of cubes and could be recovered geometrically by starting from cubes whose edges were all truncated by planes equally inclined to the edges, and that the same held for a rhombododecahedron obtained by truncating the edges of a cube by other planes. The geometric work carried out by Haüy, who at this point went further than Romé, was to enumerate all solids capable of tiling space (pentagonal prisms or those with more than six sides could not: thus, according to Haüy, only all parallelepipeds, tetrahedra, octahedra and square or isosceles triangular prisms were suitable) and to show that the observed faces of the crystals studied were recovered geometrically by intersecting all their edges or all their angles with planes equally inclined to one or the other. He named the faces according to these truncations and the inclination of the planes cutting the vertices and edges.

Haüy refused to study how crystals actually formed: they resulted from the supreme will of the creator, still inaccessible. To hope to reach it, he undertook the meticulous study of the other properties of minerals: their electrical properties (1785), magnetic properties (1787), double refraction (1788, 1792, 1793) and the appearance of electrical charges at the ends of a heated tourmaline (pyroelectricity, 1792); he also studied caloric, and revisited all of mineralogy through physics experiments in the style he had learned from his teacher Brisson and in which he became a virtuoso; he studied the chemical properties of minerals following Lavoisier's approach.

All these studies allowed him to take up Tournefort's idea and to impose the concept of a "crystalline species", to study 105 different ones, 82 of them bearing names connected to chemistry, and to create 23 new names. His precise studies of the physical properties of crystals, his connections and his integration into Laplace's circle led to his being appointed – although he had no mathematical training and made errors by using plane geometry formulas where spatial geometry was required – as professor of physics at the École normale of year III at a time when "natural" physics was becoming "mathematical". But Haüy possessed in depth the practices of experiment, observation, experimentation and theorization: and was this not above all what future science teachers needed to be trained in?

4.4.2. *From the courses of year III to the* Traité de minéralogie *of 1801: crystallography and chemistry, bases of classification*

From his first lesson on crystallography onward, Haüy, as a good teacher, told stories and led his students to see crystallization with the eyes of theory, he who until then had refused to make conjectures on this subject:

> Saline molecules, held in solution in a fluid, draw closer together as evaporation occurs [...] through their mutual attractions, they join together preferentially by certain faces; they form groups that are at first imperceptible, but that soon increase to the point of becoming visible to the eye. Then one sees appear [...] a multitude of regular bodies. (Haüy 1795, pp. 69–87)

And Haüy traced the history of interpretations of crystalline forms, from von Linné to Bergman and to Romé de l'Isle, to whom, 4 years after his death, he paid a vibrant and public homage: for his determination of the primitive form, the notions of integrant molecules and geometric truncations and his law of the constancy of angles. "In a word, his crystallography is the result of immense work, almost entirely new in its subject, and very valuable in its usefulness", added Haüy, sentences that he reproduced word for word in his *Traité de minéralogie* (Haüy 1801).

Haüy now developed his own theory: all crystals of the same substance had, whatever their varieties and their figures, a constant nucleus. They were a compact stacking of integrant molecules: their assembly could include decreases. To explain all forms, he showed that one could start from the simple form (e.g. a cube) to obtain secondary forms (e.g. an octahedron or a dodecahedron) by carrying out geometric truncations on all the vertices or on all the angles.

These intellectual truncations made it possible to predict the angles of the faces: measurement confirmed the deductions. He could state the "law of rational truncations": the passage from a simple form to the various multiple forms of a crystalline species occurred through truncations whose angles represented the steps of the decreases. The great possible variety of secondary forms depended on the deposits: any change in the fluid (in density, temperature, composition, affinity with respect to the molecules it contained, etc.) caused modifications of the crystalline forms. Haüy now distinguished only three possible categories of primitive forms: the parallelepiped, the tetrahedron, and the triangular prism: he was bothered, for example, by the observation that fluorite could cleave into a tetrahedron starting from a cube; the latter was not necessarily a primitive form.

The mineral species still had to be classified: to do so, Haüy noted that all the reasoning he carried out concerned forms, not composition or affinity: this last property was due to caloric, which moved molecules away or brought them closer. Crystallography and chemistry were not of the same nature and did not provide the same information about bodies: they "must join hands, advance with the same step: let the compass be associated with the balance". It was therefore necessary to classify in two ways: according to the forms of the integrant molecules and according to composition. Unity was not found.

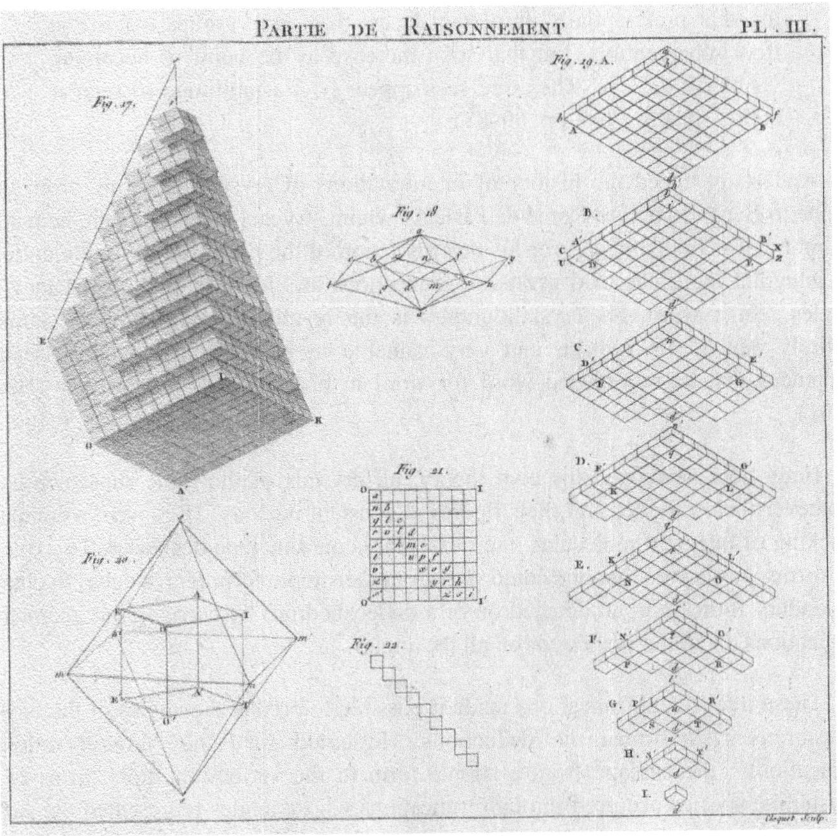

Figure 4.6. *The scalenohedron of calcite resulted from the juxtaposition of layers of rhombohedral integrant molecules; an octahedron could be obtained by truncating all the vertices of its simple form, a cube (Haüy 1801, atlas, pl. III); source: gallica.bnf.fr/ Bibliothèque nationale de France)*

The *Traité de minéralogie* of 1801 was Haüy's major work. Physics, chemistry and geometry there "joined hands" for mineralogical analysis:

> It belongs to chemistry to determine the species [...], but let geometry [...] be associated with the balance in order to bring together all known minerals under the same point of view, to compare them with one another, to study their characters, and to question in turn experiment and theory on the different phenomena of which they are susceptible. (Haüy 1801)

We were far from the statement contained in the *Essai*: "One will never be able to make crystallography the basis of a systematic distribution of minerals" (Haüy 1784). A classification had to make it possible to determine a mineral and to give the series of its properties. It was therefore necessary to determine first its chemical composition, which gave the species, and then the integrant molecule, which gave the type of the species, called "mineralogical species". For each of these, one had to determine:

1) measurable specific quantities (specific gravity, hardness, optical, electrical, magnetic properties, etc. – properties such as colors, too variable and linked to deposits, were excluded);

2) the forms that defined the varieties, with the values of their angles and the names assigned to them;

3) "annotations that [...] contain something like the history" of the minerals: deposits, associations.

It was a precise panorama of the mineral realm that Haüy outlined, proposing a new classification and rational nomenclatures of species and forms. For this last point, he had superb plates made by collaborators whom he cited: they were far more demonstrative of the three-dimensional stacking that led to the observed forms than earlier ones (see Figure 4.6). He also showed, in 240 pages in the first geometric part of the *Traité*, how this geometry made it possible to account for the secondary forms observed once the geometry of the integrant molecule had been determined (see Figure 4.7). For chemistry, he pointed out that integrant molecules were composed of "elementary molecules" (these perhaps also had a geometry, he specified), chemical compounds that assembled (as acid and soda in rock salt). This distinction allowed him to produce, in five volumes, tables for determining minerals, classified according to chemistry, which reformed those of Romé. The crystallographic part, which fascinated mathematicians and physicists, was generally not very accessible to naturalists of the time. However, the latter used the second, descriptive part.

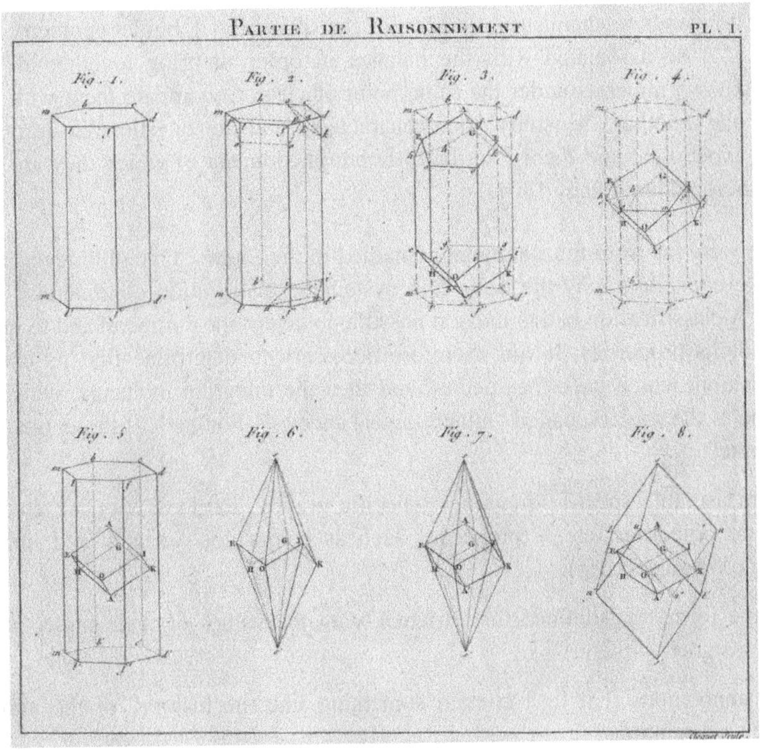

Figure 4.7. *From a primitive hexagonal prismatic form, truncations on the sides made it possible to obtain a rhombohedron. This could also be obtained from a scalenohedron or from another rhombohedron: different forms could have the same integrant molecule ((Haüy 1801, atlas, pl. I); source: gallica.bnf.fr/Bibliothèque nationale de France)*

4.4.3. *The evolution of Haüy's thought up to the* Traité de cristallographie*: a purely geometric classification*

Berthollet and other chemists pointed out that Haüy gave, as mineral species, chemical compositions that could not be those of their integrant molecules, which were nonetheless supposed to possess all the properties of the crystal they characterized. Haüy acknowledged the pertinence of this remark and resumed his studies. In his *Comparative Table of the Results of Crystallography and of Chemical Analysis, Relative to the Composition of Minerals*, he divided this table into two parts: the first arranged minerals methodically according to crystallography, and the second consisted of notes connected to the different species and mostly chemical. There one saw that Haüy indicated that an integrant molecule had one particular chemical composition among others possible, but that several compositions of molecules could be represented in the same crystal. "One easily feels how much this doctrine restricts the importance of Chemistry in distinguishing species, and how much, on the contrary, it extends that of Crystallography," he concluded (Haüy 1809).

A shift occurred: chemistry now belonged, in classification, to the "accidental principles", on the same level as hardness, density and color, whereas the geometry of the integrant molecule was constant: this geometry constituted the fixed point around which "the results of affinity only oscillate [...] there is only Geometry for which all minerals are pure" (Haüy 1809).

The *Traité de cristallographie* (1822) was the culmination of this radical change in Haüy's position. The plates it contained were identical to those of the *Traité de minéralogie*, but the demonstrations were no longer the same. Crystals were distributed into six systems, depending on the geometry of the parallelepiped of the integrant molecule (the primitive form) that characterized them: for geometric convenience in better performing truncations, a seventh system could be defined, the right prism with a hexagonal base. And Haüy concluded:

> I prove that it is to Crystallography, rather than to chemistry, that the distinction of species belongs [...] I add that Crystallography has the advantage of employing for the determination of species a means that is accessible and, so to speak, palpable [...] Mineralogy thus shares with zoology and with botany the merit of making the characteristics it employs speak to the eyes. (Haüy 1822)

We thus found von Linné again, in another way. And he insisted: "If this way of seeing is not the true one [...] [it offers] the most thoroughly reasoned mineralogical method and the one most consistent with the philosophy of science". This conclusion would not be shared by all, on two essential points of Haüy's crystallographic theory that we have not yet addressed.

4.4.4. *The remaining problems*

The first concerned the "law of rational decreases". Haüy had supported it on the fact that the "steps" of the decreases on the faces always occurred by integer multiples of the dimension of an integrant molecule and had confirmed it by the results obtained for the measurement of angles. But as early as 1802, William Hyde Wollaston (1766–1828) showed, with a very precise goniometer of his invention (Haüy refused to use it because he was English – Napoleonic France and England were at war) that the values of the inclination of the faces of calcite given by Haüy were wrong, sometimes by as much as 30 min of arc. A controversy followed, during which Haüy changed tactics and made his measurements no longer exact data but – like those made in astronomy – series of values that had to be "smoothed" to determine regularities: these were "the true mark of nature". From empirical, the law of decrease became a constructed law, and Haüy noted that it was analogous to the "principle of definite proportions" that the chemist Dalton had just formulated (in 1803).

The *Traité de Cristallographie* restated this position:

> Mechanical measurements of the angles formed by the faces [...] cannot be rigorous [...] the ratios that derive from them [...] will be represented by large numbers, whose use will harm the elegance of the theory [...] I show how I arrived at deducing from these approximate ratios [...] a limit that is recognized by the simplicity [...] of nature. (Haüy 1822).

Haüy bent experiment to the requirements of theory and specified: "One recognizes here what generally characterizes the laws emanating from the power and wisdom of God. Economy and simplicity in the means and inexhaustible fecundity in the results". Not everyone accepted this way of seeing.

The second difficulty concerned the "law of symmetry", taken up again in the *Traité* of 1822 but initially formulated in a 1815 paper, the "Memoir on a Crystallization Law Called the Law of Symmetry", in the form: "If the edge of a primitive parallelepiped undergoes a certain decrease or remains identical to itself, the other analogous edges will undergo the same decrease or remain identical. The same is true for the angles" (Haüy 1815).

For example, if one edge of a cube underwent a decrease equally inclined with respect to the two other edges, all 12 edges underwent the same decrease: the form obtained was a rhombododecahedron; if one angle underwent a decrease equally inclined with respect to three edges, the eight edges also underwent it: one obtained an octahedron. But Haüy noted that certain crystals made exceptions: cobaltite or fluorite presented tetrahedral forms, which meant that their primitive form underwent only one decrease out of two with respect to the vertices (see Figure 4.8).

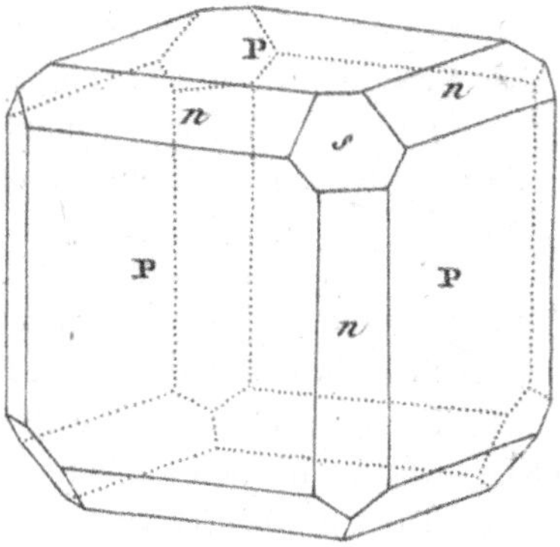

Figure 4.8. *The forms of certain crystals could be obtained by truncations of one vertex out of two: these were hemihedries, "marvelous exceptions" to the law of symmetry ((Haüy 1822); source: gallica.bnf.fr/Bibliothèque nationale de France)*

In the *Traité de cristallographie*, he added other exceptions: certain dodecahedral crystals of pyrite could be obtained by decreases of one edge out of two in a cube; a rhombohedron inscribed in a cube could not be obtained by decreases starting from that cube. Haüy named "holohedra" the crystals that could be deduced from his law of symmetry, "hemihedra" those whose decreases resulted from truncations affecting one vertex out of two or one angle out of two, "tartohedra" those whose decreases occurred on one-fourth of the identical elements (these last two terms being able to be grouped as "meriedres"). He made these crystals, as well as the pyroelectricity he discovered in tourmaline (when heated, the two ends of its opposite prismatic faces became charged with electricities of opposite signs), the results of an unknown phenomenon during crystallization. Since the cause of crystallization was itself unknown, he went no further and described these forms or phenomena as "marvelous exceptions". Not all his contemporaries or successors would be satisfied with these limitations of the subject, nor would they want to explain them through "elegance" or the "marvelous", but would devote themselves, as we shall see, to explaining these exceptions.

4.4.5. *The "polarization" of light*

There was another debate that animated the community of physicists at the beginning of the 19th century: it concerned light and crystals. In 1802, Wollaston, who was interested in mineralogy, chemistry and physics, had the idea of repeating the experiments on the birefringence of calcite, a century after Newton had contested the results obtained by Huygens. To his great surprise, he found that

Huygens' results were correct. To verify this, he developed his reflection goniometer: it was clear that Newton was wrong. The news caused a scandal. In France, the Institut submitted the problem to Haüy: using only a simple protractor, he confirmed in 1804 Newton's gross errors. From disbelief, the scientific world became perplexed.

In France, the Institut wanted the question settled and promised a prize to whoever could provide "a mathematical theory, verified by experiment, of the double refraction experienced by light when passing through various crystalline substances" (January 1808). It was the memoir submitted by Étienne Louis Malus (1775–1812) that was awarded. He confirmed Huygens' results and noted "Newton [...] substitutes for Huygens' law a law [...] absolutely contrary to the phenomena" (Malus 1810). With this point settled, Malus tried to explain the variations in intensity of rays refracted by a calcite that was rotated in its plane after passing through a first calcite. He multiplied experiments and found that a simple mirror produced the same "emotion" of the light as the first calcite. This phenomenon had never been observed before: Malus became passionate about it, observed that two reflections could in some cases "extinguish" the light, and that this phenomenon occurred through reflections, refractions and double refractions; he gave their analytical laws.

A supporter of Newton's corpuscular theory of light, against Huygens' wave theory, he explained the phenomenon by conjecture: the corpuscles would have shapes that could be referred to three axes. Passing through a medium or a simple reflection would allow only corpuscles with particular orientations to pass. The light was then "polarized", stated Malus. He explained all observed facts by considering the geometry of devices composed of several media where a polarized ray experienced variations in intensity or was extinguished. The new phenomenon was thus called "polarization", from the name of the explanation given. Malus then took up Huygens' deductions, who had linked optical properties and the structure of a crystal. He confirmed them and derived a new problem: "to find the axis of crystallization and of refraction in crystals that no longer retain traces of their primitive forms" (Malus 1811). By illuminating any sufficiently thick crystalline plate with polarized light and rotating it in its plane, he observed at its exit the alternate extinctions of the transmitted rays, at 90° to each other: these were the projection of the sought axis on the plate and its perpendicular.

In 1811, François Arago (1786–1843) passed the two polarized rays transmitted by a calcite through a thin mica plate and observed that the rays transmitted by the mica were colored and that their tints were complementary (green for one, purple for the other, the area common to both spots being white). He called the phenomenon "colored polarization". He also discovered that polarized light falling on a colored crystalline plate could produce variations in tints when the plate was rotated in its plane: he called this phenomenon "pleochroism". Jean-Baptiste Biot (1774–1862) was able to explain these two properties through difficult analytical calculations based on the corpuscular theory of light: the corpuscles would oscillate periodically while passing through crystalline media; the observed colors depended on the thicknesses traversed and the orientation of the crystalline plate. He received the support of David Brewster (1781–1868). A new relation between physical properties and crystallization had thus been revealed, after a new field of investigation had been opened.

4.5. The early 19th century: the roles of physicists and philosophies of nature

It is on the three terms of the optical properties of crystals, the "marvelous exceptions" – treated differently in Germany and France – and the laws of decrement that physicists of the first half of the 19th century focused. For clarity, we will treat them separately.

4.5.1. *Crystals and the wave theory of light*

4.5.1.1. *Interference and diffraction*

Wollaston was not the only one in England to revisit Huygens' optical work at the beginning of the 19th century. Thomas Young (1773–1829), a physician specializing in vision and hearing, studied the vibratory nature of sounds and wondered if an analogy could be made between sounds and colors: the explanation of the colors appearing in soap bubbles was indeed cumbersome and implausible in Newton. Leonhard Euler (1707–1783) had accounted for it much more elegantly with a wave theory of light, which Young knew. Young, who was ignorant of mathematics, approached the question experimentally. As two waves propagating in water superpose when they meet, creating zones of agitation while others remain at rest, he had the idea of performing an analogous experiment with a light brush passing through two holes, producing two beams that could superpose: in this zone, he observed alternating dark and bright bands (Young 1802).

"Light added to light can produce darkness": light "interferes", it can only be wave-like (Young 1802). From 1802 to 1807, he multiplied experiments, obtained results allowing him to explain the colors of thin plates, and concluded that light travels slower in water than in air, as Huygens and Euler had conjectured. His works and results, not mathematized, challenged all of Newton's optics: they were poorly received at the Royal Society, caused Young's withdrawal, but attracted the attention of Malus and Arago in France.

In 1814, Augustin Fresnel (1788–1827), trained in Descartes' optics, who had never understood Newton's corpuscular theory nor read his predecessors or Young, wished in his spare time to demonstrate that "as regards light, Newton rambles". Skillfully performing numerous experiments and virtuously handling analytical calculation, he observed that light could bend around objects and the shadow of a hair is bordered by colored fringes: he sent his observations to Arago, who encouraged him, helped refine his demonstrations and read them at the Académie des Sciences. The Academy decided to award a prize to whoever would explain "the inflection of light rays near bodies". Fresnel obtained it from an Academy that was largely Newtonian: his *Mémoire couronné sur la diffraction* resisted all experimental and mathematical objections and overturned Newton's corpuscular theory (Fresnel 1819).

4.5.1.2. *Crystals and light polarization*

Nothing in this work concerned light polarization. Yet, as early as 1816, Arago and Fresnel worked together on this phenomenon in crystals. They repeated the experiments on chromatic polarization previously performed by Arago, multiplied them, refined them and observed that the colors displayed by plates between crossed polarizers depended on the thickness traversed and birefringence, linking them to the difference in speeds of the two waves refracted by the crystals. They showed that two ordinary rays can interfere, two extraordinary rays as well, but not an ordinary ray with an extraordinary ray. They concluded that Huygens had been correct to represent the double refraction of calcite and quartz by a wavefront with two sheets, one spherical, the other ellipsoidal, tangent at two opposite points, whose orientation depends on the geometric characteristics (faces, cleavages and inclusions) of these crystals, but they specified that for quartz, the extraordinary index is greater than the ordinary (they called crystals of this type "positive"), whereas in calcite it is the ordinary index that is greater ("negative"). They further determined, for thin plates placed between crossed polarizers, through the extinctions observed when rotated in their plane, the projection of the isotropy axis, the optical axis, oriented along the two tangent points.

All these results were obtained but remained unexplained by a theory of light: "The laws have the advantage of being stated independently of any hypothesis", they wrote (Arago and Fresnel 1819). Biot and Brewster worked in the same domain, obtaining the same results, which Biot managed to justify through a complex version of his corpuscular theory. For 6 years, Arago and Fresnel pursued their investigations. They noticed that in certain minerals (micas, gypsum, etc.), both indices were “extraordinary”: their wavefront thus consisted of two ellipsoidal sheets intersecting at four points, opposite two by two; these two directions were directions of isotropy: they called these directions “biaxes”, quartz and calcite being “uniaxes”. They observed no birefringence in blende or garnets: these were isotropic. In 1824, they showed that cubic minerals are all isotropic, that those with a rhombohedral or square-based parallelepipedal integrant molecule according to Haüy are uniaxes, with the isotropy axis oriented along a precise crystallographic direction, and that others are biaxes. New experiments, illuminating thin plates between polarizers but with convergent beams of light (instead of parallel), allowed them to observe axis figures (black crosses for uniaxes, equilateral hyperbolas for biaxes), enabling them to define the crystallographic systems to which the minerals belonged and to determine, still without theoretical interpretation, their crystallographic orientations in the absence of any observable external form.

Arago observes in 1810 a very curious phenomenon: when he looks at a thin quartz plate between crossed polarizers along its birefringence isotropy direction, colors appear; they vary with the thickness of the plate. Biot revisits this study, refines it, has the idea of illuminating the plate with monochromatic light and observes that each color can be extinguished by rotating the second polarizer by a certain angle: the quartz along its birefringence isotropy direction transmits a linearly polarized vibration that has rotated relative to the incident polarization by an angle varying with color and thickness traversed. He calls this effect “rotatory polarization”. He also notes that certain plates cut from quartz rotate the polarization direction to the right, others to the left (he calls them “dextrogyre” or “levogyre”): this property is linked to the possible existence of facets opposite each other on the crystal pyramids (see Figure 4.9) (Biot 1820). All these optical observations allow Fresnel to show that the phenomenon occurs only for certain meridional crystals in Haüy’s sense. With Arago, he decisively extends the relationships between physical properties and crystal forms (Arago and Fresnel 1824).

Fresnel attempts to explain the observed phenomena by refining his wave theory of light. As early as 1819, he thinks that polarization can be explained by considering the vibration direction associated with light waves. Since Huygens, these vibrations have been implicitly or explicitly assumed to be longitudinal, that is, oriented along the direction of the light ray. Yet a calcite rotated in the plane perpendicular to normal incident rays exhibits the phenomenon, even though this direction is the only one remaining constant during rotation. If the vibrations were longitudinal, nothing could affect them during rotation: they are therefore transverse, which fits well to explain all observed phenomena, including those showing four extinctions per rotation.

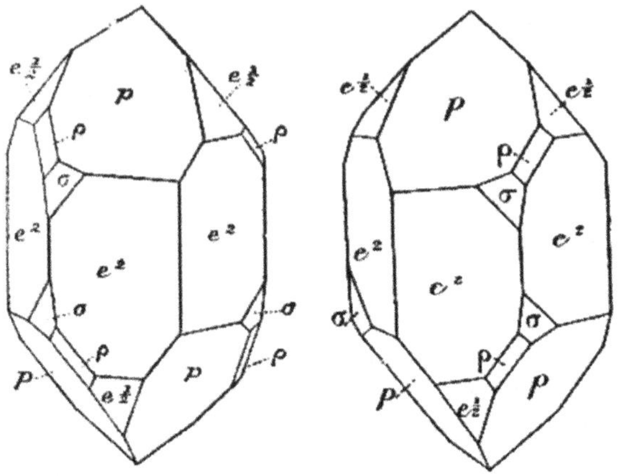

Figure 4.9. *Right- and left-handed quartz crystals*

Arago opposes this explanation: mechanically speaking, transverse vibrations can only propagate in media analogous to viscous jelly; yet the interstellar ether allows the Earth to move freely in its orbit around the Sun! Fresnel takes the leap: without being able to explain the curious properties of the ether, he asserts the transverse nature of the vibrations and provides the explanation for all observed phenomena (Fresnel 1819). In this respect, Arago no longer follows Fresnel "in (his) acrobatics". Biot and Brewster were convinced that the implausible properties attributed to ether would eventually validate their corpuscular theory and oscillating, rotating molecules. It would take the efforts of Arago, prompting Fizeau and Foucault independently in 1849, to establish that the speed of light in water is lower than in air, as predicted by wave theory against corpuscular theory, for transverse vibrations to be accepted. The continuity of waves replaces the discontinuity of corpuscles. This is not an isolated case in early 19th-century physics: electrical, magnetic and calorific properties are also progressively described by continuous values, undermining the worldview of Newton and Laplace.

4.5.1.3. *Neumann's principle of symmetry*

Development of crystal faces, cleavages, inclusions, pyroelectricity, chromatic polarization, pleochroism, rotatory polarization, differential hardness of certain minerals (kyanite), thermal expansion, electrical or thermal conductivity, elasticity, etc.: all these physical properties can be anisotropic in crystals. Romé, Haüy, Mohs, Arago, Biot, Brewster and Fresnel demonstrated this in numerous works. Generalizing the first approach made by Arago and Fresnel on the relationships between physical properties and symmetry, Franz Ernst Neumann (1798–1895) dared to generalize them: "The symmetry elements of any physical property of a crystal must include the symmetry elements of the crystal's point group[1]". This "principle of symmetry" associates all physical properties with holohedra and hemihedra: knowing the properties, one can determine the possible symmetries of the crystal; conversely, knowing the symmetry, one can attempt to reveal a particular property. Neumann's principle indicates the necessary conditions for the appearance of a property, not the sufficient ones. Only impossibilities are clearly defined. It remains to understand what "point groups" are for Neumann.

4.5.2. *Crystals in Naturphilosophie: symmetry axes, classes and lattice types*

Against the "molecular" conception of matter, a "polar" current had emerged describing it as divisible to infinity. In the 17th and 18th centuries, this way of seeing was shared by Leibniz, Boscovich and Kant: from the mid-18th century in German-speaking countries, it produced a romantic vision of the world, a context in which German sciences were embedded until around 1830: the *Naturphilosophie*. Without going into details, for these thinkers, matter is continuous, and they emphasize the description of attractive or repulsive forces exerted at mathematical points. For them, the cohesion of matter results from the dynamic equilibrium of the actions of these forces: when the distance exceeds a certain limit, the forces are attractive, and they become repulsive below it.

During crystallization, these forces are not isotropic, which explains the appearance of faces. From his studies, Friedrich Schelling (1775–1854) deduces that the universe is an organism possessing a universal formative power: it produces individual things with determined roles, such as crystals, composed of matter and form. Crystallization obeys a positive principle to realize form, but a negative principle also operates, creating differences among them. "It is a clever game that Haüy plays when he regards all crystallizations as produced by different aggregations of unalterable primitive forms", he wrote (Schelling 1798). Christian Samuel Weiss (1770–1850) took up and extended these theoretical ideas. In 1805, he publishes a translation of Haüy's *Traité de Minéralogie*, followed by *Considérations dynamiques sur la cristallisation* (Weiss 1804). For him, in a fluid, attractive and repulsive forces are in equilibrium. Crystallization occurs when

1 Neumann's course (1833) on the optical properties of crystals, according to his student Woldemar Voigt, *Die fundamentalen physikalischen Eigenschaften der Krystalle in elementarer Darstellung*, Leipzig, Veit & Co., 1898. The principle was first published by another student of Neumann, Bernhard Minnigerode, *Neues Jahrb. Mineral Geol. Paleontol*, 1887, vol. 5, p. 145.

repulsive forces become dominant. Because they are not isotropic, the resulting forms are variable, depend on the directionality of the forces and have characteristic angles. Equivalent faces appear perpendicular to the direction in which the repulsive forces have the same value: they are flat and not stepped as described by Haüy. These are therefore directions characteristic of the crystal type, and it is their distribution that must be studied.

To do this, Weiss introduced the concept of a "symmetry axis" (Weiss 1810), giving a vectorial description of crystallization in which all faces are represented by their perpendiculars converging at a point. He called "faces in a zone" those that are equivalent by rotation around a common symmetry axis, the "zone axis". This axis is assigned an "order" equal to the fraction of a full angle through which one must rotate around the axis to pass from one half-line to the next equivalent one (e.g. if two analogous half-lines are 180° apart, their axis is of order 2; order 3 if they are 120°, etc.). Applying his method to calcite, he shows that considering symmetry axes is not only geometric but has physical significance: faces, cleavages, birefringence, inclusions, etc., repeat according to these axes.

With Weiss, the notion of crystal symmetry becomes clear, which had remained implicit in Haüy. In 1815, he classified crystals into different categories according to the symmetry axes they possess (he demonstrated that only axes of order 1, 2, 3, 4 and 6, mirrors, and centers of symmetry can exist), calculated the number of arbitrary faces generated by their action (what we call the "degree of symmetry"), used spherical trigonometry for better agreement than Haüy between measurements and calculations, corrected the "law of decrement" and provided a new notation for faces, no longer descriptive as in Haüy, but based on the ratios determined by intersections of each crystal face with reference axes.

Most importantly, by combining the axes, he deduces the resulting possibilities or impossibilities (e.g. if a crystal has only one axis of order 4 – noted A4 – it can exist alone. If an axis of order 2 – A2 – is added, it can only be perpendicular to A4 to avoid repeating it: this necessarily determines 4 A2 in the perpendicular plane, forming angles of 180°/4). By combining all axes he enumerated, Weiss determined symmetry classes (e.g. all crystals with only one A4) and subclasses (A4; A4 + 4A2; etc.), noting that these coincide with the hemihedries defined by Haüy: these were no longer "marvelous exceptions" but were deduced rationally. In 1823, he proposed a general system for distributing crystallizations into classes and subclasses (their hemihedries), but he admitted that he could not incorporate all known crystals into his classification. This enumeration, even incomplete, brings great clarification and is well received in Germany, being in symbiosis with *Naturphilosophie* and providing a mathematical rationalization that Haüy lacked.

In 1823, Friedrich Mohs (1773–1839) published a *Traité de cristallographie* in which he sought, like Weiss but independently, to transform the science of crystals into a geometric science. He completed Weiss's categories by adding what we now call monoclinic and triclinic, distinguishing six (or seven) "crystal systems" (Mohs 1823).

In 1830, Johann Hessel (1796–1872) resumed the work of Weiss and Mohs. He had the idea of combining "direct axes" (already determined) with what he called "inverse axes" (entailing both rotation and inversion) and enumerated 32 symmetry classes distributed among 6 (or 7) systems. He placed all known crystals in these classes; some remain empty: the task of mineralogists will be to fill them, he wrote. The enumeration was then complete… but his book is not discovered until 1897 (Hessel 1830).

In 1842, the Englishman William Hallowes Miller (1801–1880), in his *A Treatise on Crystallography*, relied on Haüy's law of rational truncations and defined the intersections of planes (faces, cleavages) with reference triads of the seven crystal systems. He represented these faces in stereographic projection and used the coordinates of the projected points to replace spherical trigonometry. In doing so, he introduced a simple, elegant notation system, enabling all crystallographic calculations and replacing, among British and German crystallographers, Haüy's cumbersome descriptive notation, which persists only in France (Miller 1842). We will now see how all this work was received in that country.

4.5.3. *The irreverent continuators of Haüy: Delafosse, Bravais and crystal lattices*

Here is what Gabriel Delafosse (1796–1878), a student of Haüy, wrote in a biography of his professor, compiled late with Charles Durozoir, when all work on axes was completed in Germany:

> [These works rely on] idealist philosophy: this type of metaphysical consideration preoccupies all German intellectuals. Some ambiguities are borrowed from the Greeks, some sophisms based on Kant's famous antinomies, have led German physicists to prefer, in the study and interpretation of current phenomena, the kind of vague and obscure explanations they call "dynamics" rather than the simple, clear, and positive notions we derive from atomistic hypotheses. (Delafosse and Durozoir 1857)

The French did not read the Germans, so the concepts of symmetry axes and classes remained long foreign to them. They pursued their studies in another direction. In 1840, Delafosse published a text in which he said he could not accept the labeling of hemihedries as “marvelous exceptions”. He paid tribute to Romé for the law of constancy of angles and to Haüy for addressing the internal assembly of crystals, but criticized him for having “neglected all properties other than cleavage” and for not having “given his study the character of a physical theory”. He noted that, for Haüy, the integrant molecule was the smallest parallelepiped having all the properties of a crystal, that it filled space, and that if broken it gave the “principle molecules” which did not fill space. Why not study these? Haüy limited himself too much by focusing only on stacking and its geometry. Physical and chemical properties depend on principle molecules. They can have any shape not constrained to fill space. “Geometrically similar crystals can correspond to different molecular structures”, formed of “atomic polyhedra” arranged at the “nodes” of a three-periodic lattice of virtual points, defining “lattice rows” and “lattice planes”. These define, in two dimensions, parallelograms, and in three dimensions, parallelepipeds, which are the “cells” of the lattice and can be matched to Haüy’s six primitives. These atomic polyhedra do not necessarily have the same symmetries as the cells: they can be “physically identical only to half or a quarter of the primitives [...] with the lattice lines remaining the same” (see Figure 4.10) (Delafosse 1840).

With this remark, Delafosse initiated the explanation of hemihedries and properties such as rotary polarization and pyroelectricity. Starting from all physical properties of a crystal, he traced back to the molecular structure, somewhat as Huygens did for calcite from optical anisotropy. He showed that crystal systems “divide when one considers the physical modifications that can result from a change in the shape or structure of crystals”. Remaining strictly a physicist, he did not undertake the study of all possible subdivisions, describing and classifying only those revealed by experiment. Aware of this limitation, he concluded that he only wished to “show the importance of changes brought to accepted ideas and demonstrate the usefulness of new research” (Delafosse 1840).

These new, exclusively geometric studies are the work of Auguste Bravais. In an 1848 study, he took up the notions of nodes, lattice rows and cells, and generalized them. He reduced polyhedral molecules to simple points; undertook formal mathematical study; studied translation lattices in one, two and three dimensions; introduced the notions of “simple cells” (with a single proper node) and “multiple cells” (with several nodes – facilitating calculations); determined their surfaces and volumes; demonstrated the impossibility of pentagonal, octagonal, etc., symmetries in a lattice; enumerated five two-dimensional and 14 three-dimensional “lattice types”, which he assigned to Haüy’s seven systems (see Figure 4.11). Wanting to protect his study from atomistic speculation, which was being challenged in physics, he limited it to "pure geometric speculation" (Bravais 1848), daring to write a little later: "Observation of crystallized bodies [...] proves *a posteriori* [...] what my theoretical study reveals" (Bravais 1851). Nevertheless, by focusing exclusively on lattice study, Bravais did not quite carry his study of possible hemihedries to its conclusion.

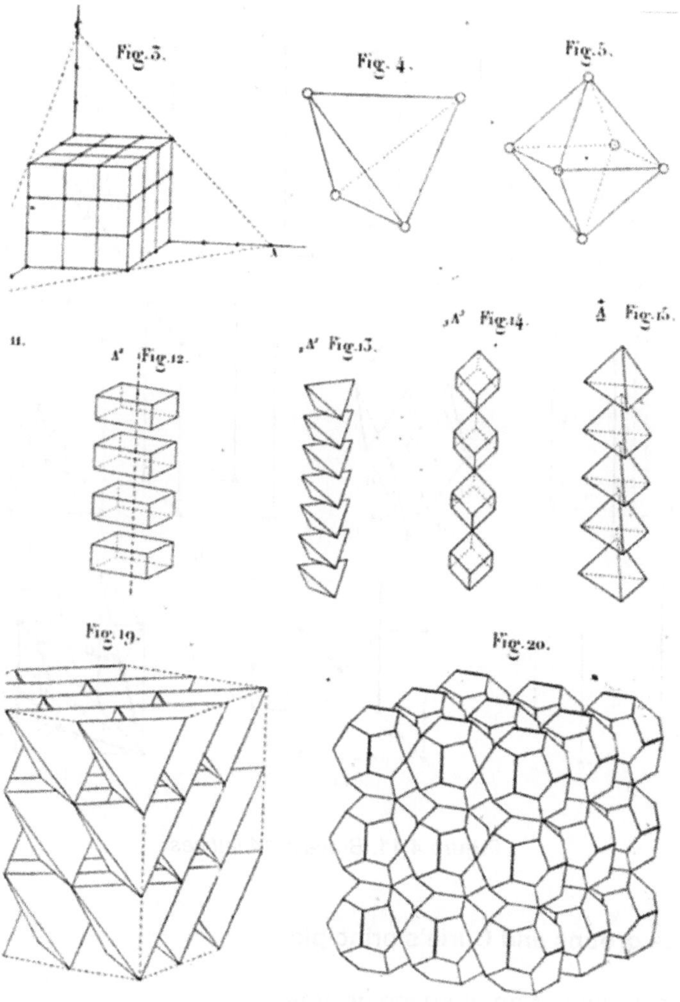

Figure 4.10. *Lattices and polyhedral molecules that do not fill space. Their assembly can have lower symmetry than the lattice ((Delafosse 1840); source: gallica.bnf.fr/Bibliothèque nationale de France)*

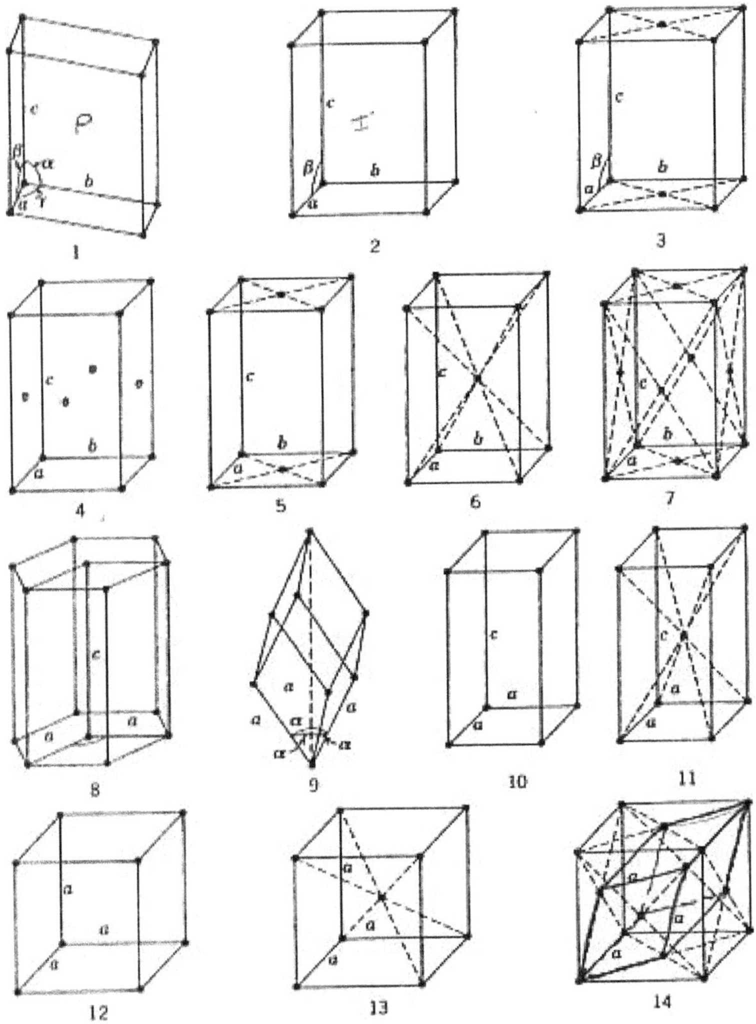

Figure 4.11. *Bravais' 14 lattices*

4.6. Space groups and Curie's principle

4.6.1. *The enumeration of space groups*

4.6.1.1. *Sohncke's groups*

We have said that, to explain hemihedries, German crystallographers were led to subdivide the seven crystal systems into 32 symmetry classes, while the French used

12 lattice types. Since the Germans read the French, Leonard Sohncke had the idea of converging the two approaches. For this, he combined the symmetry classes he knew (those of Mohs) with Bravais' lattices. Each class of a given system can present all the lattice types possible in that system: in this way, he obtained subdivisions (Sohncke 1879). He completed this study by addressing the problem of rotary polarization. This seemed to him to result from a molecular dissymmetry that did not affect the lattice (the influence of *Naturphilosophie* waned, and the Germans came to assume the existence of molecules), as Delafosse and Bravais had noted: he conjectured that this dissymmetry was helical and had no effect on the periodicity of the cells. He therefore added "helical axes" of symmetry to the axes defined by Weiss: in quartz, for example, where the phenomenon manifests along the threefold symmetry axis, this helical axis would rotate a molecular motif by 360°/3 while simultaneously translating it by c/3, c being the nodal distance along the axis (see Figure 4.12).

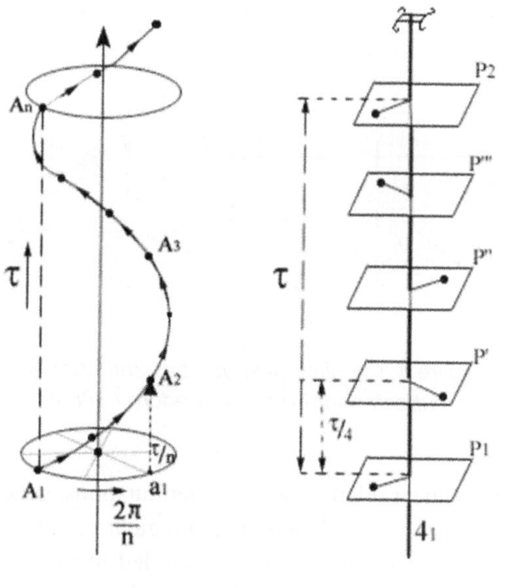

Figure 4.12. *A helical axis, here of order 4*

More generally, a helical axis of order n rotates by 360/n and translates by p/n, where p is an integer: he thus added 12 helical symmetry axes to Weiss' axes, which allowed him, starting from Mohs' incomplete enumeration of symmetry classes, to define 54 different space groups distributed among the 24 classes known at the time and the seven systems. Once Hessel rediscovered and completed the enumeration of classes, 65 groups could be counted, which are called today "Sohncke groups".

4.6.1.2. *Fedorov's and Schönflies' groups*

Six years later, Eugraph Stepanowitsch Fedorov (1853–1919) published in Saint Petersburg his *Éléments de la théorie des Figures*, in which he posed the following problem: "to find all the different arrangements of small congruent spheres such that each sphere is surrounded by all the others in an identical or inverse manner" (Fedorov 1885). The problem of the regular partitioning of space and systems filled with spheres was thus posed and studied for the first time in full generality. To solve it, Fedorov considered a small arbitrary sphere inside a pyramid and repeated it using symmetry elements: he obtained a regular system of spheres. When these are connected by line segments, octahedra and cuboctahedra "reciprocal" to each other are formed, which together tile space (see Figure 4.13).

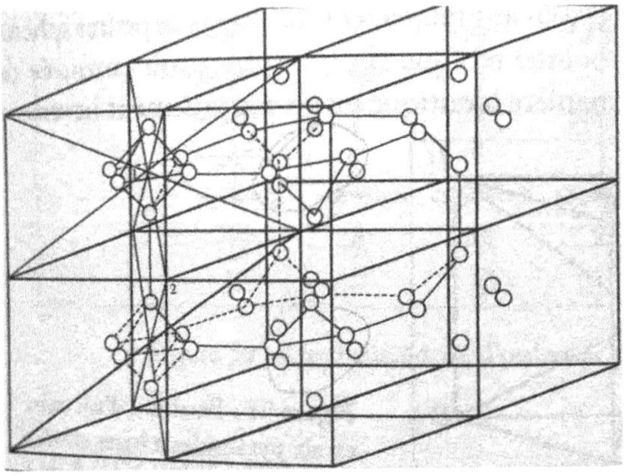

Figure 4.13. *Tiling of space by octahedra and cuboctahedra (source: Fedorov 1885)*

Fedorov gave their linear relations and, by permutations, generalized the possible solutions. He realized that he could not obtain a regular partition of space using only direct, inverse or helical symmetry axes: he was led to introduce other symmetry elements, non-concurrent, which repeat the groupings of points inside the cells without modifying the lattice, the "glide planes". These invert the groupings relative to a mirror while simultaneously performing a translation of module a/2, a being the lattice parameter in the considered direction. Six years of labor were necessary for him to solve the problem he had posed: he then published his *Simetriia Pravil'nykh Sistem Figur*, in which he enumerated 230 different space groups, including 17 plane groups (see Figure 4.14), integrating Sohncke's groups and grouping them into the 32 symmetry classes and the 6 (7) crystal systems (Fedorov 1891).

It was realized in the 20th century that all 17 plane groups exist in the Alhambra of Granada in Islamic art tilings. In the same year, independently of Fedorov, Arthur Moritz Schönflies (1853–1928) obtained the same results: like Fedorov, he approached the question from a general point of view, considered only the symmetry of the configuration, and ignored the figures (Schönflies 1891). These works, leading to no practical application, went unnoticed, and Fedorov's remained completely ignored for a long time. They nevertheless mark the complete systematization of the enumeration work anticipated by Huygens and initiated by Haüy.

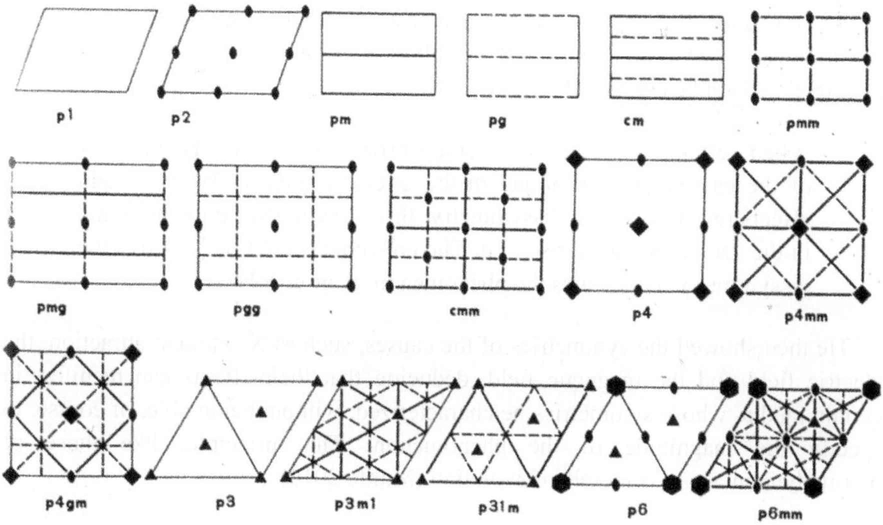

Figure 4.14. *The 17 plane groups*

4.6.2. *The physical properties of crystals and Curie's principle*

Throughout this history, we have emphasized the links between the physical properties of crystals and their symmetry, systematized by Neumann. It is the study of one of these properties, the pyroelectricity of tourmaline, discovered by Haüy, to which Pierre Curie (1859–1906) and his brother Jacques (1856–1941) devoted themselves in 1880. They then established the symmetry conditions necessary for its production in crystals; their generalization allowed them to predict that this tourmaline, compressed along its "hemihedry axis", would generate opposite electric charges at its ends. They experimentally verified this conjecture, showing that the sign of the charges is reversed when the crystal is expanded, and that, conversely, if the crystal is electrified along this direction, it elongates, while it contracts when electrified in the opposite sense. They called this phenomenon "piezoelectricity", showing that crystals exhibiting the same dissymmetries possess the same property, such as quartz and topaz, and that only crystals belonging to 20 symmetry classes can present it, but not necessarily. They did not consider the reason for this restriction (Curie and Curie 1880).

Pierre Curie, alone, undertook the study of its causes and, for this, reversed the problem. He started from Neumann's principle, which allows one to indicate the necessary conditions for the appearance of a property, not the sufficient conditions, expressing only the impossibilities and leaving experience to determine the possibilities. In 1894, he published the result of his reflections, known as the "Curie principle", which he expressed as follows:

> When certain causes produce certain effects, the symmetry elements of the causes must be found in the effects produced. When certain effects reveal a certain dissymmetry, this dissymmetry must be found in the causes that gave rise to it. The converse is not true [...] It is the dissymmetry that creates the phenomenon. (Curie 1894)

He then showed the symmetries of the causes, such as Newtonian attraction, the electric field and the magnetic field, deducing that their effects can manifest in certain media, whose symmetries he characterized, without being able, of course, to specify the magnitude of the phenomenon. This principle, like those of thermodynamics, is a principle of evolution (Maitte 2014).

With Pierre Curie, symmetry loses its form, no longer applying to the study of shapes, but becomes a reasoning tool, somewhat like in Archimedes, where the balance allows reasoning in which shape is mentally inscribed and produces ideas. With symmetry, a single structure describes a set of varied phenomena that seemed completely unrelated. The fecundity of this approach is connected to the search for perfection to the pursuit of beauty.

4.7. Mineralogy at the end of the century and its development prospects

We have said how, with Romé de l'Isle and Haüy, the number of described mineral species increased considerably: this trend continued throughout the 19th century. The description of their characteristics also became more precise: to purely macroscopic and sensory aspects (forms, crystal systems or classes, cleavages, twins, inclusions, deposits, associations, colors, hardness, tongue tests, flame tests) were gradually added chemical compositions, optical properties under the polarizing microscope, calorific, electric, or magnetic properties, etc., not to mention industrial applications, which became increasingly important over the century.

These quantitative and qualitative progressions were accompanied by the discovery of new chemical elements, revealed by the study of mineral composition, such as radium and polonium in 1898 by Pierre and Marie Curie (1867–1934) from pitchblende, or by the discovery of new physical properties, like piezoelectricity, as we have said, and in 1892, from the study of radiation emitted by thin crystalline layers of potassium uranyl sulfate, radioactivity by Henri Becquerel (1852–1908).

Mineralogists also had increasingly precise and complete works based on systematic studies, whether general manuals or identification tables. But the problems of classification remained: we have followed them through the hesitations and developments of ideas in Haüy's work. They persisted at the end of the century.

A work such as that of Félix Pisani (1831–1920) – later taken up in France by A. des Cloiseaux and A. de Lapparent, then revisited and masterfully completed by Alfred Lacroix (1863–1948) in describing precisely all the deposits in France and its colonies – well illustrates the difficulties. Pisani, after a brief description of crystallographic laws and notations, proceeded to the study of each system and its holohedral or hemihedral forms, twins and crystal groupings. He then described all the physical properties crystals could manifest and the ways to identify and measure them; he concluded with the methods to determine their "primitive forms".

At his institution, crystallographic analysis according to Haüy remained predominant (many mineralogists struggled to follow his latest geometric developments, even more so those introduced by Anglo-Saxon crystallographers). Pisani then analyzed, quickly, the chemical properties of minerals and the ways to characterize them, macroscopically, by qualitative analysis (primarily), but also quantitatively and by spectroscope.

Moving to classifications, he noted that there had always been two ways to classify, by physical properties or by chemical properties: the first was that of naturalists and for the second there was no unified method: it could be done "according to the bases, or according to the acids": the one by bases is preferable when studying industrial applications or metals, but it is unusable for silicates, which constitute the majority of minerals (Pisani 1883).

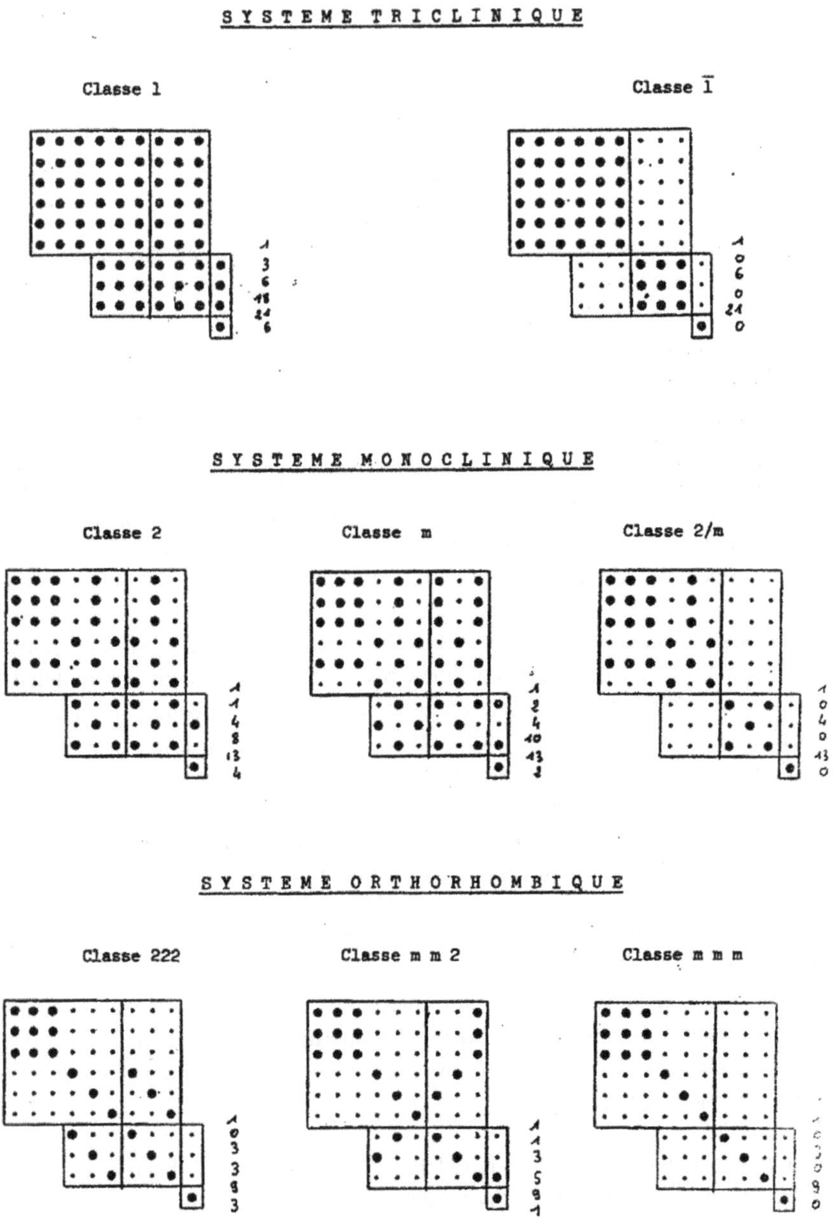

Figure 4.15. *Physical properties of crystals (source: Nye 1961)*

COMMENTS ON FIGURE 4.15.– *All physical properties of all crystals can be summarized in a table of representative tensors of these properties. Here, they are for the classes of the triclinic, monoclinic and orthorhombic systems. A point represents a zero component, and a black circle a non-zero component. In the table, the 6 × 6 tensor (top left) represents elasticities, 3 × 6 piezoelectricity, 3 × 3 permittivities, 1 × 3 pyroelectricity, 1 the density and all scalar quantities.*

In mineralogy, it was therefore necessary to adopt the second, according to their acids, "more logical" and allowing for "entire series such as sulfides, carbonates, etc., whose different members often have greater analogy among themselves", both from a crystallographic point of view and in terms of physical properties. He concluded by describing more than 2,000 mineral species, divided into 41 families (hydrides, silicides, borides, carbides, etc.): for each, he gave the mineral names, their equivalents in German or English, the crystal systems with descriptions of forms noted according to Haüy, the macroscopic physical properties, then optical macro- and microscopic properties, electrical, magnetic, chemical properties, chemical compositions and industrial uses of the mineral, and its varieties. English-language works were of a different nature: they also remained descriptive but adopted Miller's notation and were much more precise regarding the descriptions of crystallographic classes, chemical compositions and uses. At the end of the 19th century, mineralogy was mature regarding the descriptions of crystallographic forms and the characteristics allowing identification of minerals and their associations, but far from unified, both in terms of nomenclature and classifications. Each practitioner worked "within their own system" and according to their national bases.

The work carried out in physics and chemistry in the 19th century prepared and allowed a complete transformation of the situation of mineralogy and its applications in the 20th century. We have mentioned the discovery of radioactivity: it experienced the developments we know. Even more in crystallography and mineralogy, the discovery in 1895 by Wilhelm Röntgen (1845–1923) of X-rays caused a complete upheaval. In 1912, the following were gathered in the laboratory of the University of Munich: the theorist Sommerfeld, studying light/matter interaction and hiring Max von Laue to assist him; the mineralogist von Groth, studying crystals; the experimentalist Röntgen, aiming to demonstrate that X-rays are corpuscular; the physicist Ewald, calculating the variation of the refractive index of light in a triperiodic lattice and discussing it with von Laue; Friedrich Walter and Paul Knipping had the idea to pass X-rays through crystals and record the results on a photographic plate: they obtained spots. Von Laue interpreted them as resulting from diffraction. Contrary to Röntgen's postulate, the wave nature of X-rays and the existence of crystal lattices (which until then had only a conjectural existence) were demonstrated, X-rays being linked to Maxwell's electromagnetic waves, already confirmed by Hertz in the discovery of radio waves.

From then on, the systematic study of the lattices of all minerals and their structures began, interpreted through "reciprocal lattices" noted using Miller indices. It became apparent that these structures fell within the theoretically defined space groups, which now found practical application.

Structural classifications gradually became predominant; a separation of the two disciplines, mineralogy and crystallography, occurred, even as, thanks to the conjunction of Curie's principle, the invention of matrices and structural analysis, all physical properties of all crystals could be represented, class by class, to scale, in highly practical tables that allowed one to dispense with tedious species-by-species descriptions (see Figure 4.15).

A new era began.

4.8. References

4.8.1. *Primary sources*

Arago, F. and Fresnel, A. (eds) (1819). Mémoire sur les rayons polarisés. In *Œuvres complètes de Fresnel*. Imprimerie Impériale, Paris.

Arago, F. and Fresnel, A. (eds) (1824). Notice sur la polarisation de la lumière. In *Œuvres de François Arago*. Gide, Paris.

Bartholin, E. (1669). Experimenta crystalli Islandici disdiaclastici. *Rev. Hist. Sc.*, 1977(XXX/3), 193–224.

Bergman, T. (1784). *Manuel du minéralogiste ou sciagraphie du règne minéral distribuée d'après l'analyse chimique*. Cuchet, Paris.

Biot, J.-B. (1820). Sur de nouveaux rapports qui existent entre la réflexion et la polarisation de la lumière des corps cristallisés. *Mémoires de l'Académie royale des sciences de Paris*, III, 240–257.

Bravais, A. (1848). Les polyèdres symétriques de la géométrie. *Journal de mathématiques pures et appliquées*, 1(XIV), 137–140.

Bravais, A. (1851). Études cristallographiques. *Journal de l'École Polytechnique*, 20, 101–267.

Bravais, A. (1866). *Études cristallographiques*. Gauthier-Villars, Paris.

Cronstedt, A.F. (1771). *Essai d'une nouvelle minéralogie*, Wiedman, M. and Dreux, M. (trans.). Didot, Paris.

Curie, P. (1894). Sur la symétrie dans les phénomènes physiques, symétrie d'un champ électrique et d'un champ magnétique. *Journal de physique*, 3(III), 393–416.

Curie, J. and Curie, P. (1880). Développement par compression de l'électricité polaire dans les cristaux hémièdres à faces inclinées. *Bulletin de minéralogie*, III–IV, 90–93.

Cuvier, G. (1823). Éloge historique de M. Haüy. Éloges historiques des membres de l'Académie royale des sciences, lus dans les séances de l'Institut royal de France par M. Cuvier, 3.

Delafosse, G. (1840). Recherches relatives à la cristallisation considérée sous les rapports physiques et mathématiques. Académie des sciences. *Comptes-rendus*, XI, 394–400.

Delafosse, G. and Durozoir, C. (1857 [1843]). *René-Just Haüy*. Biographie universelle, Michaud.

Delamétherie, J.C. (1792). *Bergman Torbern's Manuel du minéralogiste; ou sciagraphie du règne minéral, distribuée d'après l'analyse chimique, mise à jour par Ferber, trad. augm. de notes par Mongez le Jeune, nouvelle édition considérablement augmentée par J.C. Delamétherie*, vol. 2. Cuchet, Paris.

Delamétherie, J.-C. (1811). *Leçons de minéralogie*. Courcier, Paris.

Fedorov, E.S. (1885). Elemente der Lehre von den Figuren. *Verhandlungen der russisch-Kaiserlichen mineralogischen Gesellschaft zu St Petersburg*, 21, 1–279.

Fedorov, E.S. (1891) Симметрія правильныхъ системъ фигуръ (The symmetry of regular systems of figures). *Записки Императорского С.-Петербургского Минералогического Общества* (Proceedings of the Imperial St. Petersburg Mineralogical Society), 28, 1–146; 346–390.

Fresnel, A. (1819 [1866]). Mémoire couronné sur la diffraction de la lumière. In *Œuvres complètes*. Imprimerie impériale, Paris.

Haüy, R.J. (1784). *Essai d'une théorie sur la structure des cristaux, appliquée à plusieurs genres de substances cristallisées*. Goguée, Paris.

Haüy, R.J. (1792). *Exposition abrégée de la théorie sur la structure des crystaux. Extrait du Journal d'histoire naturelle*. Imprimerie du Cercle Social, Paris.

Haüy, R.J. (1795). Leçons de physique d'Haüy. In *L'École normale de l'an III. Leçons de physique, de chimie et d'histoire naturelle*, Guyon, E. (ed.). ENS, Paris.

Haüy, R.J. (1801). *Traité de minéralogie*. Imprimerie Delance, Paris.

Haüy, R.J. (1809). *Tableau comparatif des résultats de la cristallographie et de l'analyse chimique, relativement à la classification des minéraux*. Courcier, Paris.

Haüy, R.J. (1815). Mémoire sur une loi de cristallisation appelée loi de symétrie. *Mémoires du Muséum d'histoire naturelle*, 1, 81–101.

Haüy, R.J. (1822). *Traité de cristallographie*. Librairie Bachelier et Huzard, Paris.

Hessel, J.F.C. (1830). Krystallometrie oder Krystallonomie und Krystallographie, auf eigenthümliche Weise und mit Zugrundelegung neuer allgemeiner Lehren der reinen Gestaltenkunde. Schwickert, Leipzig.

Hildegarde de Bingen (1993). *Le livre des subtilités des créatures divines. T. 1 ; les plantes, les éléments, les pierres, les métaux*. Jérôme Millon, Paris.

Hooke, R. (1665). *Micrographia*. Martyn and Allestry, London.

Huygens, C. (1690). *Traité de la lumière où sont expliquées les causes de ce qui lui arrive dans la réflexion, et dans la réfraction. Et particulièrement dans l'étrange réfraction du cristal d'Islande*. Pierre Vander, Leiden.

Kepler, J. (1609). *L'étrenne ou la neige sexangulaire*. Vrin, Paris.

Locke, J. (1690). *Essai philosophique concernant l'entendement humain*, Coste, P. (trans.), Henri Schelte. Amsterdam.

Malebranche, N. (1842). *De la recherche de la vérité. L. 3, De l'entendement*. Charpentier, Paris.

Malus, E.L. (1810). *Théorie de la double réfraction de la lumière*. Baudouin, Paris.

Malus, E.L. (1811). Sur l'axe de réfraction des cristaux et des substances organisées. *Mémoires de la classe des sciences mathématiques et physiques*, XI, 142.

Miller, W.H. (1842). *A Treatise on Cristallographie*. Bachelier, Paris.

Mohs, F. and Weiss, C. (1823). *On the Crystallographic Discoveries and Systems of Mohs and Weiss*. Constable, Edinburgh.

Newton, I. (1704). *Traité d'optique*. Royal Collection Trust, London.

Nye, J.F. (1961). *Propriétés physiques des cristaux*. Dunod, Paris.

Pisani, F. (1883). *Traité élémentaire de Minéralogie*. Masson, Paris.

René Descartes (1637). *Discours de la méthode pour bien conduire sa raison et chercher la vérité dans les sciences, plus La dioptrique, Les météores et La géométrie qui sont des essais de cette méthode*. Jan Maire, Leyde.

Romé de l'Isle, J.-B. (1772). *Essai de cristallographie ou description des figures géométriques propres à différens corps du règne minéral connus vulgairement sous le nom de cristaux*. Didot, Paris.

Romé de l'Isle, J.-B. (1783). *Cristallographie ou description des formes propres à tous les corps du règne minéral*. Imprimerie de Monsieur, Paris.

von Schelling, F.W. (1798). *Von der Weltseele, eine hypothese der höhem Physik zur Erklärung des allgemeinen Organismus*. F. Perthes, Hamburg.

Schönflies, A. (1891). *Crystal Systems and Crystal Structures*. Teubner, Leipzig.

Sohncke, L. (1879). *Entwickelung einer Theorie der Krystallstruktur*. Teubner, Leipzig.

Stenon, N. (1669). *Dissertationis Prodomus*. Stelle, Florence.

Weiss, C.S. (1804). Dynamische Ansicht de Krystallisation. In *Lehrbuch der Mineralogie*, Karsten, D.L.G. and Weiss, C.S. (eds). Teubner, Leipzig.

Weiss, C.S. (1810). Dynamische Ansicht de Krystallisation. In *Traité de minéralogie*, Haüy R.J. (ed.). Chez Louis, Paris.

Young, T. (1802). The Bakerian lecture. On the theory of light and colours. *Philosophical Transactions Roy. Soc.*, 92, 12–48.

4.8.2. *Secondary sources*

Authier, A. (2013). *Early Days of X-Ray Crystallography*. Oxford University Press, Oxford.

Bishop, A.C. (1967). *An Outline of Crystal Morphology*. Hutchinson Scientific and Technical Press, London.

Burke, J.G. (1966). *Origin of the Science of Cristals*. University California Press, Berkeley.

Groth, P. (1926). *Entwicklungsgeschichte des Mineralogischen Wissenschaften*. Springer, Berlin.

Halleux, R. (1974). *Le problème des métaux dans la science antique*. Les Belles Lettres, Paris.

Maitte, B. (2014). *Histoire des cristaux*. Hermann, Paris.

Metzger, H. (1930). *La genèse de la science des cristaux*. Albert Blanchard, Paris.

5

Geology and Visual Discourse: The Birth of the Geological Map

Pierre SAVATON[1,2]
[1]Université Caen Normandie, France
[2]Laboratoire CFV, Université de Nantes, France

The current detailed geological mapping of France at a scale of 1:50,000 (1 cm corresponds to 0.5 km) officially began in 1913, but in 1967 only about 100 of the 1,064 sheets planned to cover metropolitan France had been published. It is now more than 99% complete. The Bureau de la Recherche Géologique et Minière[1] (French Geological Survey) has been responsible for this program since 1968, the date of the official completion of the geological mapping project of France at a scale of 1:80,000 launched a century earlier with the imperial creation of the Service de la Carte Géologique (Geological Map Service). This first standardized cartographic coverage had been preceded in the 19th century by geological maps of the departments. Published from the 1820s to the 1880s, these first geological maps were individual works, spontaneous or commissioned, codified but without prior harmonization. They marked the birth of French geological cartography and bear witness to the simultaneous emergence of the invention of a new science, geology and the invention of a visual language to express it.

What is a geological map? Let us first clarify what object we are discussing. A geological map is a representation of a geographic space that highlights the

1 This cartography is available online via the BRGM website: https://www.brgm.fr/fr.

History of Earth Sciences,
coordinated by Pierre SAVATON. © ISTE Ltd 2026.

identification of the rocks composing its subsurface. The adjective geological refers to a science that for more than 200 years has set itself the objective of reconstructing the history of the Earth (historical geology) and explaining the processes that shape it and cause the appearance of its surface to evolve over time (physical geology). Rocks are archives. The nature and arrangement of the minerals that compose rocks are determined by their conditions of formation and transformation. Since the processes involved are considered uniform through time (the principle of uniformitarianism), the study of their present manifestations makes it possible to explain and reconstruct their history. Some of these rocks contain fossils, that is, traces left by living organisms. These traces testify to the evolution of species over time and can be used as temporal markers in the history of life and therefore of the Earth. Determining the mineralogical and chemical nature of a rock, identifying fossils and determining a relative age, characterizing the current spatial arrangement of former sedimentary depositional planes that have become rocks, dating minerals by radiochronology and so on allow the geologist to extract from these archives the elements necessary to reconstruct a history. The rock found in a geographic space contains elements of its history. By indicating on a map the location of a rock, the geologist writes the history of that space.

Representing on a geographic map the age and/or nature of subsurface rocks, reducing the diversity of rock layers to a few groupings, classifying rocks by their nature or by their age, indicating the inclination of a layer, abstracting away the soil that covers the subsurface, all result from choices guided by intentions in the service of a thesis. The geological map conveys historical choices determined by theses and by intentions of discourse. Yet what underpins the discourse of the map is the discourse of geology itself. The map visually expresses a specific discourse on the history of the Earth. The history of geological cartography reflects the history of ideas about the Earth. The emergence of this visual language (Rudwick 1976) reflects the birth of a new discourse on the Earth founded on the interpretation of rocks accessible to observation. By studying the history of geological maps, one studies the historical constitution of geological discourse.

The historical evolution of geological maps was determined by an evolution of discourses on the Earth that constituted this new science, which emerged at the junction between the 18th century and the very beginning of the 19th and was then called geology (Rudwick 1976, 2005, 2008; Porter 1977; Ellenberger 1983, 1994; Taylor 1985; Oldroyd 2013). The history of geological cartography is the history of a discourse under construction that detached itself from mineralogy, physical geography and geognosy[2] in order to reconstruct a history of the Earth from the archives constituted by rocks, fossils, the arrangement of layers and mineral masses. The history of geological cartography is the history of the construction of a language to express on the two dimensions of a sheet, the three dimensions of geographic space and the fourth dimension that is time.

5.1. Statistics and inventories plotted on maps

The first projects of maps depicting information on the nature of mineral objects (minerals, fossils, ores, rocks) found by digging into the ground date from the end of the 17th century. The aim was not to map in order to establish and display a history, but to produce maps to show the locations of sites of mineralogical interest.

5.1.1. *Soil maps and maps of fossil mines*

From the end of the 17th century onward, some authors wishing to establish an inventory of exploited rock deposits (mines and quarries) proposed supplementing lists of localities with maps locating them geographically. The period, in England as in France, was one of inventories, generally constructed from surveys. The aim was to inform patrons about the riches of a territory and to do so in a manner conducive to rapid communication. Mapping could accelerate this communication. It was therefore necessary to make things visible.

The Royal Society of London, created in 1660, had set itself the mission of encouraging empirical studies. By launching surveys, it sought to initiate and stimulate these studies and to collect, critique and disseminate all information likely to increase knowledge. Natural history lay at the heart of its concerns. Deposits of ores, coal, sand, building stone, crystals and fossils triggered surveys and debates from the very origins of the society (Porter 1977). Projects were proposed and debated there, but realizations were rare.

2 Nicolas Desmarest proposed in 1757 to define physical geography as a science charged with the empirical description of the major phenomena of the Earth, far removed from the speculative discourses of theories of the Earth and distinct from mineralogy, then reduced to the sole description of specimens. The geognosy defended by Abraham Gottlob Werner from 1776 onward differed from it by its mining origins and purposes, which fundamentally aimed to reconstruct the superposition of rock layers and, in so doing, the chronological succession of their formation.

In 1683, Martin Lister (1638–1712) presented to the Royal Society trials of maps produced over the previous 10 years in northern England. None of these maps seems to have been preserved, and we can therefore only refer to his communication (Lister 1684). These soil or mineral maps, as he proposed to call them, were to be constructed from a geographic map of England and indicate cities and rivers, the extent of rock and terrain outcrops such as clays, sands and coals, but also iron ore, pyrite, gypsum, chalk or sandstones, specifying as far as possible their limits of extent and their contacts. These maps were intended to allow the observer to encompass at a glance the history of the soils of England. For Lister, sands originally constituted a layer covering the entire surface of the globe before being eroded by water runoff and accumulated in valleys and along seashores. By plotting the distribution of sands on a map, one should have been able to observe the result of a history.

John Aubrey's (1626–1697) map project was quite close to that of Lister, without it being known which influenced the other. His study of Wiltshire, submitted to the Royal Society in 1675, was not published until 1691 by a local learned society (Aubrey 1691), which has led several historians to say that it probably did not benefit at the time from wide circulation. It nevertheless came to complete a series of regional studies on the history and physical and human geography of the regions of Stafford, Oxford and Leicester. The work, although it was scarcely read at the time, was partially republished in 1847 by the Wiltshire Topographical Society for the 150th anniversary of Aubrey's death, but the geographical and mineralogical aspects of his writing were scarcely taken up, as they were then judged too obsolete (Briton 1847). Aubrey defended the idea, still common one to two centuries later in many studies of regional history and geography, of a link between the soil, landscapes and the characteristics of the inhabitants.

Like many other scholars of the 17th century, he developed a personal theory on the formation and structure of the Earth explicitly borrowed from that of his friends Robert Hooke and John Ray. Observations and descriptions of soils and outcrops led him to wish that soil maps were established with the indication of the locations of fossil and mineral deposits. By soil, he essentially meant the substratum of cultivation, but also the subsoil, in the modern sense. The term fossil is to be understood in the 17th century, as still at the beginning of the 18th century, in the broad sense of "any object found by digging into the ground", that is, not only mineralized remains or traces of living organisms, but also minerals, and even objects of anthropogenic origin. Aubrey did not depart from this usage, but reserved the term minerals for metallic ores and used the term *formed stones* (*pierres figurées*) to designate more specifically what we now call fossils. The maps he called for would therefore have been essentially maps of the extent of types of soils or subsoils on which the locations of deposits of all kinds would also be indicated. Aubrey, unlike Lister, did not envisage these maps as capable of producing a discourse on history and the relationships between terrains. Nevertheless, his discourse was not devoid of geological remarks, since he noted, for example, the absence of coal in limestone terrains or the frequent association of certain soils and ores, but without drawing conclusions or applications from them.

In France, the Académie Royale des Sciences, created in 1666, also intended to establish maps from naturalist data. Bernard de Fontenelle (1657–1757), referring to René-Antoine Ferchault de Réaumur's (1683–1757) work on the faluns of Touraine, envisaged in 1720 the production of maps on which the locations of all deposits of shells buried in the ground would be plotted (Fontenelle 1720).

This idea postulated that such a representation would be able to make less speculative the attempts to reconstruct the ancient landscapes of the Earth that flourished in the wake of the theories of the Earth:

> M. de Réaumur imagines how the Gulf of Touraine was connected to the Ocean and what the current was that carried the shells there, but this is only a simple conjecture offered in place of the true unknown fact, which will always be something approximate. To speak with certainty on this matter, one would need to have kinds of geographical maps drawn up according to all the deposits of shells buried in the ground. How many observations would be needed, and what time to obtain them! Who knows, however, whether the sciences will not one day go that far, at least in part! (Fontenelle 1720, p. 9)

Fontenelle's wish, taken literally, was for maps that one could today qualify as paleontological. They were to make it possible to reconstruct ancient geographies (paleogeographies) by, for example, shifting the present shoreline either inland or, on the contrary, farther out to sea, in order to explain the present-day presence of deposits of marine shells or freshwater shells on lands that are now cultivated.

This idea of maps proposed by Fontenelle was taken up by the chemist Guillaume François Rouelle (1703–1770) in his courses at the Jardin Royal (1750–1760) to visually translate, by plotting these locations, the existence of ancient marine territories that had until then remained invisible to observers. Cartographic expression was to make it possible to highlight fossiliferous provinces. Geographic space became a paleontological space.

The idea of soil maps was also present in German-speaking countries, where the intersection between mining tradition and geographic cartography of territories was to lead, at the beginning of the 18th century, to a few projects and trials of plans.

5.1.2. *Mineralogical geography*

The map of the Danube basin by Luigi Ferdinando Marsigli (1658–1730), published in 1726, was an inventory of mining deposits depicted by symbols identifying their mineralogical nature (Marsigli 1744). A general of the Habsburg Empire, Marsigli traveled across Europe over the course of his military and diplomatic career (Stoye 1994). A cartographer, through his travels and campaigns he took an interest as much in mineralogy, botany and zoology as in ruins and objects bearing witness to ancient societies. A diplomat, his gaze also lingered on economic and industrial aspects. His work on the Danube valley formed part of a vast statistical project. It brought together in six richly illustrated volumes a set of geographical, astronomical, hydrographic, historical and physical studies. There he described the mining activity of Hungary and complemented his texts with representations of the subsurface in plans and cross-sections constructed on the model of Georgius Agricola's *De Re Metallica* (1556). His mineralographic map of the mines of Hungary (*Mappa mineralographica fodinas in Hungaria*) located mines and outcrops of useful substances on a geographic base indicating cities, the course of rivers and mountain chains. The originality of his cartographic locations lay in his choice to designate minerals and rocks by symbols. He used classic chemical and metallurgical pictograms, but also invented some to designate, for example, opal, garnet or asbestos.

This symbolization, while it accentuated the diagrammatic dimension of his maps, still expressed only a statistical concern. This choice of a scatter of signs characterized maps that can be described as mineralogical. They geographically located deposits, but refrained from proposing vanished or unknown continuities between two adjacent similar deposits. Marsigli's map did not specify the nature of the soils or rocks that contained these deposits. For the historian David Oldroyd (Oldroyd 2013), Marsigli nevertheless took a new step in the representation of the subsurface, because his handwritten notes contain sketches, never published, of terrain cross-sections clearly showing the author's thinking in terms of rock layers, distinct strata, sometimes inclined or even folded or unconformable (Marsigli 1744). Marsigli may perhaps have questioned the spatial relationships between point deposits and the limits of geographic extension of rocks and terrains visible at outcrop. In his cartographic sketch of a gypsum and sulfide deposit at Cesena near

Bologna, Oldroyd believes he perceives the emergence of a geological discourse through the drawing of a boundary in the field between gypsum and sulfides.

Marsigli's sketches nevertheless remained figurative, even if they reflected a new analytical acuity, and they cannot suffice to prove the existence of a geological intention. The historian François Ellenberger also notes with interest that in his map of the mines of the Chemnitz region, Marsigli indicated a preferred orientation of veins (Ellenberger 1983, p. 15). It would nevertheless be premature here as well to see in his maps a symbolic discourse on the geology of a place. Marsigli's interest in the description of volumes is doubtless to be related to the mining tradition of the Habsburg territories and to the first projects of underground cartographies of mines, but it does not translate clearly into his cartographic representations.

A further step toward a geological discourse was, however, taken a few years later in France by Jean-Étienne Guettard (1715–1786). First a correspondent and then a collaborator of Réaumur, Guettard likewise situated himself within a statistical approach to the inventory of mineral deposits. The establishment of networks of correspondents ensured for some academicians the collection of numerous and geographically dispersed point observations. There is no doubt that this mode of data collection predisposed such work to cartographic expression. Botanical, zoological and mineralogical inventories, which continue even today by calling upon networks of amateurs, increasingly lead in a similarly automated manner to the publication of location atlases.

The *Mémoire et carte minéralogique sur la nature & la situation des terreins qui traversent la France et l'Angleterre*, which Guettard submitted to the Academy in 1746, marked the birth of a new type of map, because it is associated with the depiction of the location of point deposits by symbols information on the nature of the terrains that contain them. This small-format map indicated a presence more than it truly made it possible to locate. Guettard distinguished there 51 different resources, which he located in three fundamental types of terrain: schists, marls and sands. These terrains drew on the map three "bands", sandy, marly and schistose (or metallic), which sketched out an initial mapping of geological surfaces. The marly band brought together beneath the English Channel the Paris and London basins. For this first project, Guettard benefited from the geographic maps established by Philippe Buache (1700–1771), and in particular from his *Carte physique ou géographie naturelle de la France* of 1744 (Buache 1752).

Buache was at the origin of a thesis on the structuring of the geographic space of the globe by mountain chains, which he extended beneath seas and oceans, making islands the emergent summits of submarine chains. It is difficult to specify the meaning to be given at that time to these three bands, which Buache also indicated on his map of the Middle East of North America (Kish 1976). Guettard seems to have made them a global characteristic, a general rule, by indicating the existence of these three bands on his maps of Egypt, Palestine, Switzerland, Canada and Poland.

These mineralogical maps did not, by their scale, make it possible to achieve the precision that Guettard wished to give to his mineralogical inventory of the subsurface. He therefore continued his cartography through detailed surveys that allowed him to establish a mineralogical map of Paris, and then of Étampes in 1757. His cartographic enterprise having found financial support from the minister Henri Bertin as early as 1763, he embarked in 1766 on an ambitious project to publish a *Atlas minéralogique de la France* in 214 sheets. From 1766 to 1770, Guettard, in collaboration with Antoine Lavoisier (1743–1794), surveyed, in 16 sheets, the Île-de-France, a large part of the Brie and Champagne regions, and part of Alsace, Lorraine and Franche-Comté. The geographic base of these maps was drawn by reduction from the geographic maps of César-François Cassini by the king's geographic engineer, Jean-Louis Dupain-Triel. On each sheet, deposits of rocks and minerals were indicated. A column representing the superposition of terrains, as one might ideally observe it by means of a pit or an embankment, was drawn in the margin.

The thicknesses and altitudes of the layers were determined by barometer by Lavoisier, who, out of a concern for precision, intended to indicate the depth at which the deposits were found. Lavoisier's cartographic project was more ambitious than that of Guettard (Rappaport 1969). He was concerned with the arrangement of layers in the three dimensions of space. The material ambition of the project, by its scale, required numerous field surveys and a corresponding budget. The budget allocated by Bertin was insufficient for such an undertaking. As early as 1773, Lavoisier unsuccessfully attempted to convince Bertin to revise the project by drastically reducing the scale and thus the number of sheets to be produced, and by relying on surveys already carried out throughout France by local scholars. In 1777, Bertin appointed Antoine Grimoald Monnet (1734–1817), an inspector of mines trained at the Freiberg School in Saxony, Inspector General of the mines and mining works of the kingdom, and charged him with directing the surveys necessary for this atlas by relying solely on the reports and surveys of mining engineers. Guettard and Lavoisier were administratively dispossessed of their projects. Bertin and Monnet had a very utilitarian vision of this cartography, which was intended above all to serve the development of the French mining industry. Collecting and mapping information other than the geographic locations of deposits lost their interest. Monnet published, from 1778 to 1780, 17 of the sheets prepared by Guettard and Lavoisier. In the end, 31 sheets were published.

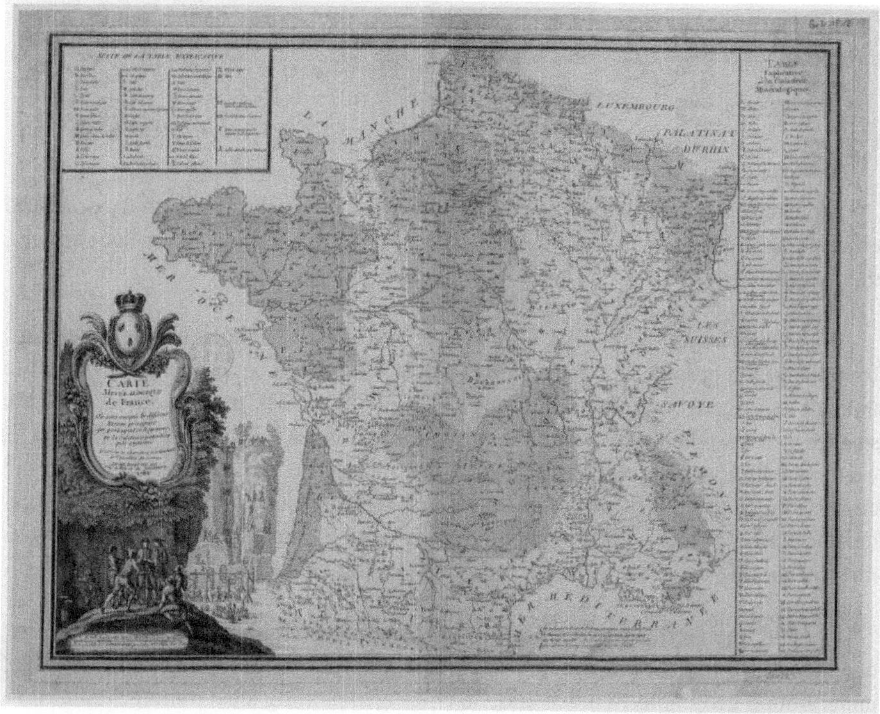

Figure 5.1. *Mineralogical map of France by Guettard and Buache (source: Guettard 1781[3])*

Guettard published in 1781 a *Carte minéralogique de France, où sont marqués les différens terreins principaux qui partagent ce Royaume et les substances particulières qu'il renferme*, in continuity with his 1746 map, but with 210 symbols to identify the diversity of deposits of stones, minerals and ores, which essentially had the effect of making an overall reading impossible and a detailed reading very complex, due in particular to the figurative proximity of certain symbols (Guettard 1781). The author nevertheless seemed to have sensed that cartographic representation needed to free itself from point-based inventory in order to move toward the production of a more prospective and speculative discourse. The three bands were once again drawn, but the map was still not conceived as a representation of surfaces. Guettard's maps were only cartographic inventories of observations. Opposed to the speculative discourses of the authors of theories of the Earth, he refused to inscribe on his maps interpretations that would fall under speculation.

3 See: https://gallica.bnf.fr/ark:/12148/btv1b8440697j/f1.item.

Drawing a continuity between contiguous deposits of the same nature seemed to him too speculative.

5.2. Physical geography

Alongside these early attempts at mineralogical maps, the 18th century produced other experiments seeking to translate underground geography or, on the contrary, the form of the ground at the surface, or geomorphology.

In 1738, Daniel Tilas (1712–1772), a Swedish geographer and mineralogist, formulated the project of a geographic map of the subsoil of Sweden. He wished to indicate on a geographic map not only the location of rock masses observed at the surface, but also their extension at depth, information useful for their exploitation. It was an ambitious project that proposed, in a sense, to move from the plans and cross-sections used at the scale of mines to a representation at the scale of a larger territory. Tilas produced two mineralogical maps of Finnish provinces and numerous maps of mining deposits, but none truly fulfilled this project.

In 1746, the German mineralogist Carl Friedrich Zimmermann (1713–1747) set out his wish for a new science whose object would be specifically the study of underground geography. To that end, he envisaged plotting local mining studies on a geographic map of a territory. This project served a discourse on the geographic availability of mining resources. Zimmermann already at that time defended the need to create in Saxony an academic institution to develop and disseminate mining knowledge. He may be considered one of the initiators of the project of the Freiberg School of Mines founded in 1765.

That same year, 1746, Abbé Pierre Augustin Boissier de Sauvages (1710–1795) presented before the Société Royale des Sciences de Montpellier a lithological study intended to serve a natural history of Languedoc. His memoir was also read at the Académie Royale des Sciences in 1749 and 1750 (de Sauvages 1751, 1752). The region he had just explored presented layers of terrains that were folded, fractured and cut into blocks. It was difficult to follow their lateral continuity. De Sauvages, faced with this chaos, thought that a topographic map of the terrains could make it possible to translate and set out this complexity:

> It would be especially desirable that those who take an interest in the progress of natural history should work on a topographic map of the terrains on which one would mark the continuities, the interruptions, the different grains, the nature and the properties. The execution of this map would have another great advantage, in that it would

> influence economics, and that it would show, as if at a glance, the crops of which an entire province is capable, the extent and the quality of the harvests that can be obtained from it; it is a work that could be done at little cost, as one draws up the geographic maps of the kingdom. (de Sauvages 1751, p. 723)

This topographic map of the terrains served both concerns that can be described as agronomic and a discourse seeking to reconstruct a tectonic history.

This idea of a topography of terrains is found again in the *Carte d'une partie de l'Auvergne où sont figurés les courants de laves, dans lesquels se trouve le Basalte en Prismes*, published in 1774 by Nicolas Desmarest (1725–1815) and François Pasumot (1733–1804). This map accompanied a *Mémoire sur l'origine et la nature des basaltes à grandes colonnes* that claimed the volcanic origin of these rocks (Desmarest 1771). On the occasion of observations made in the field, Desmarest recognized in a dark columnar rock a resemblance to the rocks of the Giant's Causeway in Ireland. These rocks were not yet associated with a volcanic phenomenon, but their strange morphology had attracted the attention of many observers. Desmarest saw in the landscapes he observed in Auvergne the result of the processes of alteration and erosion to which all rocks that outcrop at the surface of the globe are subjected. He intended to develop a physical geography, or a geography of the Earth's physical surface. The *Carte topographique et minéralogique d'une partie du département du Puy de Dôme*, published in 1823 by his son Anselme Gaëtan Desmarest (1784–1838), continued his study. His prior observations on erosion allowed him to bring together distant deposits and to morphologically reconstruct ancient lava flows. He thus identified cartographically, through the local relief, an ancient volcano and its flows.

The *Carte géographique de la Nature ou disposition naturelle des minéraux, végétaux, etc.*, observés en Vivarais, published in 1780 by Abbé Jean-Louis Giraud-Soulavie (1751–1813), is to be related to Desmarest's work by the attention it pays to morphology and to the disposition of layers at outcrop. His mineralogical discourse was constructed on the idea of a superposition of terrains over time. He observed in the Vivarais basalt flows resting on fluvial conglomerates, superposed on limestones with granites at the base. He also distinguished ancient basalts forming reliefs and eroded, and recent basalts occupying the bottoms of valleys. Morphology as much as lithology led him also to identify the existence of ancient volcanism. His map delimited by colors the extent of these different rocks. He added to it mineralogical indications by a scatter of symbols. His map is close to the German petrographic maps that will be discussed later, but from reading the accompanying text (Soulavie 1780–1783) one discovers a discourse that links the nature, age and disposition of rocks, and leads him to group under the same symbol rocks (limestones) distinguished in the field by their fossil contents.

The recognition of ancient volcanic rocks, in France and in Europe, quickly led some observers to defend the idea of the existence of ancient volcanoes against the general opinion that envisaged a relief or rocks as being of volcanic origin only on condition that volcanic activity had been historically observed. A learned controversy over the origin of rocks then opposed proponents of a fundamentally aqueous origin of all rocks (rocks issued from precipitation and deposits at the bottom of a sea or an ocean) to proponents of the existence of two types of rocks, some formed by the action of water and others by the action of fire. History later categorized them as Neptunists and Vulcanists. This controversy, by centering mineralogical discussions on the origin of rocks, contributed to the evolution of a geographic discourse toward a historical discourse. The mineralogical map was centered on the location of objects whose origin was at least secondary. The map could produce only a geographic discourse. By taking an interest in the origin of rocks or in the origin of a morphology, objects became products of events, became archives. The map could lay claim to a historical discourse.

The *Carte minéralogique des Monts-Pyrénées* by Abbé Pierre-Bernard Palassou (1745–1830), published in 1781 as an accompaniment to his *Essai sur la minéralogie des Monts-Pyrénées*, differs from the maps of Guettard, Lavoisier and Monnet, contrary to what its title might suggest (Palassou 1781). It certainly indicates on a geographic base a few mineralogical deposits by symbols, but it adds lines to indicate the preferred geographic orientation of the bands of "schists or clayey stones" and of "limestones" that alternate regularly. The map represents contiguous surfaces presenting a particular disposition in plan and at depth. The Pyrenees range would be characterized by the spatial arrangement of its terrains. Palassou went beyond the mineralogical cartography of his contemporaries in favor of a map indicating the disposition of a few terrains that seemed to him to structure the landscape. This mineralogical map was therefore also a tectonic map.

5.3. Geognosy and geognostic maps

Johann Gottlob Lehmann (1719–1767), in a work published in 1756, laid the foundations of a relative chronology of terrain layers (Lehmann 1756). In the tradition of German miners, the prospecting and exploitation of copper relied on the recognition of a succession of layers framing the copper-bearing horizon. Lehmann generalized this observation by proposing to distinguish, on the one hand, deep rocks (*Gangsgebirge*), which contain metallic veins, and, on the other hand, stratified rocks and terrains (*Flötzgebirge*), which cover them and sometimes contain metallic levels, but in the form of veins (within the plane of the layers) and not of lodes (cutting across the layers). These rocks characterized two types of mountains that formed successively in the history of the Earth: primitive mountains and secondary mountains. These two types were covered by terrains that he judged to be recent. Stratified masses could be ordered from the deepest to the most superficial, that is, from the oldest to the most recent. By thus chronologically classifying terrains (*Gebirge*) on geometric arguments of superposition, Lehmann gave them a historical dimension and opened the way to a description of the stack of layers succeeding one another through time.

Georg Christian Füchsel (1722–1773) then established in 1761 a map of Thuringia in which he described a succession of terrains (*Gebirge*) laid one upon another (Füchsel 1761). According to his thesis, each rock layer is derived from a deposit or precipitation. Several successive and similar layers constituted a "formation" that translated a moment in the history of the Earth. Since layers formed horizontally, any observed inclination of them translated a later deformation according to principles already set out in 1669 by Nicolas Steno (Steno 1669). Füchsel then described the succession of the fourteen formations that he observed in the territory of Thuringia. At the base, he named as "basement" (*Grundgebirge*) massive rocks on which rested upturned schistose layers, then a series of less inclined schistose layers, then sandy layers, and finally, at the top, limestone layers rich in shells (*Muschelkalk*). Füchsel wished to develop a *scientia geognostica* (science of geognosy) whose aim would be to describe the succession of terrains and to reconstruct the history of lands and seas. It was no longer a matter of describing a layer in order to recognize it elsewhere so as to locate an ore deposit, but of describing it in order to reconstruct a history. It was no longer a mineralogical map with a scatter of symbols disseminated on a geographic base, but a map of the extent of terrains that outcrop at the surface. These surfaces, identified by a number, were difficult to distinguish on the map, but if colors were added, they became visible and a geologist immediately saw a spatial arrangement of the layers. The addition of color was not merely a question of aesthetics; it contributed to the clarity of cartographic discourse by making reading more immediate.

This map by Füchsel reflected new objectives. The disposition of layers and rock masses had to be characterized, because this disposition translated into a chronology. This geognostic science, which Abraham Gottlob Werner formalized 15 years later, differed from physical geography by its focus on the disposition of rocks. To translate the diversity of mineral deposits, the authors of mineralogical maps resorted to an ever greater diversity of symbols, but this inflation could not be followed by maps that aimed to identify superposed surfaces. The layers observed in the field could seem unique, since their superficial aspect could be changeable: the weathering of a layer modified its color and cohesion; erosion determined its morphology at outcrop. Füchsel therefore had to go beyond immediate appearances in order to determine a general aspect; he had to group the successive layers that he judged similar in order to plot their outlines on a map. One cannot account for all the diversity and complexity of terrain observations; one must extract the simple from the complex. One must reduce immediate complexity to a simplified cartographic expression: one must group. Yet these groupings did not impose themselves and therefore required the construction of prior rules. Observation was guided and constructed as principles were gradually established.

Friedrich Gottlob Gläser (1749–1804), a mining engineer of the very young Freiberg School, published in 1774 a work on the Henneberg region in western Thuringia (Gläser 1774). It contained a color map on a very simple topographic base. One clearly sensed the influence of Füchsel in the choice to represent the extent of three types of rocks and in the choice of grouping and designation. Gläser distinguished a surface of granites, sandstones (*Buntsandstein*) and shelly limestones (*Muschelkalk*). He also depicted ore deposits by mineralogical symbols (iron, gold, silver, copper, etc.).

In 1778, Johann Friedrich Wilhelm Charpentier (1738–1805), professor of mathematics and physics at the Freiberg School, published a map representing by distinct colors the geographic extent of the different types of rocks. He called it a petrographic map (Charpentier 1778). Included in a work on the mineralogical and petrographic geography of Saxony, it is distinguished by colors of eight types of terrains arranged in an order compatible with that of Füchsel and Gläser: granites, gneisses, schists, limestones, gypsums, sandstones, sands, clays and silts. It is also indicated by symbols for the location of deposits of a few minerals, and thus doubled the color coding of these lithological units with a symbolic coding.

The Freiberg School of Mines, the first of its kind, could already at that time be considered a center of cartographers. In addition to the preceding authors, it also trained Johann Phillip Becher (1752–1831) and Johann Wilhelm Voigt (1752–1821), authors of petrographic maps in the 1780s. But the school was to become known throughout Europe and to radiate beyond it thanks to another of its professors, in charge of the mineralogical collections from 1775 onward, Abraham Gottlob Werner. Although he did not produce a map, he nevertheless formalized and developed Füchsel's idea of a geognostic science and established a framework for thinking about rocks as resulting from successive formation through time. Behind the diversity of rocks observed, it was therefore necessary to focus on specifying their relative disposition with respect to one another, because this was the necessary condition for placing them back in the order of their formation. This geognosy, which developed especially in the years 1800–1810, truly carried a chronological discourse that foreshadowed the historical discourse of the geology to come.

In 1785, James Hutton (1726–1797) presented before the Royal Society of Edinburgh his *Theory of the Earth*. In it, he defended the thesis of a recycling of stacked layers by fusion at the base of the pile. This fusion was, according to him, at the origin of the formation of granites and of veins that cut across stratified rocks. These molten rocks and the granites that result from them caused, through buoyant rise, a thrust on the overlying stratified rocks and led to their tilting and deformation.

Hutton defended the existence of rocks of magmatic origin, granites and basalts in particular. For him, granites displayed crystals visible to the naked eye because they resulted from slow cooling at depth. To the claim by the vulcanists of a volcanic origin for ancient basalts was added a claim later described as Huttonian or plutonist of a magmatic origin for granites.

That same year, 1785, Werner published his *Short Classification and Description of the Different Types of Rocky Terrains* (Kurze Klassifikation und Beschreibung der verschiedenen Gebirgsarten). The ideas contained in this work were major and marked the history of geology (Werner 1787). Werner wrote very little, and the diffusion of his ideas occurred through his students and through this small 28-page work. As its title suggested, the work proposed a new classification of rocks. He distinguished 31 types of rocks. But the work was not a simple petrographic or mineralogical classification, because, associated with the idea of a chronological succession of these rocks, it determined a project of reconstructing a history of deposits that was to mark nascent geology and cartography up to the 1820s. Rocks were classified there from the angle of their origin and separately from minerals. A rock was an assemblage of minerals, and it was this combination that became central for describing terrains. It was a petrographic and not a mineralogical classification; petrography emancipated itself from mineralogy and in doing so freed itself from a confused heritage little suited to the birth of a new science.

Werner defended the idea that all rocks derived from the precipitation or transformation of a deposit in an ocean: a vision described as Neptunist. The Earth at its origin was covered by an ocean in which the first rocks of the present-day Earth's crust precipitated and were deposited. By gradually retreating, this primitive ocean allowed its deposits to emerge, stacked from the oldest to the youngest. These successive deposits, by extracting minerals from the water, progressively modified its composition, which translated into an evolution in the nature of the deposits. Thus, during the first period of the history of the Earth's crust, or primitive epoch, the deep and calm ocean was conducive to the precipitation of crystals at the origin of crystallized rocks known as granites. These rocks were at the base of the pile of deposits and could be observed today both at the center of mountains and at the base of stratified terrains. Subsequently, deposits continued to differentiate, with schistose deposits rich in minerals, then well-stratified deposits rich in fossils. The schistose deposits showed deformations linked, according to Werner, to the agitation that marked the ocean at that period. Werner later divided this second period, or *Flötzgebirge*, by naming as "transition terrains" the terrains situated at the base, at the contact with granites.

The pile of deposits, of formations, formed a sequence that translated the evolution of the ocean since the origins. By reconstructing the pile, geognosy could reconstruct the history of deposits, that is, the history of the Earth. According to Werner, primitive, transition and stratified rocks all derived from one and the same ocean. The alluvial and volcanic rocks that covered them derived from recent local histories. The molten rocks emitted by volcanoes required a combustion according to a common idea. This combustion could only be envisaged once coal layers had formed. Volcanoes therefore could only be recent events in the history of the Earth.

For Werner, a succession of layers accounted for a moment in the sequential history of deposits. He took from Füchsel the term "Formation" to designate these ensembles. To describe the sequence was to describe the succession of formations. To identify a formation was to place it in the chronology. The primordial ocean covering the entire Earth, there was only one pile of layers, with possible regional variations. If the pile was conceived as universal, its reconstruction was not easy. The observer who traversed the field seeking to identify the formations had to distinguish rocks that were similar but formed at different periods from rocks that were dissimilar but formed simultaneously. Werner supported his thesis with his own observations in Saxony and invited his students to develop their local studies so as to reconstruct the global pile by correlating observations in Europe and beyond.

Field observations were to lead to the production of maps of terrain layers, or geognostic maps, which fundamentally aimed to reconstruct a history of the Earth's surface. Werner's classification of rocks was rapidly adopted and disseminated even by opponents of his Neptunist vision, because they saw in it a possibility for harmonizing vocabulary recognized as necessary for exchanges and for establishing correlations. The multiple proposals for classifications since the middle of the 17th century had led to multiple confusions among authors. At least a common descriptive language was required in order to build collectively.

The Freiberg School – and the geognostic cartography that it promoted – was carried by a theoretical framework. This was probably what was lacking at that time in Great Britain and in France to give meaning to their accumulation of empirical observations.

The enterprise of mineralogical cartography, conceived as an inventory on a geographic base, was exhausted in France during the 1780s. In vain did Lavoisier defend the interest of describing the superposition of layers to understand the long history of deposits and the interest of associating vertical sections with maps. From his observations in the Paris Basin, he had recognized as early as 1766 the existence of an alternation through time of limestone layers with pelagic faunas and detrital layers with littoral faunas, translating the succession in the same place of the sea and then emergent lands. This distinction, already used by Guillaume-François Rouelle and by John Woodward, took on a new significance with Lavoisier. He showed that the succession of littoral terrains and pelagic terrains indicated a "rising" sea (transgression), whereas the opposite succession translated a "falling" sea (regression). The succession determined a cycle. But Lavoisier did not go further in exploiting the historical significance of this sequence. His sections still had a lithographic purpose; they served to define facies. Although he was interested in their fossil contents as witnesses of ancient living environments, he did not make them medals, markers of an era.

The mineralogical cartography of Guettard, Lavoisier and Monnet came to an end, unfinished, but the political will to possess an inventory of mineralogical and mining resources remained. The Royal School of Mines, founded in 1783 with the aim of training men capable of directing the mines of the kingdom, was abolished in 1791 and then reorganized in 1794. It left Paris to settle in Peisey-Nancroix in Savoy, on the site of the Peisey mine. Teaching thus sought to move closer to the field and to develop practical knowledge and skills on the model of the Freiberg School.

The Conseil des Mines announced in 1796 its intention to establish a mineralogical map of France presenting both the nature of terrains and the location of mineral deposits and processing sites (furnaces, smelters, tile works, etc.). This map was to rely on the collection of information by state mining engineers during their tours of their mineralogical districts. Memoirs and local and departmental maps accumulated on the model of a statistical inventory of resources, but the general map was not produced. The Peisey School of Mines was closed in 1814 and reorganized in Paris in 1816.

Source gallica.bnf.fr / Bibliothèque nationale de France

Figure 5.2. *Geological map of France, the Netherlands, and several neighboring regions, by Omalius d'Halloy, 1822*[4]*. For a color version of this figure, see www.iste.co.uk/savaton/history.zip*

In 1810, Charles Coquebert de Monbret (1755–1831), director of statistics at the Ministry of the Interior, entrusted Jean Baptiste d'Omalius d'Halloy (1783–1875) with assembling the materials necessary for the geological description of the French Empire. This map project was above all economic in nature. It sought to identify the relationships between the nature of the subsoil and agricultural production and was part of a political context aimed at describing and rationalizing production in order

4 See: https://gallica.bnf.fr/ark:/12148/btv1b84944732.

to increase economic development. Coquebert de Monbret, also editor-in-chief of the *Journal des mines*, had accumulated data to this end for years, but to produce a general map of the Empire, substantial fieldwork still remained to be done. Omalius d'Halloy, who had made himself known to the scholarly community in 1808 through a remarkable work on the geology of northern France, then traveled throughout the Empire and submitted a map in 1813 to the Conseil des Mines. This map, set aside during this politically unstable period, was published only in 1822 under the title *Geological Map of France, the Netherlands, and Several Neighboring Regions* (Omalius d'Halloy 1822). Meanwhile, Omalius d'Halloy published in 1816 in the *Annales des mines* an *Outline of a Geological Map of the Paris Basin and Several Neighboring Regions* (Omalius d'Halloy 1816). His map of the former French Empire published in 1822 chronologically grouped all terrains into six main units: Primordial (Hercynian basement, Alps, Pyrenees), Pennean (Permian), Ammonian (Triassic, Jurassic), Cretaceous, Mastozootic (Tertiary and Quaternary) and Pyroid.

The enormous synthetic work of Omalius d'Halloy, which in 1811 represented considerable progress, was published too late to receive the praise it would have deserved. In 1822, it was overtaken by the evolution of ideas carried by mineralogical maps and geognostic maps. The geognostic description of terrains had already been reoriented by the determining chronological role of fossils.

Werner's geognostic cartography was also exported to the United States, where William Maclure (1763–1840), trained in Wernerian petrology in Europe, published as early as 1809 a geological map of the United States limited to the East Coast. Maclure recognized there four of Werner's terrains (primitive, transition, secondary and alluvial) and noted the absence of volcanic terrains. His notes show a lack of interest in the question of the origin of rocks or of their temporal succession. If he borrowed from Werner the term "Formation", it was in an acceptance limited to the idea of a mass of rocks similar in appearance and disposition. The strictly lithological aim of this map thus brought it closer to the German petrographic maps of the 1770s. His numerous geological travels in Europe (1810–1814) led him to adopt the idea then defended by Cuvier and Brongniart: that layers could be placed in a chronology solely through the determination of the fossils they contained. The revised second edition of his map, published in 1817, incorporated this dimension. It no longer referred to Werner's Formations, but relied on the distinction of "Classes of terrains". In 1818, he published a new classification proposal that broke with both the Neptunists and the Vulcanists. Maclure served as president from 1819 to 1828 of the first Geological Society of America, a learned society dedicated to the study of geology. The present-day Geological Society of America was created 60 years later (1888) by James Hall, James Dwight Dana and Alexander Winchell.

At the end of the 18th century, observation of the disposition of rocks and terrains, the presence or absence of fossils, and the question of the origin of rocks – particularly the challenge to an exclusively aqueous origin (controversy between Neptunists and Vulcanists, then Neptunists and Plutonists) – fed a broad reflection on the history of the Earth. Mineralogy, physical geography and geognosy were overwhelmed by new questions and concerns: a geological science was in the process of emerging. It both fed and was fed by cartographic work.

5.4. Geological maps

Werner's death in 1815 freed the voices of his former students and disciples. Taking distance from a model judged too limited or erroneous fractured the inheritance and made the construction of new shared principles all the more necessary. While a large number of rocks derived from the transformation of precipitations and deposits in water, others had been produced by fusion followed by cooling and crystallization at depth or at the surface. The opposition between Neptunists and Vulcanists, Wernerians and Huttonians, was obsolete.

The description of stratified terrains supplemented lithology with a description of fossils. Witnesses of a form of life and a living environment, the presence of fossils became, at the very end of the 18th century, a reason to question their origins and their links with present-day species. Naturalists, by renewing the question of the history of life, determined a new perspective on fossils, on the layer that contained them, and by extension on the history of rocks.

5.4.1. *Lithostratigraphic and geological maps*

In 1815, William Smith (1769–1839) published a stratigraphic map of England, Wales, and part of Scotland (*A Delineation of the Strata of England and Wales with part of Scotland*) and a memoir explaining his method of identifying terrains (*A Memoir to the Map and Delineation of the Strata of England and Wales, with part of Scotland*). He designated layers by the term stratum and distinguished them, often lithologically similar, by their fossil content (Smith 1815, 1816, 1817). To construct his map, Smith traveled in detail across England and Wales, as no one before him had done (Torrens 2003). He was a man of the field, a surveyor with solid mining experience, a prospector, a collector of fossils, an observer who since the 1790s had sought to find an order to arrange layers that were deformed, broken or discontinuous in the field. Smith sought to order terrains both vertically and horizontally. In 1799, he had already published a map of the surroundings of Bath, and in 1801, a first attempt at a map of England and Wales. His cartographic enterprise was a personal and individual work constructed from his very numerous field campaigns.

His originality compared with the geognostic project and maps was linked to the importance he gave to fossils as descriptors and markers of his strata, capable of establishing correlations between distant outcrops. He also distinguished himself by postulating, incorrectly, that the disposition of the terrains he observed was similar to their original disposition. His use of fossils was nevertheless limited to simple elements of description of strata. He did not consider them as chronological markers and thus missed the central role they could play in directly placing a stratum within a chronology. This idea emerged in France at the very beginning of the 19th century through the writings and work of Jean-André De Luc, but above all of Georges Cuvier and Alexandre Brongniart. Smith's map, published in 1815, did not take this thesis into account, even though it had been published as early as 1810 in the *Philosophical Magazine.* The geological description of the Isle of Wight (*Description of the Isle of Wight*) by Thomas Webster (1773–1844), published in 1816, did integrate the work of Cuvier and Brongniart and thus acknowledged its adoption in British cartography.

To Smith's 1815 map was added in 1820 a new geological map of England and Wales directed by the founding president of the *Geological Society*, George Bellas Greenough. Although Smith's thesis of strata defined by their fossils was criticized by this nascent community of British geologists, it also inspired them.

George Bellas Greenough (1778–1855) was trained in mineralogy by attending the lectures of Johann Friedrich Blumenbach at the University of Göttingen, then by exploring the field in the Harz, France, Italy and Scotland. In 1807, he founded in London with a few friends the Geological Society, the first learned society to claim this designation. It defended an empirical approach to the field that sought to free itself from the earlier conflicts of the Wernerian and Huttonian schools. The aim was to develop a new discourse on the Earth, far from speculative theses. As early as 1809, the Geological Society set itself the objective of establishing a mineralogical map of England and Wales by collecting all the information provided by a network of correspondents, in the tradition of inventories and statistics. Greenough, assisted by William Buckland, William Conybeare, Thomas Webster, and many others, organized these data into a map. A first realization in 1812 did not satisfy them, and the project was delayed. The map (*Geological Map of England and Wales*), in six sheets, was finally published only in 1819 (Greenough 1819). The idea defended by Smith of grounding the definition of strata in their fossils was not taken up, and the map grouped successive terrains on lithological arguments. The limits of the terrain groupings were nevertheless sufficiently close to those of the terrains on Smith's map that one could sometimes envisage simple reprises. These two maps and the disputes associated with their production and publication marked the birth of the Geological Society and of English geology. At the same time, they laid the foundations of geological cartography, which Great Britain immediately undertook at a more detailed scale.

5.4.2. *Biostratigraphic and geological maps*

Jean-André Deluc (1727–1817), then Alexandre Brongniart (1770–1847) and Georges Cuvier (1769–1832), by centering their attention on the significance of fossils, reoriented Werner's geognostic project and Smith's stratigraphic project. Layers became archives in themselves, dated by fossils and a source of a narrative of the history of the Earth.

In 1811, Georges Cuvier and Alexandre Brongniart published an *Essai sur la géographie minéralogique des environs de Paris* with a geognostic map and terrain cross-sections (Cuvier and Brongniart 1811). The authors referred to Werner and established a "geographical review of the various kinds of terrains or formations that constitute the soil of the environs of Paris, and of the places where each of them can be observed". The rocks of the Paris region were grouped into formations, mapped with seven colors as "transport and alluvial terrain, upper lacustrine terrain and millstone, upper sandstone and sand, gypsum, middle lacustrine terrain and siliceous limestone, coarse marine limestone, chalk". All these terrains lay spatially above the chalk formation and plastic clays, considered by the Wernerian school as the most recent formation of the secondary. The sections accompanying the map were intellectual constructions and not natural sections. They sought, like the text and the map, to convey two strong ideas: the alternation of terrains of marine origin and of freshwater origin, and the renewal of faunas from one formation to another. The thicknesses of the terrains relied on barometric leveling measurements. The principal fossils found in these formations were drawn in plates.

The care given to the description of layers and their contents recalled the work of the German geognosts, but by pursuing Lavoisier's observations and developing Deluc's paleontological stratigraphy, Cuvier and Brongniart went further. It was the succession of distinct fossil faunas through time that justified grouping layers into successive terrains. The study of strata had to be paleontological in order to establish a historical chronology. If Cuvier and Brongniart were not the first to use fossils to establish correlations between layers, they were the first, however, to make them the criterion of stratigraphic correlation (Rudwick 2008). To the lithological description of strata, they added a description of the fossils they might contain, not to refine their description as William Smith did, but because they defended the idea of a succession of fossil faunas and floras over the course of Earth history. In seeking the lateral limits of extension of the terrains they described, they observed that lithology could change without fossils varying and conversely thus proposed relying on fossils rather than lithology. The temporal marker to be favored was the fossil.

The geognostic map of Cuvier and Brongniart was the expression of a new discourse that conceived fossils as ancient species specific to a given epoch and made their presence the Ariadne's thread of the reconstruction of the narrative. Wernerian geognosy, by making lithology the criterion of correlation, posed many problems as observations expanded across Europe and then the American continent. The work of Cuvier and Brongniart also departed from that of German geognosy, which was oriented toward reconstructing a universal sequence of strata through its insistence on local aspects. The aim was first to reconstruct the local history of the environs of Paris and then by extension to correlate it with other local histories to be established. Cartography historically progressed within this tension between local history and general history, detailed description of the diversity of layers and proposals for grouping into formations.

The Wernerian chronological classification of terrains, which rested fundamentally on the spatial disposition of layers, was replaced in the project of Cuvier and Brongniart by a chronological classification based on fossil criteria. The object of the map was a discourse on the geological history of the represented territory. But this attempt at a map at the scale of the environs of Paris was slow to translate into a national project of scope. In 1820, Great Britain had two geological maps covering most of its insular territory, whereas France still had none.

Louis Becquey (1760–1849), director general of Bridges and Roads and of Mines, received Greenough's map in 1820 and immediately forwarded it to the Conseil des Mines, which then decided to support the map project that André Brochant de Villiers (1772–1840) had been defending since 1811. A student of Werner at Freiberg, Brochant joined the École des mines in 1794 before being appointed professor of mineralogy and geology in 1802. He developed teaching in the spirit of Wernerian geognosy and submitted to the Conseil des Mines in 1811 a project for a geognostic map of France. The project was not accepted. The royal ordinance of 1816 relating to the organization and administration of the École des mines nevertheless recalled among the missions of its Council that of collecting the materials necessary for the mineralogical description of France. But the administration, which already possessed the unpublished map of Omalius d'Halloy, did not then seem concerned with this cartography more focused on the designation of terrains than on the statistics of deposits. Greenough's map may well have triggered a nationalist reflex. Becquey then entrusted Brochant de Villiers, in 1822, with producing a mineralogical description of France. The project was particularly ambitious intellectually and materially. It aimed to make up for a delay and to surpass its predecessors.

A program of geological cartography of French territory was launched in a few lines. It comprised two components: a general geological map of France including all indications of the nature of terrains, and departmental geological and mineralogical maps. It was necessary to produce maps that met economic and industrial expectations, but also more academic expectations. The differing, even opposing, constraints of these expectations could be overcome by producing cartography at two scales. Brochant realized that the extent of the administrative tasks of mining engineers left them little time and energy to carry out these more scholarly cartographic surveys. French academic delay may perhaps be related to this hold of administrative missions. The engineer could certainly not take on these two missions without one suffering greatly. The geological effervescence of the very beginning of the 19th century thus disrupted statistical and administrative priorities. Academic scholars and local notables took an interest in this project of positive geology that sought to construct a new discourse on the history of the Earth by fundamentally relying on the rational study of outcrops. Economic interests and scholarly interests converged and pushed toward the establishment of geological maps.

Brochant entrusted the general geological map project to Léonce Élie de Beaumont (1798–1874) and Pierre Armand Dufrénoy (1792–1857), two talented young engineers from the École des mines de Paris. The three men first traveled to England in 1823 to familiarize themselves with English Jurassic and Cretaceous formations. They met Smith and Greenough there. The British example was recognized. The aim was to draw inspiration from their remarkable work on secondary terrains, which complemented the work of German geognosts on primary terrains.

This joint campaign played a significant role in harmonizing future surveys, because it allowed the authors to agree on how to read the terrain and to write its formations. This was all the more important since the prospecting work was subsequently carried out separately. Instead of relying on a survey of all mining engineers, in continuity with statistical mineralogical work, the project relied on the surveys of two engineers, who soon became professors at the École. Geological surveys were no longer plotted on Cassini's geographical map, but on Chanlaire's map of France. The map was surveyed from 1825 to 1829, then the two men returned together to the field from 1829 to 1834 to correct and harmonize their work conducted individually up to that point.

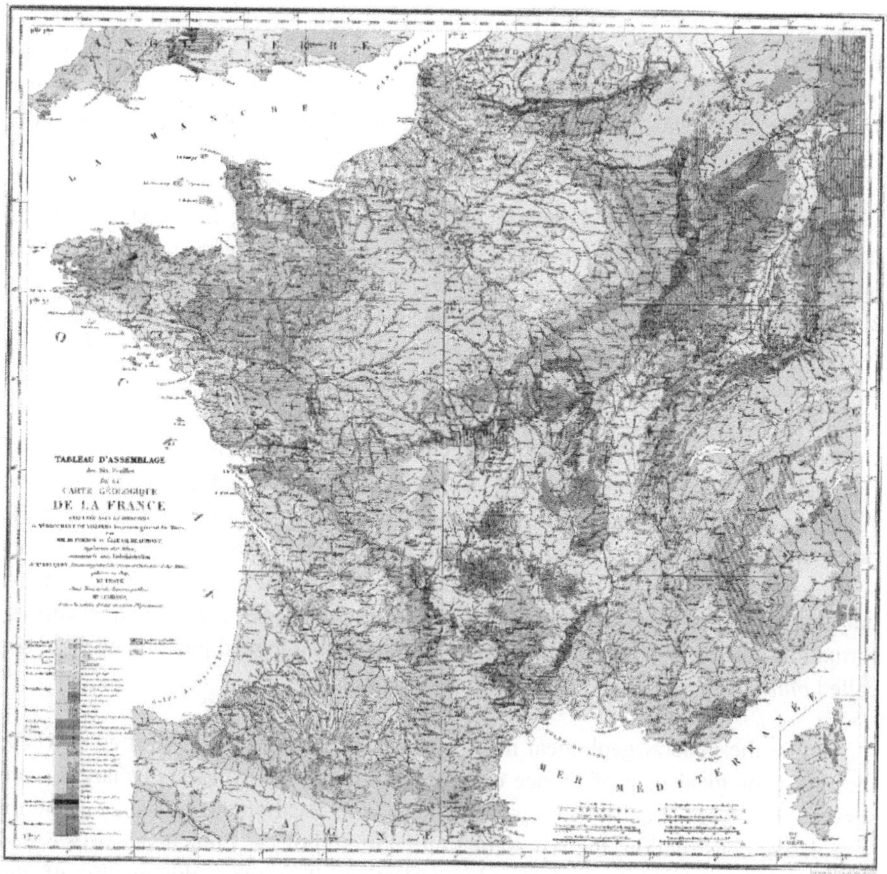

Figure 5.3. *Assembly table of the six sheets of the geological map of France, executed under the direction of Mr. Brochant de Villiers by Messrs. Dufrénoy and Élie de Beaumont, published in 1841. For a color version of this figure, see www.iste.co.uk/savaton/history.zip*

The drafting of the map was completed in 1835, and a first proof, without the engraving of the relief, was immediately presented to the Academy of Sciences and then exhibited in a room of the École des mines. The engraving of the topographic base was completed in 1840 and the map was printed in 1841. The geological observations carried out for this map were published in the form of memoirs in the *Annales des mines* starting in 1827 and then grouped into four volumes titled *Mémoires pour servir à l'explication de la carte géologique de la France*, published from 1830 to 1838. A new series of memoirs was published starting in 1841 under the title *Geological Map of France* (*Explication de la carte géologique de France*).

These works spread throughout France through learned societies and by engineers and fueled departmental cartographic projects. Announced for 1836, the map was presented to the Academy of Sciences on December 20, 1841, and put on sale in 1842. The map was delayed for financial reasons. It arrived 30 years after Brochant's first project; he had died in 1840.

The geological map of France at the scale of 1:500,000, in six sheets, was a document nearly 2 m on a side. All of the geology of France was summarized there in 37 distinct terrains, represented in colors, labeled by an alphanumeric code, and classified chronologically. The chronological groupings of stratified rocks were based on the chronology established from fossils, supplemented by the study of the relative disposition of terrains. The designations were lithological and took up those already established in Germany (those of the *Carte géognostique des pays du Rhin, entre Bâle et Mayence* by Karl von Oeynhausen, Heinrich von De la Roche and Heinrich Karl von Dechen from 1825) or in Great Britain, or indicated equivalences. A scatter of initials indicated the principal deposits of ores, salts and coal and the processing plants. The *Primitive Terrains*, renamed *Crystalline Terrains* (mica schists, gneiss, granite, syenite), were placed at the base of the chronological stack, but above the *Plutonic Rocks* nevertheless indicated as intercalated within various formations and above the *Volcanic Terrains*. The chronological position of these so-called *plutonic* or *volcanic* rocks was in fact not known.

Unlike the general geological maps of France that followed it starting in 1889, the first geological map of France did not result from the synthesis of larger scale surveys, but from specific surveys correlated with local studies. Despite its more speculative construction, the representation it offered of the French subsurface was very close to more recent maps, demonstrating, if needed, the quality of the observations and interpretations of its authors. The national project of 1822, by placing the production of the general map first, inverted the logic of inventory and synthesis that had prevailed until then. A framework, a global thesis, was needed to guide subsequent more detailed work. The example of Omalius d'Halloy's map undoubtedly played a role in this change of approach.

The general geological framework was built, and France could now carry out its detailed cartography. This general map, exceptional in view of the documents already established at that time and of its speed of execution, was decisive in the institutional impetus given to the surveying of departmental geological maps in 1835, but local initiatives had not waited, and departmental geological maps were already being surveyed at that date.

5.5. From map to map: geological cartography in 19th-century France

The French Revolution, by creating the departments, had created a new geographical and administrative division of the territory of France. The law of February 17, 1800, on the division and administration of the territory of the Republic, drafted by the Minister of the Interior Jean-Antoine Chaptal, specified the subdivisions of the department into arrondissements, cantons and communes and distributed administrative responsibilities among the prefect, the general council and the prefectural council. While the prefect was responsible for administration, the council was responsible for budgetary and fiscal matters. The department thus became an entity concerned with its economic, industrial, agronomic, social and cultural development.

This new framework accompanied, if it did not determine, a new concern at the local level for knowledge of the department's natural resources. Under the restoration, local notables, sensitized by a revival of learned societies, paid increasing attention to mining prospecting and exploitation, quarry openings, the development of mineral soil amendments, and the establishment of manufactories and plants for processing mineral materials. This essentially economic interest converged with a scholarly interest developed by many local erudites.

The diffusion of geognostic ideas, then of Cuvier's thesis and the first works claiming to be geological, favored the emergence of geological map projects at the departmental scale as early as the 1820s (Savaton 2011). These individual projects originated both from local erudites, often founders or members of local learned societies, and from mining engineers posted in the provinces over a mineralogical arrondissement. Specific motivations, theoretical frameworks, scholarly skills, available time, and material means differed greatly from one project to another, but individual determination and scholarly, political or administrative support led to the start of most of these projects. Departmental geological maps began to be surveyed at the very moment when the Conseil général des mines supported and launched Brochant's general geological map of France project.

Notables without a profession or physicians, pharmacists, notaries, teachers, retired military officers, industrialists or engineers, all naturalists and passionate about this positive geology, then traveled through their department, sampled rocks and fossils, identified them or had them identified by a few Parisian specialists and drew their geological map. From the 1820s and early 1830s onward, Arcisse de Caumont published a geological map of the departments of Calvados and Manche; Pierre Marcel de Serres produced a geognostic map of the departments of Pyrénées-Orientales and Hérault; Henri Lecoq and Jean-Baptiste Bouillet took charge of Puy-de-Dôme; François-René Dubuisson of Loire-Inférieure; Émile Gueymard of Isère and Hautes-Alpes; Scipion Gras of Basses-Alpes; and so on.

In 1835, the general map project of France was sufficiently advanced for Victor Legrand, undersecretary of State for public works and director general of Bridges and Roads and of Mines, to write to the prefects inviting them to have a geological map of their administrative territory established at departmental expense. He emphasized the economic benefits that geological knowledge of their subsurface could bring. In 1836 and 1837, two special instructions and a ten-point regulation specified the conditions for establishing these maps.

As early as 1836, 36 general councils voted the necessary subsidies. Most entrusted the surveying work to a mining engineer already paid by the State, with skills recognized by his title and better able to take into account the state of progress of geological cartography overseen by the Conseil des Mines. The mining engineers responsible for surveying a departmental map received two extracts of Cassini's geographical map as support for their surveys, and an extract of the geological map of France established by Dufrénoy and Élie de Beaumont. Each year they had to report on the progress of their work by presenting their draft and their field notebook. The Conseil des Mines, through national and local administration, thus took control of its project again.

This national organization provided that a map would be surveyed on average in 6 years. Some were completed in 4 years, but many were not completed. In 1840, 30 departments officially had a map out of the 48 that had undertaken surveying. The other departments were slow to vote the budget. Some, after voting for the budget for surveys, did not vote for the funding necessary for printing and publishing the completed map. The problem of financing drawing, engraving and printing discouraged new private projects.

The financial issue was not the only weakness of this cartography. As years passed and geological science evolved, certain cartographic choices determining the surveys or the intended final drawing quickly became obsolete. Individual works, the maps reflected individual cartographic choices that led to a lack of harmonization of productions. They differed in scale, in their geographical or topographical base, in the choices of grouping terrains into formations, in the sometimes chronological attribution of certain layers or crystallized rocks, in the colors chosen for printing, in the legend or in the scope of the explanatory work that generally accompanied them. The initial project of completing the general geological map of France with detailed departmental maps capable of specifying geology at another scale could not be achieved.

As early as 1852, Dufrénoy and Élie de Beaumont attempted to harmonize the geological maps of northern France by plotting them on the topographic base of the 1:80,000 sheets of the military staff map being produced since 1827 by the army service. This harmonized geological cartography was presented at the Paris Universal Exhibition of 1855[5]. In 1867, a provisional map commission, composed exclusively of members of the École or the Conseil des mines, was created. In 20 months, it organized the transfer of available departmental geological maps onto 62 of the 267 sheets at 1:80,000. At the Paris Universal Exhibition of 1867, geological maps were presented with geographical maps and no longer with plans of mines and quarries. Statistical cartography was replaced by explanatory and interpretive cartography based on the state of advancement of academic discourse. The geological map left the domain of mines and industry for that of education.

In 1868, an imperial decree established a Service of the Detailed Geological Map of France at the scale of 1:80,000, placed under the responsibility of Élie de Beaumont. It relied exclusively on mining engineers. To each sheet were to be attached, as appendices, plates containing cross-sections and longitudinal projections, vertical sections, and photographic perspectives. The technical legend specified subsurface resources and the diversity of enterprises associated with their exploitation through a multitude of symbols. The maps were supplemented by the publication of *Mémoires pour servir à l'explication de la carte géologique détaillée de la France.*

Departmental geological maps, too diverse or even disparate, thus had to be replaced by unified and centralized cartography. Excessive standardization and codification immediately proved unfavorable to the rapid completion of the long-awaited cartography.

The first 12 sheets at 1:80,000 (the sheets of Paris, Beauvais, Meaux, Melun, Soissons, Rouen, Provins, Évreux, Chartres, Sens, Fontainebleau and Châteaudun) were not published until 1874, the year of Élie de Beaumont's death. The service was reorganized in 1875 and henceforth associated new collaborators, professional geologists and university professors of geology who were sometimes already authors of departmental geological maps. This evolution was not secondary, because the geological map was now an academic object studied in universities.

5 The first World's Fair, The Great Exhibition, took place in London in 1851. Paris hosted the second in 1855, then the fourth in 1867, the seventh in 1878, and then the tenth in 1889, marked by the construction of the Eiffel Tower.

5.6. Conclusion

The modern geological map was built over the course of the 19th century by visually translating the evolution of discourses on the surface of the Earth. The mineralogical maps of the 18th century sought to facilitate the reading of mineralogical statistics and inventories by plotting on a geographical map the location of the different types of mineral deposits, mines and quarries, and processing sites. They were nothing more than plots of points in a geographical space whose substratum was devoid of importance for thinking about this distribution. Only Guettard envisaged the host rock of these deposits in the form of three types of terrain: schistose, marly and sandy. The geognostic maps of the Werner school, by centering their discourse on the reconstruction of the chronological succession of rock layers, established the layer, the surface and not the point, the deposit, as the entity to be mapped. The reading of the map was centered on the distribution and disposition of terrains inscribed within a chronology of the history of the Earth.

Cuvier's thesis, by directing attention to the fossils contained in the layers and elevating them to the rank of markers of this chronology, gave terrains the dimension of archives of the history of the Earth. Geology, through the study of present-day phenomena and through their interpretation placed at the service of understanding these archives, determined a new positive discourse in the service of reconstructing history. The geological map, by representing through colors and symbols the boundaries, groupings, nature, disposition, chronology and history of the origin of terrains, brought together on a limited paper space all the information necessary for producing this discourse. The birth of the geological map is inseparable from the emergence of geological science at the very beginning of the 20th century. As a complex medium for a scholarly discourse, reading the map now required training. The map would become in the 20th century an object of study, training and evaluation of geological knowledge and skills. Revised and corrected, it became a privileged mode for presenting the state of knowledge.

5.7. References

5.7.1. *Edited sources*

Aubrey, J. (1691). *The Natural History of Wiltshire*. Wiltshire Topographical Society, Salisbury.

Britton, J. (1847). *The Natural History of Wiltshire by John Aubrey, Wiltshire Topographical Society*. Nichols and Sons, London.

Buache, P. (1752). Essai de géographie physique où l'on propose des vues générales sur l'espèce de charpente du globe, composée des chaînes de montagnes qui traversent les mers comme les terres, avec quelques considérations particulières sur les bassins de la mer et sa configuration intérieure. *Mémoires Acad. Roy. Sciences, Paris*, 399–416.

Charpentier, J.F.W. (ed.) (1778). Petrographische Karte des Churfürstenthums Sachsen und der Incorporirten Lande. In *Mineralogische Geographie der Chursächsischen Lande*. Siegfried Lebrecht Crusius, Leipzig.

Cuvier, G. and Brongniart, A. (1811). *Essai sur la géographique minéralogique des environs de Paris, avec une carte géognostique et des coupes de terrain*. Baudouin, Paris.

Desmarest, N. (1771). Mémoire sur l'origine & la nature du basalte à grandes colonnes polygones, déterminées par l'histoire naturelle de cette pierre, observée en Auvergne. *Histoire Acad. Roy. Sci.*, 1771, 705–775.

Fontenelle, B. (1720). Sur les coquilles fossiles de quelques cantons de la Touraine. *Histoire Acad. Roy. Sciences, Paris*, 1720, 5–9.

Füchsel, G.C. (1761). Historia terrae et maris : ex historia Thuringiae, per montium descriptionem. *Acta Academiae Electoralis Moguntinae scientiarum utilium quae Erfordiae est*, 2, 47–209.

Gläser, F.G. (ed.) (1774). Geographischer Plan der Gefuersteten Graffchaft Henneberg. In *Versuch einer mineralogischen Beschreibung der gefürsteten Graffschaft Henneberg Chursächsischen Antheils, nebst einer kurzen Geschichte des ehemaligen und ietzigen Bergbaues derselben*. Siegfried Lebrecht Crusius, Leipzig.

Greenough, G.B. (1819). *Geological Map of England and Wales. Accompanied by an Explanatory Memoir*. Geological Society of London, London.

Guettard, J.-E. (1746). Mémoire et carte minéralogique sur la nature & la situation des terrains qui traversent la France & l'Angleterre. *Histoire Acad. Roy. Sciences, Paris*, 1746, 363–392.

Guettard, J.-E. (1781). *Carte minéralogique de France, où sont marqués les différents terrains principaux qui partagent ce royaume et les substances particulières qu'il renferme, dressée sur les observations de M. Guettard... par M. Dupain-Triel père, Paris*. Gallica, Paris.

Guettard, J.-E. and Monnet, A.-G. (1780). *Atlas et description minéralogique de la France, entrepris par ordre du Roi*. Didot, Paris.

Lehmann, J.-G. (1756). *Versuch einer Geschichte von Flötz-Gebürgen: betreffend deren Entstehung, Lage*. Springer, Berlin.

Lister, M. (1684). An ingenious proposal for a new sort of maps of countrys, together with tables of sands and clays. *Philosophical Transactions*, XIV, 739–746.

Marsigli, L.F. (1744). *Description du Danube, depuis la montagne de Kalenberg en Autriche, jusqu'au confluent de la rivière Jantra dans la Bulgarie, contenant des observations géographiques, astronomiques, hydrographiques, historiques et physiques*. J. Swart, The Hague.

Omalius d'Halloy, J.B. (1816). Mémoire sur l'étendue géographique du terrain des environs de Paris, accompagnée d'une Esquisse d'une carte géologique du Bassin de Paris et de quelques contrées voisines (pl. III). *Annales des mines*, I, 231–266.

Omalius d'Halloy, J.B. (1822). Observation sur un essai de carte géologique de France, des Pays-Bas et des contrées voisines, notice accompagnant l'Essai d'une carte géologique de la France, des Pays-Bas et de quelques contrées voisines. *Annales des mines*, VII, 353–376.

Palassou, P.-B. (1781). *Essai sur la minéralogie des Monts-Pyrénées*. Didot, Paris.

de Sauvages, P.-A. (1751). Mémoire contenant des observations de Lithologie, pour servir à l'Histoire Naturelle du Languedoc, & à la théorie de la Terre. *Mém. Acad. Roy. Sciences*, 713–758.

de Sauvages, P.-A. (1752). Suite du Mémoire contenant des observations de Lithologie, pour servir à l'Histoire Naturelle du Languedoc, & à la théorie de la Terre. *Mém. Acad. Roy. Sciences*, 1752, 629–743.

Smith, W. (1815). Map and delineation of the strata of England and Wales with part of Scotland. London.

Smith, W. (1816). *Strata Identified by Organized Fossils*. W. Arding, London.

Smith, W. (1817). *Stratigraphical System of Organized Fossils*. E. Williams, London.

Soulavie, J.-L. (1783). *Histoire naturelle de la France méridionale, ou recherche sur la minéralogie du Vivarais, du Viennois, du Forez, de l'Auvergne, du Velai*. Belin, Paris.

Steno, N. (1669). *De solido intra solidum naturaliter contento disertationis prodromus*. Accademia della Crusca, Florence.

Werner, A.G. (1787). *Kurze Klassifikation und Breschreibung des verschiedenen Gebürgen*. Waltherifchen Hofbuchhandlung, Dresden.

5.7.2. *Studies*

Ellenberger, F. (1983). Recherches et réflexions sur la naissance de la cartographie géologique, en Europe et plus particulièrement en France. *Histoire et nature*, 22(23), 3–54.

Ellenberger, F. (1994). *Histoire de la géologie 2 : la grande éclosion et ses prémices, 1660-1810*. Tec & Doc, Paris.

Kish, G. (1976). Early thematic mapping: The work of Philippe Buache. *Imago mundi*, 2(28), 129–136.

Lagarde, L. (1987). Philippe Buache, ou le premier géographe français, 1700-1773. *Mappe monde*, 87(2), 26–30.

Oldroyd, D. (2013). Maps as pictures or diagrams: The early development of geological maps, In *Rethinking the Fabric of Geology*, Baker, V.R. (ed.). Geological Society of America, Boulder.

Porter, R. (1977). *The Making of Geology. Earth Science in Britain, 1660-1815*. Cambridge University Press, Cambridge.

Rappaport, R. (1969). The geological atlas of Guettard, Lavoisier, and Monnet: Conflicting views on the nature of geology. In *Toward a History of Geology*, Schneer, C.J. (ed.). MIT Press, London.

Rudwick, M.J.S. (1976). The emergence of a visual language for geological science, 1760-1840. *History of Science*, 14, 149–195.

Rudwick, M.J.S. (2005). *Bursting the Limits of Time. The Reconstruction of Geohistory in the Age of Revolution*. University of Chicago Press, London.

Rudwick, M.J.S. (2008). *Worlds Before Adam*. University of Chicago Press, London.

Savaton, P. (2007). The first detailed geological maps of France: Contributions of local scientists and mining engineers. *Earth Science History*, 26(1), 55–73.

Savaton, P. (2011). Les cartes géologiques départementales : entre savants et ingénieurs des mines. *Cahiers François Viète*, II(1), 103–119.

Stoye, J. (1994). *Marsigli's Europe, 1680-1730. The Life and Times of Luigi Ferdinando Marsigli, Soldier and Virtuoso*. Yale University Press, New Haven.

Taylor, K.J. (1985). Early geoscience mapping, 1700-1830. *Proceedings of the Geoscience Information Society*, 15, 15–49.

Torrens, H. (2003). *Memoirs of William Smith LL. D. by John Philipps. An Introduction to the Life and Times of William Smith and the William Smith Lecture 2000 by Hugh Torrens*. Bath Royal Literary and Scientific Institution, Bath.

6

Birth and Development of Stratigraphy

Philippe GRANDCHAMP
COFRHIGEO, Paris, France

Stratigraphy (from the Latin *stratum* "bed covering", "bed", "paving", and from the Greek *graphein* "to inscribe", "to write") is the branch of geology that studies the layers or *strata* that were deposited at the surface of the Globe and their distribution in space and time, with the aim of reconstructing the history of the Earth[1].

6.1. The birth certificate of stratigraphy

This concern does not go back much beyond the 17th century. Indeed, it is found clearly expressed for the first time only in 1669, in a small treatise of barely 78 pages, the *Prodromus* of Nicolas Steno (1638–1684). The most important and best-known passage of this founding text, whose complete title is *Prodromus of a dissertation concerning a solid body enclosed by process of nature within a solid*, is

1 "The true individuality of Stratigraphy within the vast ensemble of the geological sciences results from the fact that it always brings into play the 'geographical point of view', that is to say, the distribution in space of the facts that these sciences study" (Gignoux 1926, pp. 1–2).

History of Earth Sciences,
coordinated by Pierre SAVATON.

in fact the one in which the basic principles of the method for investigating the layers or *strata* (the term is from Steno) of the Earth were clearly formulated for the first time.

De solido intra solidum naturaliter contento dissertationis prodromus, pp. 29–31	Prodromus of a dissertation concerning a solid body enclosed by process of nature within a solid, pp. 29–31
De loco stratorum sequentia certis annumeraripoterunt: *1. Quo tempore formabatur quodlibet stratum, fuit sub eodem strato aliud corpus, quod materiæ puluelruentæ viteriorem descendum impediebat, adeoque quo tempore formabatur stratorum infimum, fuit sub eo, vel corpus aliud solidum, vel si quod fluidum ibi extiterit, fuit illud tum diversæ naturæ à fluido superiori, tum grauius solido sedimentofluidi superioris.* *2. Quo tempore formabatur vnum è stratis superioribus, stratum inferius iam tum solidam consistentiam acquifiuerat.* *3. Quo tempore formabaturstratum quodlibet, vel alio corpore solido à lateribus cinctum fuit, vel totum terræ globum obduxit. Hinc sequitur quocumque in loco stratorum nuda latera conspiciuntur, vel corumdem stratorum continuationem quærendum esse, vel innenìendum esse aliud corpus solidum, quod materiam stratorum retinuit, ne disilueret.* *4. Quo tempore formabatur quodlibet stratum, materiam illi superincumberis totam fluidam fuisse, adcoque, quo tempore infimum stratum formabatur nullum è superioribus stratis extitisse.* *Quod figuram spectat, certum est, quo tempore formabatur stratum quodlibet, superficiem eius inferiorem, ut & laterum eius superficies, inferioris corporis, & corporam lateralium superficiebus respondisse, superiorem verò superficiem horionti, quantum superficiem horionti, quantum maximè licuerit, parallelam extitisse ; adcoque strata omnia præter infimum, duobus planis horizonti paralleliscontineri. Hinc sequitur, strata, vel perpendicularia ad horizontzm, vel ad illum inclinata, alio tempore horizonti parallela existisse.*	Concerning the position of strata, the following can be considered as certain: 1. At the time when a given stratum was being formed, there was beneath it another substance which prevented the further descent of the comminuted matter; and so at the time when the lowest stratum was being formed either another solid substance was beneath it, or if some fluid existed there, then it was not only of a different character from the upper fluid, but also heavier than the solid sediment of the upper fluid. 2. At the time when one of the upper strata was being formed, the lower stratum had already gained the consistency of a solid. 3. At the time when any given stratum was being formed, it was either encompassed on its sides by another solid substance, or it covered the entire spherical surface of the earth. Hence it follows that in whatever place the bared sides of strata are seen, either a continuation of the same strata must be sought, or another solid substance must be found which kept the matter of the strata from dispersion. 4. At the time when any given stratum was being formed, all the matter resting upon it was fluid, and, therefore, at the time when the lowest stratum was being formed, none of the upper strata existed. As regards form, it is certain that, at the time when any given stratum was being produced, its lower surface, as also its lateral surfaces, corresponded to the surfaces of the lower substance and lateral substances, but that the upper surface was parallel to the horizon, so far as possible; and that all strata, therefore, except the lowest, were bounded by two planes parallel to the horizon. Hence it follows that strata either perpendicular to the horizon, or inclined toward it, were at one time parallel to the horizon.

Table 6.1. *Extract from Steno's* Prodromus *in which the two fundamental principles of stratigraphy that we now call the superposition principle and the principle of continuity are set out for the first time (translation into English by John Garrett Winter)*

Drawing the consequences of the meticulous observations he made in Tuscany, the author admits, in a section entitled "strata terræ" (the strata of the Earth), that the layers are the product of a *sedimentation* that took place in a fluid[2] (Sténon 1669, 26–27). He then moved on (Sténon 1669, pp. 29–31) to the study "de loco stratorum sequentia" (of the position of the strata) and concluded (see Table 6.1) that they were deposited horizontally and successively from *bottom to top* (so that their sequence tells us a story); this is the formulation of what we now call the *superposition principle*; he also deduced that a single layer has the same age throughout its entire extent (layers have universal or limited extent), which corresponds to our *principle of continuity*. If these two common-sense principles, which are the foundation of modern stratigraphy, seem self-evident to any geologist, they nevertheless had to be stated explicitly at a given moment.

The influence of Steno's study on the subsequent development of geology has been called into question by claiming that it fell into oblivion[3] (Ellenberger 1988, p. 245). But the thesis of a Steno, an unrecognized genius, does not hold up in the face of a serious examination of the geological literature, which clearly shows that the work of the Dane was known very early in England: among others, Martin Lister referred to it explicitly in 1671, John Ray in 1673, John Woodward in 1695 and William Whiston in 1696 (Ellenberger 1988, p. 246). One must therefore admit that Steno's *Prodromus* represented, from the time of its publication, a definitive achievement for science.

However, the application of only the principles stated by Steno could scarcely allow anything more than the establishment of local chronologies, limited to the deposits studied. To succeed in reconstructing a history of the Earth, it was necessary to find a means of synchronizing the series observed in the various regions of the Globe with one another.

6.2. The impasse of Wernerian lithostratigraphy

The first syntheses consisted of finding, at points increasingly far apart, the same succession of rocks, the most important of these rocks being called *formations*[4]. This

---

2 "Ad sedimenta fluidi terræ strata pertinent" ("The strata of the earth are due to the deposits of a fluid"), translation by John Garrett Winter.

3 Among the historiographers who spread this false assertion are de Lapparent, Zittel, Beringer, Lenoble, Cailleux, Fenton, Garboe, Metzger, Noe-Nygaard, etc. As Montaigne wrote: "Ce n'est rien que foiblesse particuliere qui nous faict contenter de ce que d'autres ou que nous mesmes avons trouvé en ceste chasse de cognoissance" (*Essais*, III, chap. XIII).

4 Thus, "it has long been recognized, for example, throughout all of northern Europe, the existence of a 'chalk formation', clearly recognizable everywhere" (Gignoux 1926, p. 2).

key term was introduced into the Earth sciences by Abraham Gottlob Werner (1749–1817), professor at the Freiberg Mining Academy in Saxony and founder of a new discipline, *Geognosy*. This science aimed to describe meticulously the lithological units of the subsurface, to give them a name, to determine the order of their superposition and to endeavor to trace this laterally from one place to the next. The approach, based on the study of the structural characteristics of layers of ground, was therefore purely lithostratigraphic[5]. It was rooted in the know-how of miners, among whom it had been in germ since the 16th century (Rudwick 2005, 84 sq.).

His numerous observations led Werner to take up again the structural model proposed before him by Johann Gottlob Lehmann (1713?–1767) and which, in his time, was adopted everywhere in Europe. The subsurface was then divided into two major ensembles: the lowest terrains, devoid of fossils (azoic), were called *primitive*, and the terrains that covered them, fossil-bearing, were called *secondary*. Werner completed this standard sequence by distinguishing four great lithological ensembles: primary "formations" (*Urgebirge*), from which he separated "transition formations" (*Übergangsgebirge*), secondary "formations" (*Flötzgebirge*) and finally "wash formations" (*Aufgeschwemmtegebirge*).

However, Wernerian lithology could not provide a true history of the Earth, and this for two reasons. The first, to which sufficient attention has not always been paid, lies in the fact that the practice of geognosy, inherited from that of miners, was fundamentally a three-dimensional practice[6], which encouraged geognosts to concern themselves more with the structural relations of rock masses than with the chronology of their emplacement. Determining the structural order of the "formations" of a given region was above all meant to clarify their three-dimensional relationships.

Of course, this structural order was recognized as being the result of the chronological order in which these terrains formed (by virtue of the superposition

5 Werner wrote no treatise (except for a work on the theory of veins). His teaching was entirely oral, and his doctrine is known to us only through the testimonies left by his students. During the stay he made in Paris in 1802, Werner spoke with Delamétherie, and the latter recorded the substance of the Freiberg master's remarks, notably the way in which he conceived geognosy: "Geognosy [...] is [...] that part of mineralogy which, considering minerals as portions of our globe, deals principally with their occurrence and with their relations of position with respect to one another. [...] To know [...] the structure of the solid crust of the terrestrial globe, and the reciprocal disposition of the mineral masses that compose it, that is its goal; observation, that is its means" (Delamétherie, 1802, p. 444).

6 Anyone concerned with mining questions "lives in a three-dimensional lithological world" (Ellenberger 1994, p. 247).

principle), but their geognostic order (or, if one prefers, their stratigraphic succession) was a question of structure before being a question of geohistory (Rudwick 1997, pp. 120–121; 2005, pp. 88–89, 421–422). The second reason is that when it undertook to seek the processes (crystallization, precipitation or other processes) that might have produced the structural order it strove to recover, Wernerian geognosy arrived at the simple but false idea according to which a universal ocean had deposited not only stratified and fossil-bearing rocks, but also granite and all crystalline rocks, beginning with the latter. For Werner, a convinced Neptunist, at each stage of Earth history there had been generated, beyond local variations and in a constant order of superposition, general formations that were necessary to recover and identify wherever possible. Thus, not only was the practice of geognosy not the best-suited tool for undertaking a geohistorical reading of the underground edifice, but moreover the history to which it led implicitly assumed that observations made from the subsurface of Europe were generalizable to the entire Globe.

Nevertheless, even though it had to be abandoned, the Wernerian utopia was far from having been useless. The certainty of finding the same formations in all parts of the Globe effectively stimulated the many students to whom Werner knew how to communicate his enthusiasm. And if they ultimately had to be disillusioned, the ideal that animated them, and even more the methods of field investigation they had inherited from Wernerian geognosy, largely contributed to establishing its successor, paleontological stratigraphy[7]. This is attested by the vocabulary that stratigraphers drew from the miners' repertoire and that they have commonly used ever since, such as the words *roof*, *wall* or *fault*[8].

6.3. The emergence of paleontological stratigraphy

It was, in fact, the use of fossils that was to make it possible to introduce a chronological dimension that went beyond the notion of lithological facies and made possible the correlation at a distance of lithologically different units.

7 The term *biostratigraphy*, coined by Louis Dollo, would not appear until 1909.

8 FAULT (*faille*): come from old French *faillir*, "to fail", because beyond a fault the miner no longer finds the vein or layer he was following; wall: miner's term designating the lower surface of a formation or the terrains located immediately beneath it; roof: miner's term designating the upper surface of a formation or the terrains immediately overlying it (*Dictionnaire de Géologie* by Foucault et al., 9th ed., Dunod, Paris, 2020). The miners' repertoire provided stratigraphers with these terms, which were perfectly suited to describing the three-dimensional structures of the subsurface that they study.

6.3.1. *Beginnings*

The starting point was the observation that fossil faunas and floras change from one layer to the next. This discovery had been made empirically on several occasions before 1800, but without truly being taken advantage of. Let us cite Lister (1671), Woodward (1695), de Sauvages (1751, 1752), Füchsel (1761, 1773) and above all Arduino (1774 and 1783), "the most modern of the geologists of the 18th[9] century". In 1780, Giraud Soulavie (1752–1813) formed the project of building a chronology of the limestone terrains of the Vivarais by using several converging criteria, including the use of fossil faunas. But it was Jean-André De Luc (1727–1817) who crossed the crucial step by suggesting much more explicitly that it is possible to date layers by fossils (Ellenberger and Gohau 1981).

In a letter addressed to Professor Blumenbach (published in 1793–1794 in England and in 1798 in France), he spoke of two collateral histories, that of the layers and organized beings, which amounts to postulating the existence of two independent scales for identifying layers, one lithostratigraphic and the other paleontological:

> The *monumens* from which I have drawn the details above are far more within reach of men who desire to instruct themselves than are those from which the history of ancient peoples is deduced; for the former are spread everywhere. These *monumens* concern two collateral *histories*, those of layers and of *organized beings*[10]. (De Luc, 1798, pp. 382–383)

But for it to be able to write the history of the Earth, it was necessary that the scale of faunas (and floras) free itself completely from the lithological criterion. This would be accomplished in successive stages.

6.3.2. *Cuvier, Brongniart and Smith*

The starting signal was first given in France with the publication in 1811 of the memoir by Georges Cuvier (1769–1832) and Alexandre Brongniart (1770–1847),

9 Giovanni Arduino (1714–1795) had already, in 1760, given a table of the lithological ensembles constituting the subsurface that was much more complete than that of Lehmann, distinguishing four "orders" of terrains (a prefiguration of our modern four eras). See (Ellenberger 1994, pp. 258–264; Gohau 1987, p. 108; 2003, p. 70).

10 Seventh letter to Blumenbach.

entitled *Essai sur la géographie minéralogique des environs de Paris*[11]. This book marked a major turning point for formations that were no longer described solely by their lithological characteristics, but also by the entirety of their fauna.

For example, when they dealt with the "chalk formation" with which the series of terrains they studied begins (see Figure 6.1), the two authors specified (Cuvier and Brongniart 1811, p. 11): "What essentially characterizes this formation are the fossils it contains, fossils entirely different, not only in species, but often in genera, from all those contained in the coarse limestone". And after listing about 20 species that they had collected in the chalk, they concluded: "None of these species is found in the coarse limestone. The belemnite genus is the characteristic fossil of the chalk. This formation is therefore perfectly distinct from the formation of the marine limestone that overlies it" (Cuvier and Brongniart 1811, p. 12).

It will be noted that the two authors use the word "formation" to designate the chalk and the coarse limestone. The choice of this word is not neutral: in a footnote, Cuvier and Brongniart specify that they have taken up this term "adopted by the Freiberg school to designate a set of layers of the same nature or of different nature, but formed at the same time". They were therefore still beholden to geognosy, and their true innovation here consisted of "injecting" fossils into Wernerian lithostratigraphy (Cuvier and Brongniart 1811, p. 8). This was something that Werner and his collaborators would have found very difficult to do, since they were concerned above all with the exploration and exploitation of mineral resources. Now, most ore deposits were found in the primary terrains, which were devoid of fossils[12]. This was not the case for the "formations" of the environs of Paris, as is moreover indicated by this remark: "Most [of the] formations [discussed in the essay] have until now been unknown to the geologists of the famous Freiberg school" (Cuvier and Brongniart 1811, pp. 8–9, note).

---

11 On April 11, 1808, Cuvier had presented to his colleagues at the Académie des sciences the first conclusions of the work undertaken with Brongniart. In the same year, an abridged version of the memoir was published in the *Journal des Mines* (pp. 421–458) and in the *Annales du Muséum national d'Histoire naturelle* (pp. 293–365). The definitive version of the *Essai* (dated 1810) was published in 1811 in the *Mémoires de l'Académie des sciences*, and this edition was reissued the same year by the printer Baudoin.

12 Hence, there is a lack of interest in fossils in the practice of geognosy. Even in the secondary terrains, where fossils often abound, these appeared marginal in relation to the central concerns of geognosy (Rudwick 1997, p. 121). It nevertheless seems that Werner eventually introduced the study of fossils into his courses (Guntau 1995; Rudwick 2005, pp. 422, 444).

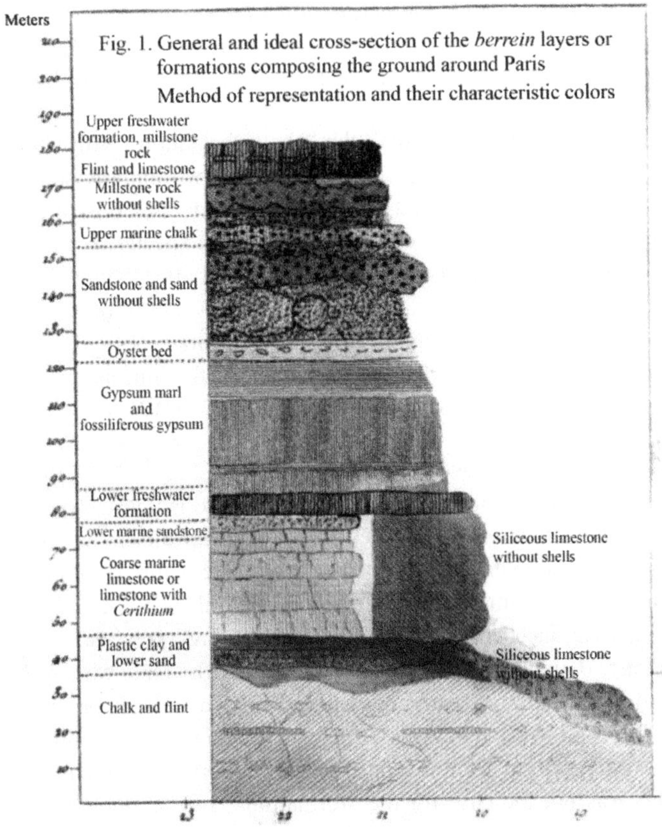

Figure 6.1. *A translation of the sequence of the "formations" recognized and described by Cuvier and Brongniart in their* Essai *of 1811. This synthetic section, which shows an alternation of marine and freshwater facies determined by fossils, can be interpreted in terms of geohistory (successive advances and retreats of the sea). For a color version of this figure, see www.iste.co.uk/savaton/history.zip*

For his part, the Englishman William Smith (1768–1839) demonstrated in the same way, in a small book that he published in 1816 and in which he dealt with *Strata identified by organized Fossils*, that each of the strata of England corresponds to a given assemblage of fossils. He provided a list of 17 levels with their characteristic fauna, a list that he completed the following year by adding 10 more levels in a new work that he entitled *Stratigraphical system of organized fossils*. This was the first time that the term *stratigraphy* (as an adjective and in connection with fossils) appeared in the geological literature. Much debate has taken place over which of the two Frenchmen or the Englishman (nicknamed *Strata Smith*) should be granted priority for the discovery, since Smith's first writings on the subject began to circulate as early as 1796[13]. But this quarrel over priority, to which national equations are not foreign, is of no interest, because it obscures an essential point that must be set out here. Smith, who was a surveyor by profession, worked as a pure geognost by seeking to determine the structural order of strata in a fossil-rich region (the secondary sedimentary terrains of England). To this end, he used fossils as an additional character allowing him to identify layers of ground with very similar lithological appearance (limestones and clays) in areas where outcrops were scattered. What is truly original in his approach is the use he made of "characteristic fossils", considered as one criterion among others for determining the geognostic position of terrains.

This approach was therefore purely structural, without any geohistorical aim (Rudwick 2005, pp. 444–445). Cuvier and Brongniart likewise applied the methods of geognosy[14]; but, being naturalists by training, they had a different perception of fossils. Cuvier, in particular, through the research he had undertaken on fossil bones, which had revealed to him the existence of extinct species, had gradually become aware that it was possible to reconstruct a true history of life on Earth. But to do so, he had to

13 In 1796, Smith wrote for the first time a note (which remained in manuscript form) on the relations between *strata* and fossils, in which he showed that it was possible to use empirically "characteristic" fossils to identify outcrops across the countryside. In 1801, he published a prospectus for a map that would trace "the natural order of the various Strata" and their outcrops. The following year, the manuscript of this map was exhibited in various places. That same year, taking advantage of the Peace of Amiens, Brongniart made a trip to London, and it is likely that on this occasion he saw the draft of Smith's map. This map may have suggested to him that it was possible to carry out in the same way a survey of the terrains of the environs of Paris and to use their fossils to follow the outcrops across this region. In the preliminary outline of their memoir, which they published in 1808, Cuvier and Brongniart wrote that "certain circumstances oblige us to present today this essay and to take date for researches as long and as laborious before the happy moment when [*sic*] we shall believe we have brought them to their conclusion". That was enough for Farey to assert in 1810 that the "circumstances" in question were nothing other than the desire of the two Frenchmen to get ahead of the Englishman Smith. Jean Gaudant has another interpretation: for him, it was Jacques-Michel Coupé (1737–1809), who had begun to decipher the succession of the layers of the Parisis in a memoir published in 1805 in the *Journal de Physique*, who was targeted by the remarks of Cuvier and Brongniart (Gaudant 2001, p. 18).

14 In addition to the explicit references to the "Freiberg school" cited above, it can also be noted that the map accompanying the essay is entitled "geognostic map".

determine the geognostic position of his samples, which had led him to integrate geognosy into the field of his research[15] (Grandchamp 2009a, 2009c) and to collaborate with Brongniart. Thus, the participation of Cuvier in the memoir on the terrains of the environs of Paris conferred on it its most important characteristic, which owes nothing to the Smith's study: the descriptive geognosy of successive terrains is systematically interpreted there *in geohistorical terms*[16] (Rudwick 1997, p. 125; 2005, pp. 476–477). For example, when they note that some of the Parisian terrains contain shells belonging to freshwater genera, Cuvier and Brongniart give this observation great significance by affirming that it indicates an alternation between marine periods and freshwater periods (Rudwick 1997, p. 125; 2005, p. 480).

The synchronisms established by Smith or by Cuvier and Brongniart, whether of a simply structural nature or endowed with a historical dimension, had a common point: all had been obtained within the limits of a sedimentary basin. Each fossil species was therefore associated with one geological formation and only one. As a result, there was no decoupling between the lithology of the strata and their paleontological content. The step that was to achieve this decoupling and make true long-distance correlations possible was to be taken by Alexandre Brongniart alone at the end of an intellectual maturation that was to take 10 years.

6.3.3. *The decisive step*

In 1811, at the time of the publication of the *Essai sur la géographie minéralogique des environs de Paris*, Brongniart, as we have seen, was still captive to the Wernerian model. He would later confide that at that time "the organic remains contained in certain rocks offered an additional means, and a very effective means, of establishing a new geological chronometer with which one could further subdivide the terrains" (Brongniart 1829b, pp. 285–286). But this did not for all that

---

15 In this regard, it is not insignificant to note that Cuvier himself declared that he was an "antiquary of a new kind" in the very first lines of the *Discours préliminaire* that he inserted at the head of his *Recherches sur les ossements fossiles* (which moreover contain a printing of the *Essai sur la géographie minéralogique des environs de Paris*).

16 It is likely that Brongniart was encouraged to introduce a historical perspective into the world of fossils and formations not only by the collaboration he maintained with Cuvier, but also by the position of director that he held at the Sèvres Manufactory, which had made him sensitive to the historical dimension of artistic styles and of the techniques *characteristic* of ceramics from different "periods" (Rudwick 2005, p. 472).

call into question the existence of the great lithological ensembles constituting the subsurface as Werner had defined them[17].

In 1813, Brongniart extended his research to the Cotentin, and the observations he made there somewhat disoriented him. In particular, the study of granite deposits, a rock regarded as the oldest of all by the Freiberg school, led him to admit that, in certain cases, it might well be younger than some fossiliferous schists. This would mean that the conditions governing the formation of granite reappeared several times at the surface of the Earth[18] (Brongniart 1814). This was a first breach in the Wernerian edifice. In the same year, Brongniart gave a course in geognosy (Grandchamp 1999, 2009b) at the Sorbonne in which he distinguished eight classes of terrains, four of which were divided into *epochs*, a term with a geohistorical connotation (see Table 6.2). But each of these epochs (or nearly each) was designated by the name of a rock (walls of a mineral), which implies that each epoch is characterized by the deposition of a particular type of rock. One recognizes in this feature the foundation of Wernerian lithostratigraphy.

Fossils were not ignored (Brongniart drew up long lists of them), but they were present solely to characterize formations, in the same way as minerals or rocks. Brongniart thus employed them here similar to Smith. However, geohistorical concerns were not entirely absent. Brongniart specified, for example, that "the presence of mammals is reported for the first time" in the upper zootic terrains, which is manifestly the mark of Cuvier's influence. One also finds this passage with a premonitory air: "In the Salzburg region, above the saline formation there is a darkish limestone very abundant in shells that appear to approach lucines; this limestone perhaps corresponds to the chalk formation and to that of the environs of Paris". But much time would still be required before Brongniart definitively freed himself from Wernerian lithostratigraphy. In a letter dated 1819, he still hesitated:

> Three very different kinds of considerations may be employed together or separately to characterize the various terrains that compose the crust of the globe: 1) The respective position and, consequently, the greater or lesser antiquity. This consideration may be open to discussion; 2) The mineralogical structure, which is a matter of fact; 3) The presence of organized bodies, which is also open to discussion,

17 And this all the more so because, as we have seen, the Tertiary terrains with which Cuvier and Brongniart had concerned themselves were unknown to the Freiberg school, which had practically studied only Paleozoic formations.

18 The fact that, in the title of this study, Brongniart used the expression "geognostic history" while at the same time speaking of "geological classification" is revealing of the unease in which he found himself.

but which has a very great and new interest [...] Which of these considerations must be chosen? Can they be combined...? (de Launay 1940, p. 126)

<table>
<tr><th>Essai of 1811</th><th colspan="2">Course of 1813</th></tr>
<tr><td>11) Formation of alluvial silt</td><td>6) Limono-clastic epoch</td><td rowspan="6">Upper zootic terrains</td></tr>
<tr><td>10) Formation of the second freshwater terrain</td><td>5) Epoch of freshwater terrains[19]</td></tr>
<tr><td>9) Formation of millstone beds without shells and of clayey sand
8) Formation of the upper marine sandstone
7) Formation of sandstone without shells and of sand</td><td>4) Sandy epoch</td></tr>
<tr><td>6) Formation of marine marls
5) Formation of gypsum with bones and of the first freshwater terrain</td><td>3) Gypsiferous epoch</td></tr>
<tr><td>4) Formation of siliceous limestone
3) Formation of coarse limestone and marine sandstone</td><td>2) Epoch of coarse limestone with cerithium and of sandstones</td></tr>
<tr><td>2) Formation of plastic clay</td><td>1) Clayey-sandy epoch</td></tr>
<tr><td>1) Formation of chalk</td><td colspan="2">Lower zootic terrains</td></tr>
</table>

Table 6.2. *Table of the formations of the environs of Paris published in the* Essai *of 1811, compared with the divisions of the class of upper zootic terrains given in the course of 1813*

COMMENTS ON TABLE 6.2.– *The term "formation", borrowed from the Freiberg school in 1811, gave way to that of "epoch", which conferred a geohistorical dimension on the classification of the 1813 course. But the latter was still entitled "course of geognosy", and the epochs were designated there by their lithological nature, which shows that Brongniart was still beholden to Werner.*

19 Brongniart announced that he preferred here to make the general history of these terrains "which had already appeared in the gypsiferous formation" (which is in agreement with the table of terrains of the *Essai* of 1811).

In 1821, Brongniart decided the issue. In an article entitled “On the zoological characters of formations”, he demonstrated (Brongniart 1821, pp. 560–570) that the fossils contained in certain limestone layers of the Fiz mountain (in the Buet range) are the same as those of the lower chalk and concluded that, despite their different lithology, he considered these two terrains to be of the same age (Brongniart 1821, p. 569). Of the two criteria, lithological or paleontological, it was the latter that ultimately appeared to him to be the more reliable[20]. He had just taken a decisive step, for the first time a synchronism was obtained, not within a sedimentary basin, but between two geographically very distant formations, one of which belonged to a sedimentary platform and the other to an orogen. It was therefore a true long-distance correlation.

The following year, William Daniel Conybeare (1787–1857) and William Phillips (1775–1828) published their *Outlines of the Geology of England and Wales*, in which they set out the principle of long-distance paleontological dating, which they had perfectly understood. And for most of the formations they described, they provided lists of specific fossils. Henceforth, making a complete inventory of the fauna (and flora) of each formation became a priority to which the geological community set itself. Paleontological stratigraphy thus progressively took possession of all the emerged formations at the surface of the Globe in order to date and correlate them.

In 1828, notably, Lardner Vanuxem (1792–1848), who had attended the courses of Cuvier, Haüy and Brongniart between 1817 and 1819, established the first continent-to-continent synchronization, as attested by his notes gathered by Samuel George Morton (1799–1851) and read on January 28, 1828, before the Academy of Natural Sciences of Philadelphia (which published them in Volume 6 of its *Journal*[21]).

20 Brongniart had begun his article with this question (Brongniart 1821, pp. 541–542): “When, in two distant terrains, the rocks are of a different nature, while the organic remains are analogous, should one, on the basis of this difference, regard these terrains as of different formation, or should one, because of the general and *suitably determined* resemblance of the fossil organized bodies, regard them as of the same *epoch of formation* when moreover the order of superposition does not obviously oppose it?” The field facts he assembled allowed him to answer: “I therefore regard the characters of epoch of formation drawn from the analogy of organized bodies as of primary value in geognosy and as having to prevail over all other differences, however great they may appear” (Brongniart 1821, p. 545).

21 *Rev. bibl. Sc. nat.* pour servir de complément aux *Ann. Sc. nat.*, 1830, Crochard, Paris, pp. 5–6; see Laffitte (1972, p. 19).

GENERAL TABLE (P. 27.)
OF DIVISIONS, CLASSES, AND ORDERS OF TERRAINS

NAMES OF CLASSES AND ORDERS	SYNONYMY	DEFINITION AND ESSENTIAL CHARACTERISTICS
JOVIAN PERIOD, that is, the present epoch, or post-diluvial terrains		
I.re Cl. ALLUVIAL TERRAINS	*alluvium*	formed by transport and sediment
II.e Cl. LYSIAN TERRAINS		dissolved and formed by chemical processes
III.e Cl. PYROGENIC TERRAINS	modern volcanic and igneous	
P. VOLCANIC		Formed by the action of heat from active volcanoes
P. *PHLOGISTIC TERRAINS*	pseudo-volcanic terrains	formed by combustion without swelling
P. ATMOSPHERIC	Meteorites	
SATURNIAN PERIOD, that is, ancient, prior to the last revolution of the globe		
1.re ***Consideration.*** **STRATIFIED** or **NEPTUNIAN TERRAINS.**		
IV.e Cl. *CLYSMIAN TERRAINS*	*diluvium*	by transport or alluvial action
V.e Cl. *IZEMIAN TERRAINS*	sedimentary	chiefly formed by sediment
Iz. THALASSIC	upper; tertiary terrains	that is, marine
Iz. PELAGIC	middle; secondary terrains	of the open sea
Iz. ABYSSAL	lower	of the ancient sea
VI.e Cl. *HEMILYSIAN TERRAINS*	compact transitional terrains	formed partly by sedimentary, partly by chemical processes
VII.e Cl. *AGALYSIAN TERRAINS*	primordial terrains	entirely dissolved and crystallized
A. EPIZOIC	crystallized transitional terrains	overlying terrains containing remains of organized bodies
A. HYPOZOIC	primitive terrains	underlying all known terrains containing organic remains
2.e ***Consideration.*** **MASSIVE** or **TYPHONIAN TERRAINS.**		
VIII.e Cl. PLUTONIC TERRAINS	effusive terrains	expelled from the Earth with evidence of liquefaction
P. GRANITOID		
P. OPHIOLITIC		
P. ENTHRITIC	porphyritic, etc.	a matrix enclosing crystals
P. TRACHYTIC		
IX.e Cl. VOLCANIC TERRAINS	ancient volcanic terrains	bearing clear signs of igneous liquefaction
V. TRAPPEAN		pasty liquefaction
V. LAVAS		fluid liquefaction

2*

Figure 6.2. *A translation of Brogniart's (1829a) general table of divisions, classes and orders of terrains. Of all the stratigraphic classifications proposed before 1830, that of Alexandre Brongniart is unquestionably the most complete. But it met with little success among geologists, mainly because of the names drawn from Greek that were used to designate the classes and orders of stratified terrains*

At the same time, the division of the Earth's framework into primitive, transition, and secondary terrains, inherited from Wernerian geognosy, gave way to finer subdivisions into *eras* and *systems*. In 1822, Conybeare coined the term *Carboniferous* to designate the coal-bearing terrain of England. In the same year, the chalk formations were called *Cretaceous* by Jean-Baptiste Julien d'Omalius d'Halloy (1783–1875). These two names still referred to the lithology of the terrains; this was the final trace left in stratigraphic nomenclature by the lithostratigraphic approach. Alexandre Brongniart introduced the name *Jurassic* in 1829[22], and Friedrich August Alberti (1795–1878) that of *Trias* in 1834. Four years earlier, the Tertiary had been divided by Gérard Paul Deshayes (1795–1875) into three systems, immediately named *Eocene*, *Miocene* and *Pliocene* by Charles Lyell[23] (1797–1875).

New general classifications appeared[24]. The most complete and most general of its kind published before 1830 was Brongniart's *Tableau des terrains qui composent l'écorce du globe* (Brongniart 1829a), in which the author proposed a nomenclature drawn from Greek to designate the classes of stratified terrains (see Figure 6.2). This terminology, which d'Archiac judged "somewhat pretentious" (d'Archiac 1864, vol. 2, p. 145), had no success and even contributed to casting discredit on this essay. It should nevertheless be noted that certain terms have survived, such as *pelagic*, *plutonic*, *granitoid*, *ophiolitic* or *trappean*.

6.4. An alternative path: purely geometric stratigraphy

Despite the accumulating observations that confirmed the validity of the paleontological criterion, the use of fossils in stratigraphy did not meet with

---

22 In England, Smith and his successors preferred to use the terms "oolitic group" (or "oolitic series") to designate the same ensemble.

23 Lyell established his divisions by taking as a criterion the number of species of the present fauna identified within the fauna of the period considered. He relied on mollusks, well represented in terrestrial sediments. He thus distinguished the *Pliocene* ("more recent"), with 49% living species; the *Miocene* ("less recent"), with 18%; and the *Eocene* ("dawn"), with 3.25%.

24 Everyone was eager to propose their own. In the French translation of Henry-Théodore de la Bèche's *Geological Manual*, the translator, after giving the author's classification of terrains, judged it useful to append to it a table "which presents, for each of the divisions or groups indicated above, equivalents in different modes of classification. Thus, those of Messrs. Conybeare, Brongniart, d'Omalius d'Halloy, and even that of Werner perfected, are arranged alongside mine in different columns, so that each of the divisions of these geological methods is placed opposite its equivalent in the one that has just been set out" (de la Bèche 1833, pp. 45–47).

unanimous approval among geologists. Ami Boué (1794–1881), for example, maintained that "geology, without fossils, possesses enough data to lead to all the major results, to all the acquired chronological divisions" (Boué 1831, p. 88, 315). And Henry Théodore de la Bèche (1796–1865) was just as doubtful when he wrote:

> It must not be supposed, as some now do, that in any accumulation of ten or twenty beds, characterized by the presence of distincts fossils, in a given district, the organic remains will be found equally characteristic of the same part of the series at remote distances. (de la Bèche 1832, p. 34; 1833, p. 41)

But the strongest reservations came in France from the Corps des mines and from its most illustrious representative, Léonce Élie de Beaumont (1798–1874). The latter had had as his professor of geology André Brochant de Villiers (1772–1840), who had followed Werner's teaching at Freiberg. The course he taught at the École des mines de Paris reflected this: unlike the lessons of geognosy that Brongniart was giving at the same time at the Sorbonne (see section 6.3.3), those of Brochant de Villiers gave no place to fossils. His course, although entitled *course of geology* in the official programs, in reality described the structural characteristics of the Earth's crust at all scales in the manner of Werner[25] (Grandchamp 2005, 2009b). In 1826, the administration of studies at the École des mines did indeed attempt to orient the teaching of geology along the path traced by Brongniart, but Brochant de Villiers turned a deaf ear[26] (Jourdy 1917, p. 5).

The following year, Élie de Beaumont took over the geology course and persisted in ignoring fossils. That same year, he discovered at Petit-Cœur, in the Alps, a schistose layer containing coal plants intercalated between two beds containing belemnites (considered Liassic), which reinforced in him the idea that fossils were not a reliable tool for dating terrains (Elie de Beaumont 1828).

In 1829–1830, Élie de Beaumont published a memoir entitled *Recherches sur quelques-unes des révolutions de la surface du globe*[27], presenting different

25 This can be understood insofar as, like Werner, Brochant de Villiers had to train miners.

26 Lessons in paleontology were instituted only in 1848, as an annex to geology, without a special examination. This development given to paleontology was not without provoking protests, because it was feared that the École des mines would be diverted from its purpose by training naturalists rather than engineers (Aguillon 1889, p. 67).

27 A first extract from this memoir (more than 300 pages long!) had been read at the Académie des sciences in April 1828 and published the following year in the *Annales des sciences naturelles*.

examples of coincidence between the uplift of the layers of certain mountain systems and the sudden changes that established the lines of demarcation observed between certain successive stages of sedimentary terrains.

In this book, the young mining engineer admitted that all mountain chains had been produced by rectilinear uplifts and explained that it is possible to date these uplifts by means of the angular unconformities observed between the uplifted strata and those that were deposited on their edges after their tilting. He thus showed that several uplifts (he counted four) had affected the subsurface of Europe at different times and in different directions. One can therefore divide the geological history of a region based not on fossils, but on epochs of uplift: to each of these epochs (whose age is given by an angular unconformity) corresponds a *mountain system* characterized by a particular direction of the tilted layers and of the chains they form; between these epochs are intercalated periods of tranquility during which the deposition of sedimentary terrains resumes its course[28].

The question thus comes down to a geometric analysis of the arrangement of the strata forming the underground framework; this is, in a certain way, a remanence of Wernerian geognosy, but pushed to the extreme of mathematization (by measurements of dip and strike angles[29]). This rigor introduced into geology seduced a great number of contemporaries to the point that François Arago (1786–1853), then perpetual secretary of the Académie des sciences, did not hesitate to proclaim[30]: "Today, geology has taken its place among the exact sciences" (Arago 1829, p. 294).

28 The *uplifts* invoked by Élie de Beaumont could easily be related to the *revolutions* spoken of by Cuvier in his *Discours sur les révolutions de la surface du globe*. Moreover, the title given by Élie de Beaumont to his memoir is perfectly explicit in this regard.

29 This is assuredly the imprint of the teaching of Brochant de Villiers who, as has been said, had attended Werner's courses. But it is important to note that, if the principle of structural analysis of underground rock masses advocated by the Freiberg master is indeed taken up here, the general model of the formation of the great lithological ensembles of the subsurface by aqueous crystallization and sediment that he proposed is, on the other hand, profoundly amended. For Werner, Neptunism was fundamentally *atectonic*, since it postulated that, as they were deposited, sedimentary layers merely molded themselves to the preexisting irregularities of the Earth's surface.

30 Alexandre Brongniart himself could not refrain from writing: "The memoir of Mr. de Beaumont certainly sets forth one of the newest, boldest, and most ingenious theses that have been proposed for a long time" (Brongniart 1830, p. 301).

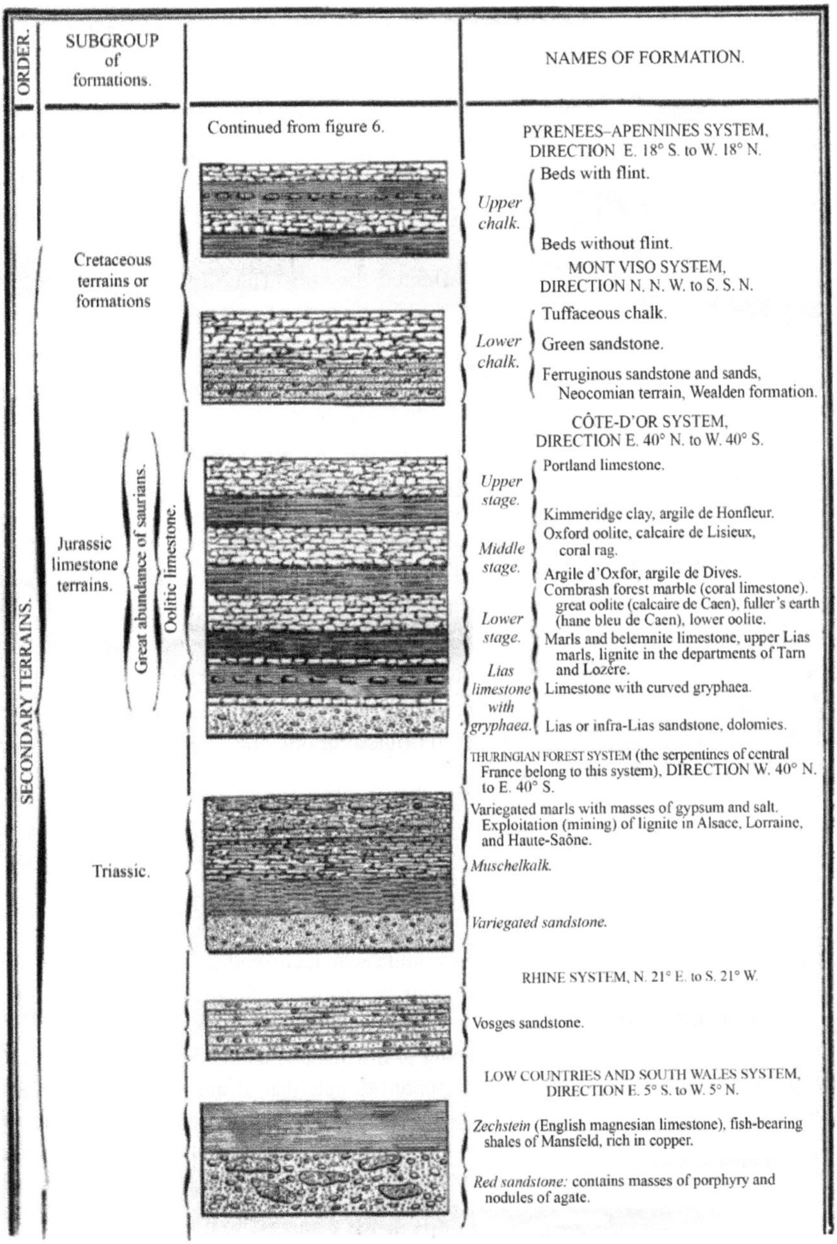

Figure 6.3. *General table of formations translated from an excerpt of the* Tableau général des formations *in Dufrénoy and de Beaumont (1841)*

COMMENTS ON FIGURE 6.3.– *The limits of the major divisions are determined by the uplifts that the Globe is supposed to have undergone. Conceived in this way, the table indicates at once the relative position of the terrains, the epoch of the uplift of mountain systems and the direction of the strata that were tilted. Fossils are mentioned only incidentally, because, in Élie de Beaumont's view, paleontological characters should serve only to "supplement stratigraphic characters" (Élie de Beaumont 1845, p. 66).*

Throughout his career, Élie de Beaumont, who was not a man to accept the slightest questioning of his ideas, envisaged no stratigraphy other than his own. When, for example, Deshayes explained to the members of the very young Société Géologique de France that the Tertiary terrains, regarded as being of a single epoch, could be divided into three major zoological periods perfectly distinct by the entirety of the species contained in each of them, and by the constant proportions between the number of analogous living species and those that are lost (see section 6.4), Élie de Beaumont replied that he had arrived at analogous results by following *a purely geological method*, and that by endeavoring to trace the deposits overlying the chalk, in France, Switzerland and Piedmont, *according to the geometric arrangement and the nature of their layers*, he had succeeded in dividing them into three stages, each of which corresponds to a period of tranquility intermediate between two mountain uplifts[31].

In 1841, in the first volume of the *Explication de la carte géologique de la France*, which he wrote with Armand Dufrénoy[32] (1792–1857), he announced the existence in Europe of "twelve uplift systems corresponding to twelve of the intervals in the series of stratified terrains" (Élie de Beaumont et Dufrénoy 1841, p. 48) and inserted a *general table of formations* in which the directions of the chains of the 12 systems and the epochs of their emplacement are indicated (see Figure 6.3). And when, almost three decades later, he wrote his voluminous *Rapport sur les progrès de la stratigraphie*[33], the same Élie de Beaumont, after recalling in the introduction that

31 *Bull. Soc. géol. Fr.*, 1, 1831, pp. 135–136 (session of May 2, 1831).

32 Dufrénoy had been entrusted, together with Élie de Beaumont and under the direction of Brochant de Villiers, with the survey of the geological map of France, between 1825 and 1835.

33 This volume formed part of the 29 reports on the progress of the sciences in France commissioned by Victor Duruy in preparation for the Universal Exhibition of 1867.

"stratigraphy is the part of geology concerned with the geometric description and the graphic representation of mineral masses[34]", continued his enumeration of mountain systems by adding 32 new examples to the list and predicted that "the number of mountain systems will rise within a few years to more than one hundred[35]" (Élie de Beaumont 1869, pp. 29–30). This entire fine intellectual construction, erected at the cost of an extraordinary expenditure of effort, collapsed after the death of its promoter, the few disciples of Élie de Beaumont not having the same advantages as their master, who held sway over a large number of institutions[36].

In Belgium, André-Hubert Dumont (1809–1857) had adopted a position analogous to that of Élie de Beaumont. This geologist had begun his career as a land surveyor and mine surveyor, and, according to d'Omalius d'Halloy, these early occupations had contributed to developing in him a practice of stratigraphy that favored the measurement of the directions and dips of formations (Omalius d'Halloy 1858, p. 14; Groessens 2008, p. 14). Correlatively, this pioneer of Belgian stratigraphy was fiercely opposed to the use of fossils to date terrains. In 1847, he had read a communication *Sur la valeur du caractère paléontologique en géologie* in which he challenged the applications of this discipline to stratigraphy, and whose text was published in the *Bulletin de la Classe des Sciences de l'Académie royale de Belgique* and in the *Bulletin de la Société géologique de France* (Groessens 2008, p. 14).

34 "For a modern reader, Élie de Beaumont redefines stratigraphy and diverts it from its meaning" (Gohau 2009a). But this was already the opinion of some of the geologist's contemporaries: "There has been the unheard-of audacity to publish at government expense, in the midst of the official reports on the progress of the sciences in France, an enormous volume under the title: *Rapport sur les progrès de la stratigraphie*, [...] in which not only is not a single word said about stratigraphy, except in the title itself, but which is its boldest and most unheard-of negation" (Marcou 1875, pp. 14–15).

35 The essential part of this report is devoted to the *pentagonal network*, a mathematical theorization of the distribution of mountain chains on the surface of the Globe, the examination of which lies outside the scope of our study.

36 Élie de Beaumont entered the Institut in 1835, became perpetual secretary of the Académie des sciences in 1853, taught at the École des mines from 1827 onward, and attained the chair of natural history at the Collège de France in 1832 (he was even a senator under the Second Empire). His two faithful assistants were Charles Sainte-Claire Deville (1814–1876), who deputized for him at the Collège de France but died barely 2 years after him, and Alexandre-Eugène Béguyer de Chancourtois (1820–1886), who was far from having the aura of his master (Touret 2006).

6.5. The invention of a key concept: the stratigraphic stage

GEOLOGICAL AGES.			NUMBER OF SPECIES	
TERRAINS.	STAGES.		BY STAGES.	BY TERRAINS.
TERTIARY .	27. Sub-Apennine		606	6,042
	22. Falunian .	Upper or Falunian .	2,754	
		Lower, or Tongrian.	428	
	25. Parisian		1,576	
	24. Suessonian		678	
CRETACEOUS	33. Danian		66	4,291
	22. Senonian		1,579	
	21. Turonian		380	
	20. Cenomanian		849	
	19. Albian		410	
	18. Aptian		156	
	17. Neocomian		851	
JURASSIC .	16. Portlandian		60	3,846
	15. Kimmeridgian		199	
	14. Coralian		655	
	13. Oxfordian		739	
	12. Callovian		281	
	11. Bathonian		546	
	10. Bajocian		603	
	9. Toarcian		288	
	8. Liasian		301	
	7. Sinemurian		174	
TRIASSIC .	6. Saliferous		792	927
	5. Conchylian		135	
PALEOZOIC .	4. Permian		91	3,180
	3. Carboniferous		1,047	
	2. Devonian		1,198	
	1. Silurian ...	upper Silurian	418	
		lower Silurian	426	
		Total	18,286	18,286

Figure 6.4. *D'Orbigny's stages*

COMMENTS ON FIGURE 6.4.– *The 27 stages proposed by d'Orbigny (*Cours de paléontologie et de géologie stratigraphiques, vol. 2, p. 250*). Most of the names created by d'Orbigny from the designation of a locality supplemented by the suffix -ian have become familiar in geology, but his classification is heterogeneous (Devonian, for example, is on the same rank as Toarcian) and it lacks balance (the Paleozoic is largely undervalued).*

The wanderings of stratigraphy along the path of a mathematization aimed at making geology an "exact" science, governed by rigorous geometric laws, did not prevent paleontological stratigraphy from continuing its advance[37].

As classifications of terrains multiplied[38], the need was felt to introduce divisions that could facilitate the establishment of correlations and synchronisms on the scale of the Earth.

The authors of the various classifications then in use in fact spoke of formations such as the *Buntsandstein*, the *Oolitic* or the *Chalk*, that is, lithological units, or else of divisions defined by a paleontological designation, such as the Ammonite-bearing unit (*Ammonéen* in French). And some geologists clearly perceived the drawback presented by these nomenclatures, as did d'Omalius, who set out the problem very clearly as early as 1828:

> We find ourselves at a loss when we must say that the formation of the Alpine limestone is composed, in that place, only of beds of sandstone and schists. There are not even names drawn from a geological circumstance, good in itself, that do not sometimes become defective: the name of the Cerithium-bearing terrain, for example, given to one of the formations of the environs of Paris, embarrasses us when it must be applied to a system of beds that contains no Cerithia. (Omalius d'Halloy 1828, pp. 21–22)

37 Proof of this is that, parallel to the publication of Élie de Beaumont's *Rapport sur les progrès de la stratigraphie*, a *Rapport sur les progrès de la paléontologie* had been entrusted to d'Archiac, and the entire first part of this report (pp. 25–228) dealt with nothing other than *stratigraphic paleontology* (Gohau 2009b).

38 In 1841, in particular, John Phillips had proposed naming the geological eras *Paleozoic*, *Mesozoic*, and *Cenozoic*.

It was to resolve this difficulty that Alcide d'Orbigny (1802–1857) proposed a new division: the *stage*[39]. His classification of terrains, begun in 1842 in the first volume of his *Paléontologie française*, was published in all its details in his *Cours de paléontologie et de géologie stratigraphiques* (1849–1852). D'Orbigny subdivided Paleozoic, Triassic, Jurassic, Cretaceous, Tertiary and recent terrains into *twenty-seven successive stages*, which he designated by adjectives recalling a type locality and ending with the suffix *-ian* (*-ien* in French) (see Figure 6.4).

> One will see, from the nomenclature adopted in the terminology of the names of stages, that I wished [...] to take names drawn from the places where the stage is best developed, in order to put an end to this confused nomenclature, drawn from local mineralogical composition, so variable from place to place, and from dominant fossils at one point, which may be absent elsewhere. (d'Orbigny 1842, p. 604)

And immediately afterward, speaking of the *Sinemurian*, he clearly defined what we today call the *stratotype*:

> I derived this name from the town of Semur (*Sinemurium*), where the best type is found, a deposit that I can regard as a standard, that is, as always capable of serving as a point of comparison. (d'Orbigny 1842, p. 604)

In reality, the true promoter of these homophonic locality names ending in *-ian* was Jules Thurmann (1804–1855), who began to use them in 1832 with *Kimmeridgian, Portlandian*, and *Oxfordian*[40], and then, in 1835, with *Neocomian*

---

39 The term stage (*étage* in French), borrowed from architecture, had been used before d'Orbigny with a certain semantic vagueness. Cuvier and Brongniart, for example, used it in their 1811 *Essai* in a rather ambiguous way: they distinguished four stages, from the Chalk to the great sandy plateau (*plateau sableux*), to speak of the stepping of the terrains. Élie de Beaumont, in his *Recherches* of 1829–1830, spoke of "successive stages of sedimentary terrains" in the sense of large lithologically distinct formations.

40 The first three names appeared for the first time in a memoir by Thurmann entitled *Essai sur les soulèvements jurassiques de Porrentruy* and published in 1832 in the *Mém. So. Hist. nat. Strasbourg*. The fourth name was proposed by the same Thurmann in 1835 (*Bull. Soc. géol. Fr.*, (1)7, p. 209). In the glossary of his *Chronographe géologique*, Renevier attributed the invention of Portlandian and Oxfordian to Alexandre Brongniart (in his *Tableau des terrains* of 1829) (Renevier 1897). But as Jules Marcou pointed out, Brongniart had only been the inspirer of those homophonic locality names by using terms ending with that suffix but unrelated to places (Marcou 1889, p. 44, note).

(from Neuchâtel, in Jura). But d'Orbigny was the first to succeed in freeing himself entirely from local lithological or paleontological nomenclatures by substituting for the notion of *formation* a chronological notion, that of the *stage*. For him, the stage designated no longer a rock, but a duration, an epoch, in other words a *slice of time* characterized by a given fauna and by facies as different as those of our present-day Globe (Lafitte 1972, p. 21). As he himself specified:

> A stage, for us, is an epoch in all respects identical to the present epoch. It is a state of repose of past nature, during which there existed, as in present nature, continents and seas, plants and marine animals; and, in the seas, pelagic animals and coastal animals across all depth zones. For a stage to be complete, it must show an assemblage of terrestrial or marine beings that can represent an entire epoch, analogous to the development that we currently see on the Earth[41]. (d'Orbigny 1852, pp. 256–257)

This is still the current meaning of the term stage[42], even if, for d'Orbigny, this division corresponded to a period of repose framed by two catastrophic renewals, which is obviously no longer accepted today. In this respect, moreover, d'Orbigny joined Élie de Beaumont in his dream of establishing universal divisions on the basis of catastrophic events: to the crises of faunal renewal invoked by the former respond the abrupt uplifts postulated by the latter, whose repercussions, felt on the scale of the entire Earth, determine total isochronous breaks in the fabric of time (Ellenberger 1972, pp. 28–29). Thus one sees that catastrophism, by making it possible to divide time into slices of worldwide extent, was more useful for establishing a geohistory than actualist theories, which proposed a temporal continuum into which it was impossible to insert sharp breaks.

6.6. Toward the international stratigraphic scale

After d'Orbigny, efforts continued to perfect and above all to unify the system, the goal being the establishment of an *international stratigraphic scale* allowing all series to be synchronized from one place to another on the Earth.

41 At the moment when d'Orbigny was inventing the *stage*, Albert Oppel (1831–1865) was isolating the *zone* in a fossil succession that he saw as continuous (Oppel 1856–1858). This was another way of expressing time. At the time, one did not yet speak of a *biozone*, a term that would be invented only in the 20th century, but rather of a *paleontological zone*.

42 A meaning admirably summarized by this formula of Gignoux: "A stage is therefore a *stratigraphic* synthesis" (Gignoux 1926, p. 5).

SYSTEMS	SERIES		STAGES		SUBSTAGES
			Pelagic Facies	Lagoonal Facies	
Cretaceous or Creatic	Upper		Danian	Garumnian	Montian
					Danian pre-diluvial
		Senonian	Aturian		Dordonian
					Campanian
			Emscherian		Santonian
					Coniacian
			Turonian		
			Cenomanian		
	Lower		Albien		
			Aptian		
			Barremian		
			Neocomian		Hauterivian
					Valanginian
Jurassic	Upper		Portlandian		Aquilonian-Purbeckian
					Bononian
			Kimmeridgian		
			Sequanian		
			Rauracian		
			Oxfordian		
			Callovian		
	Middle		Bathonian		
			Bajocian		
	Lower or Liasic	Lias	Toarcian		
			Charmouthian		
			Sinemurian		
		Infra-Lias	Hettangian		
			Rhaetian		
Triassic			Juvavian		
			Tyrolian		Carnian
					Norian
			Virglorian		
			Werfenian	Vesgian	

Figure 6.5. *Table of the divisions of the Secondary or Mesozoic group extracted from the* Note on the Nomenclature of Sedimentary Terrains *by Munier-Chalmas and de Lapparent (1893)*

COMMENTS ON FIGURE 6.5.– *In this study, which in fact is an attempt at a uniform stratigraphic scale both from the point of view of classifications and of nomenclatures, the two authors adopted -ic suffixes for systems and retained -ian suffixes for stages.*

The first attempts in this direction were the *Tableaux synchronistiques* published between 1864 and 1868 by Karl Mayer (1826–1907). In 1873 and 1874, there appeared, in numbers 70, 71 and 72 of the *Bulletin de la Société vaudoise des Sciences naturelles*, Eugène Renevier's (1831–1906) *Tableau des terrains sédimentaires*. In this study, which consists of nine tables printed on papers of different tints to make the succession of geological times visible to the eye, the author endeavored to establish a hierarchy of subdivisions in two corresponding series, one strictly chronological (*Eras, Periods, Epochs* and *Ages*), the other exclusively stratigraphic (*systems* subdivided into *stages*).

International geological congresses began to meet and made it possible to confront opinions. At the first, held in Paris in 1878, an international commission was appointed for the unification of geological nomenclature. Its secretary, Gustave Dewalque (1826–1905), was charged with drafting a report. This note was distributed during the second congress, held in Bologna in 1881. During this session, the discussions showed above all divergences on almost all questions (including the word *rock*!). The commission was nevertheless invited to continue its work and to present a second report at the third congress, held in Berlin in 1885. But Dewalque's attempts to have some general ideas and principles of stratigraphic geology based on accepted practice were in vain. Thus, the unification commission was doomed to disappear after this congress. It was only at the Saint Petersburg congress, in 1897, that the question of unifying classifications and nomenclatures was taken up again, with the appointment of a new commission composed of two categories of members: eight "official" members and 22 consultative members.

Meanwhile, Ernest Munier-Chalmas (1843–1903) and Albert de Lapparent (1839–1908), convinced "of the great advantages that the adoption of a uniform stratigraphic scale would offer", published in 1894 in the *Bulletin de la Société géologique de France* a *Note sur la nomenclature des terrains sédimentaires* in which they proposed a revised classification applicable to the whole world (see Figure 6.5). This was the most notable effort attempted since d'Orbigny. In this essay, the two authors recognized that the adoption of uniform suffixes accords poorly with the genius of each language, because it is liable to create ambiguities (e.g. that which would result from the use of the term *Neocomic*). They also opposed the idea that a nomenclature could be "the product of a vote ending a discussion of a few hours among people who do not speak the same language and often have great difficulty understanding one another" (Munier-Chalmas and de Lapparent 1893, p. 438).

That same year, 1894, Renevier presented at the Zurich congress his *Chronographe géologique*[43] – which would not be printed with its explanatory text until 1897 – a new edition of the 1874 *Tableau* enriched with numerous innovations, notably the appearance of the range of conventional colors adopted by international congresses, and accompanied by an explanatory text with a universal stratigraphic index written in several languages, in which more than 3,000 designations are listed and defined, some general, most local. In this work of compilation, the author declared that he had had recourse, for the parts in which he felt least competent, to consult various colleagues who were willing to assist him with their advice[44]. He added that he had obviously not been able to adopt everyone's views, but that he had endeavored to be "as eclectic and as objective as possible" (Renevier 1897, p. 524).

Renevier's chronograph proper consists of 13 tables printed on colored papers, in accordance with the chromatic conventions of international geological congresses. Assembled on canvas with the title, they form an ensemble 2.5 m high by 1 m wide.

Thus, it was necessary to wait until the very end of the 19th century to be in possession of a true international stratigraphic scale, which would be constantly improved as stratigraphic knowledge became more precise. But while this scale did indeed account for the succession of deposits, faunas and floras, it gave no indication of the actual duration of geological periods.

Only with the advent, in the following century, of methods based on the use of radioactive isotopes would *relative chronology*, the fruit of a collective effort unique of its kind, be complemented by an *absolute chronology*. This quantification of time began in 1904 with the dating of rock samples by Rutherford, and the first numerical scales appeared at the end of the 1920s (Holmes 1927).

In the 1980s, new mass spectrometers made it possible to refine studies, and at the turn of the century a reliable scale was achieved, which the Société géologique

43 Renevier explained that he had entitled his book *Chronographe* because "it is, in fact, a graphic representation of time, intended to serve as an international standard to which the various international stratigraphic series may be referred" (Résumé du chronographe géologique, *Bull. Soc. vaud. Sc. nat.*, 1897, 33, p. 30).

44 The names cited include Depéret for the *Tertiary*, Karpinsky for the *Carboniferous*, Kayser for the *Devonian*, Lapworth for the *Silurian*, as well as those of Barrois, Choffat, Dollfus, Douville, Du Pasquier, Fallot, Ficheur, Gosselet, Grand'Eury, de Grossouvre, Haug, Kilian, von Koenen, Pavlow, Péron, Rutot, Saggo, Schmidt and Van den Broeck.

de France widely disseminated (Rat 2009, p. 32). Its reading highlights what is called the *perspective effect*: the place accorded in the stratigraphic scale to recent periods, close to us, rich in documents and laden with events, is proportionally more important than that accorded to ancient, distant periods, with much memory lost.

With the 20th century, the methods of stratigraphy evolved further. The means of access to the object changed: drilling replaced direct contact with the outcrop or the mine gallery and aroused great interest in micropaleontology. Access to the sea made it possible to obtain, for recent periods, through coring, continuous successions, which is not possible with continental fossils.

Moreover, one is no longer satisfied with a purely descriptive stratigraphy: one speaks of *genetic stratigraphy*, with new approaches such as *sequence stratigraphy* (based on the interpretation of depositional sequences and sedimentary systems tracts). One also seeks new indicators of the most diverse events, some repetitive (magnetism, sea-level variations) and others accidental (such as the *iridium layer* at the end of the Cretaceous) (Rat 2009, p. 35).

6.7. Conclusion

Werner was convinced that the subsurface framework was everywhere like that of Saxony and that the great lithological assemblages he had recognized there should be found everywhere else in the world; we now know that the distinction between what we today call basement and cover in Hercynian Europe is not universal.

Élie de Beaumont recognized the traces of several successive orogenies in Western Europe, but he was deluded in believing that he could extend to the entire Globe the geometric laws he had derived from the study of these few examples.

D'Orbigny had good practical experience of the stratigraphy of the Jurassic, the Cretaceous and even the Tertiary of the environs of Paris, but he had only made a superficial study of, or had not seen at all, the greater part of Paleozoic terrains, hence the glaring imbalance of the classification he proposed.

In fact, all these geologists have made the same logical error: that which, in Chevreul's words, consists of "speaking on behalf of the part one knows with the confidence that one would have if one knew the whole" (Chevreul 1871, p. 19). Here, there is a lesson that every scientist should ponder, for it holds for all periods, including ours.

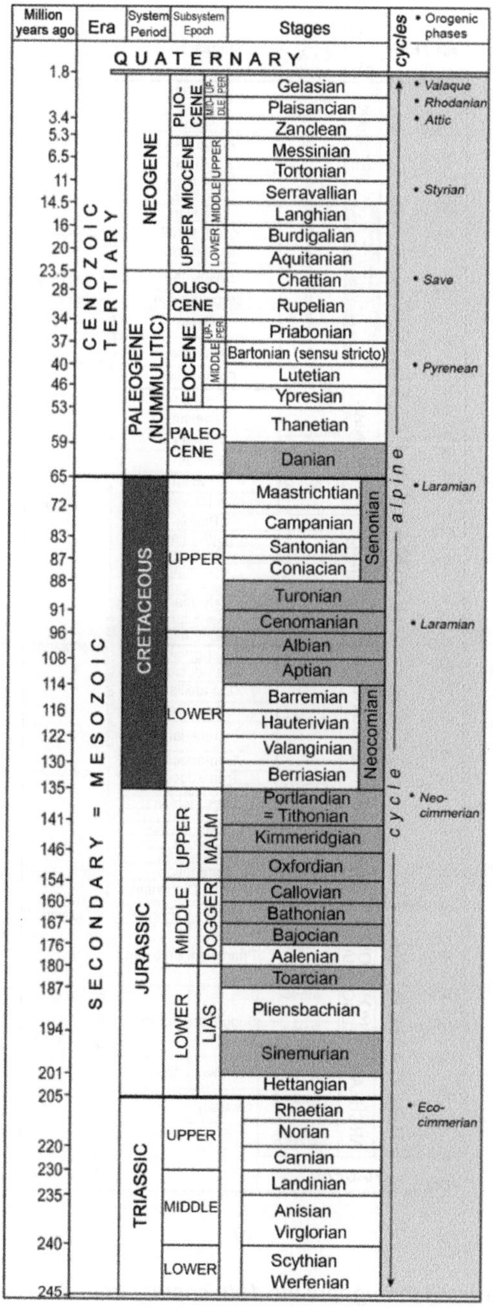
Million years ago
Era
System Period
Subsystem Epoch
Stages
cycles
• Orogenic phases
QUATERNARY
1.8
3.4
5.3
6.5
11
14.5
16
20
23.5
28
34
37
40
46
53
59
65
72
83
87
88
91
96
108
114
116
122
130
135
141
146
154
160
167
176
180
187
194
201
205
220
230
235
240
245
CENOZOIC
TERTIARY
SECONDARY = MESOZOIC
NEOGENE
PALEOGENE (NUMMULITIC)
CRETACEOUS
JURASSIC
TRIASSIC
PLIO-CENE
UPPER MIOCENE
OLIGO-CENE
EOCENE
PALEO-CENE
UPPER
LOWER
UPPER
MIDDLE
LOWER
UPPER
MIDDLE
LOWER
MALM
DOGGER
LIAS
Gelasian
Plaisancian
Zanclean
Messinian
Tortonian
Serravallian
Langhian
Burdigalian
Aquitanian
Chattian
Rupelian
Priabonian
Bartonian (sensu stricto)
Lutetian
Ypresian
Thanetian
Danian
Maastrichtian
Campanian
Santonian
Coniacian
Turonian
Cenomanian
Albian
Aptian
Barremian
Hauterivian
Valanginian
Berriasian
Portlandian = Tithonian
Kimmeridgian
Oxfordian
Callovian
Bathonian
Bajocian
Aalenian
Toarcian
Pliensbachian
Sinemurian
Hettangian
Rhaetian
Norian
Carnian
Landinian
Anisian Virglorian
Scythian Werfenian
Senonian
Neocomian
alpine cycle
• Valaque
• Rhodanian
• Attic
• Styrian
• Save
• Pyrenean
• Laramian
• Laramian
• Neo-cimmerian
• Eco-cimmerian

a)

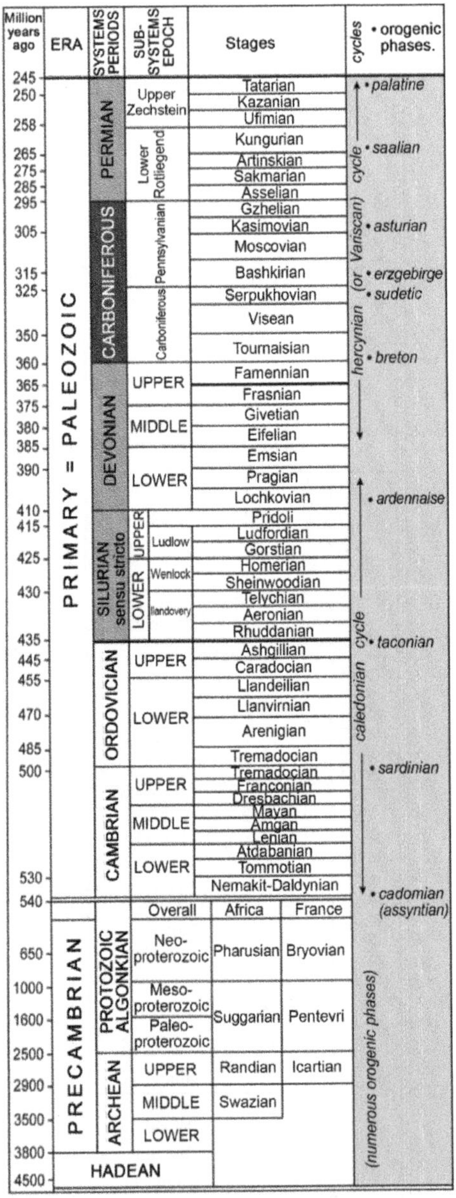

b)

Figure 6.6. *Constructed throughout the 19th century from data gathered in Western Europe by geologists whose conceptions were different from ours, the international stratigraphic scale bears the imprint of its creators*

COMMENTS ON FIGURE 6.6.– *In dark gray: nomenclature inherited from the lithostratigraphic approach; in medium gray: names of stages created by Alcide d'Orbigny (three of which now have the rank of systems); in light gray: chronological scale taking up the principles of structural stratigraphy defended by Élie de Beaumont; thick lines: major crises recalling the universal breaks invoked by catastrophists.*

The stratigraphic scale that today fixes the time frame of our representation of the history of the Earth was constructed, as we have seen, throughout the 19th century from the continental geology of Western Europe, and it retains the trace of this, both in its nomenclature and in the position of its chronological boundaries (see Figure 6.6.). Thus, the division of geological time into three great eras is an inheritance of the distinction initially established in Hercynian Europe between primary terrains (which gave rise to the *Paleozoic*) and secondary terrains (later split into the *Mesozoic* and the *Cenozoic*). The nomenclature of systems bears in two places the trace of the lithostratigraphic approach inspired by Werner, with the *Carboniferous* and the *Cretaceous*. Note that 18 of the 27 names of stages created by d'Orbigny are present in the current scale (even if three of them now designate systems). The geometric stratigraphy founded on angular unconformities, conceived by Élie de Beaumont, is restored to its proper place, in the column of orogenic cycles and phases inserted to the right of that of sub-stages and other designations.

Finally, our review would not be complete if we did not place on the stratigraphic scale the five major crises that the biosphere has experienced and that inevitably echo the abrupt breaks invoked by catastrophists to delimit their periods of Earth history[45].

Thus, the elaboration of the stratigraphic scale that we use today depended largely on people and circumstances. It is therefore more than likely that if the deciphering of stratified terrains had been carried out, not in Western Europe, but in China or on the American continent, the result obtained would have been different and the stratigraphic scale would not have the same appearance as the one we know today.

---

45 It is now accepted that the biosphere has experienced five major crises (often called the *Big Five*), which occurred at the end of the Ordovician (around 443 Ma), in the Late Devonian (the Frasnian–Famennian transition, around 372 Ma), at the Permian-Triassic boundary (the most severe, around 252 Ma), at the end of the Triassic (around 202 Ma) and at the Cretaceous-Tertiary boundary (or rather the Cretaceous–Paleocene, around 65 Ma).

6.8. References

Aguillon, L. (1889). *L'École des mines de Paris, notice historique*. Dunod, Paris.

Arago, F. (ed.) (1829). Sur l'ancienneté relative des différentes chaînes de montagnes de l'Europe. In *Annuaire pour l'an 1830, présenté au roi, par le Bureau des longitudes*. Bachelier, Paris.

d'Archiac, A. (1864). *Introduction à l'étude de la paléontologie stratigraphique*. Savy, Paris.

d'Archiac, A. (1868). *Rapport sur les progrès de la paléontologie de la France*. Imprimerie impériale, Paris.

de la Bèche, H.T. (1832). *Geological Manual*. Treuttel an Wurtz, London.

de la Bèche, H.T. (1833). *Manuel géologique*, 2nd edition, Brochant de Villiers, A.J.M. (trans.). Levrault, Paris.

Boué, A. (1831). Essai pour apprécier les avantages de la paléontologie appliquée à la géognosie et à la géologie. *Bull. Soc. géol. Fr.*, 1(2), 81, 87–91, 315–316.

Brongniart, A. (1814). Notice pour servir à l'histoire géognostique de cette partie du département de la Manche qu'on nomme le Cotentin, suivie de quelques considérations sur la classification géologique des terrains. *Journ. Mines*, 35, 109–135.

Brongniart, A. (1821). Sur les caractères oologiques des formations, avec l'application de ces caractères à la détermination de quelques terrains de craie. *Ann. Mines*, 6, 537–572.

Brongniart, A. (1829a). *Tableau des terrains qui composent l'écorce du Globe, ou essai sur la structure de la partie connue de la Terre*. Levrault, Paris.

Brongniart, A. (1829b). Rapport sur un mémoire de M. de Beaumont concernant l'ancienneté relative des différentes chaînes de montagnes d'Europe. *Ann. Chimie. Phys.*, 42, 284–302.

Chevreul, E. (1871). *D'une erreur de raisonnement très fréquente dans les sciences du ressort de la philosophie naturelle*. Didot, Paris.

Conybeare, W.D. and Phillips, W. (1822). *Outlines of the Geology of England and Wales. With an Introducting Compendium of the General Principles of that Science, and Comparating Views of the Structure of the Foreign Countries*. Phillips, London.

Coupé, J.-M. (1805). Sur l'étude du sol des environs de Paris. *Journ. Phys. Chimie Hist. Nat. Arts*, 61, 363–395.

Cuvier, G. and Brongniart, A. (1808). Essai sur la géographie minéralogique des environs de Paris, avec une carte géognostique et des coupes de terrains. *Ann. Mus. Hist. Nat.*, 11, 293–326.

Cuvier, G. and Brongniart, A. (1811). *Essai sur la géographie minéralogique des environs de Paris, avec une carte géognostique et des coupes de terrains*. Baudoin, Paris.

De Luc, J.-A. (1798). *Lettres sur l'histoire physique de la terre, adressées à M. le professeur Blumenbach, renfermant de nouvelles preuves géologiques et historiques de la Mission divine de Moyse*. Nyon, Paris.

Delamétherie, J.-C. (1802). Idées de Werner sur quelques points de la géognosie. Extrait de ses conversations. *Journ. Phys. Chimie Hist. Nat. Arts*, 50, 445–450.

Dufrénoy, A. and de Beaumont, L.É (eds) (1841). *Explication de la carte géologique de la France*, Volume 1. Imprimerie royale, Paris.

Dumont, A. (1847). Sur la valeur du caractère paléontologique en géologie. *Bull. Soc. géol. Fr.*, 2(4), 590–604.

Élie de Beaumont, L. (1828). Notice sur un gisement de végétaux fossiles et de bélemnites, situé à Petit-Cœur près Moutiers, en Tarentaise. *Ann. sci. nat.*, 14, 113–127.

Élie de Beaumont, L. (1830). *Recherches sur quelques-unes des révolutions de la surface du globe : présentant différents exemples de coïncidence entre le redressement des couches de certains systèmes de montagnes et les changements soudains qui ont établi les lignes de démarcation qu'on observe entre certains étages consécutifs des terrains de sédiment.* Crochard, Paris.

Élie de Beaumont, L. (1833). *Manuel géologique*. A.J.M., Paris.

Élie de Beaumont, L. (1845). *Leçons de géologie pratique, professées au Collège de France, pendant l'année scolaire 1843-1844*. Bertrand, Paris.

Élie de Beaumont, L. (1869). *Rapport sur les progrès de la stratigraphie*. Imprimerie impériale, Paris.

Élie de Beaumont, L. and Dufrénoy, A. (1841). *Explication de la carte géologique de la France*. Imprimerie royale, Paris.

Ellenberger, F. (1972). Quelques remarques historiques sur "la maladie infantile" de la stratigraphie. *Mém. BRGM*, 77(1), 27–30.

Ellenberger, F. (1988). *Histoire de la géologie 1 : des Anciens à la première moitié du XVIIe siècle*. Tec & Doc, Paris.

Ellenberger, F. (1994). *Histoire de la géologie 2 : la grande explosion et ses prémices 1660-1810*. Tec & Doc, Paris.

Ellenberger, F. and Gohau, G. (1981). À l'aurore de la paléontologie stratigraphique : Jean-André De Luc, son influence sur Cuvier. *Rev. Hist. Sci.*, 34, 217–257.

Farey, J. (1810). Geological remarks and queries on Messrs. Cuvier and Brogniart's [*sic*] memoir on the Mineral Geography of the Environs de Paris. *Bull. Mag.*, 35, 113–139.

Gaudant, J. (2001). Les débuts de l'exploration géologique et paléontologique du Bassin parisien. *Bull. Inf. Géol. Bass. Paris*, 38, 3–39.

Gignoux, M. (1926). *Géologie stratigraphique*. Masson, Paris.

Gohau, G. (1987). *Histoire de la géologie*. La Découverte, Paris.

Gohau, G. (2003). *Naissance de la géologie historique. La Terre, des "théories" à l'histoire.* Vuibert, Paris.

Gohau, G. (2007). Quelle idée se faisait Élie de Beaumont de la stratigraphie. In *Pour comprendre le XIXe. Histoire et philosophie des sciences de la fin du siècle*, Pont, J.-C., Freland, L., Padovani, F., Slavinskaia, L. (eds). Leo S. Olschki Editore, Florence.

Gohau, G. (2009a). Le rapport d'Élie de Beaumont sur les progrès de la stratigraphie. In *1867. L'année de tous les Rapports. Les lettres et les sciences à la fin du Second Empire*, Barbin, É., Godet, J.-L., Stenger, G. (eds). Éditions du Temps, Paris.

Gohau, G. (2009b). Le rapport d'Adolphe d'Archiac sur les progrès de la paléontologie. In *1867. L'année de tous les Rapports. Les lettres et les sciences à la fin du Second Empire*, Barbin, É., Godet, J.-L., Stenger, G. (eds). Éditions du Temps, Paris.

Grandchamp, P. (1999). Un essai inédit de classification des terrains : le cours de Géognosie professé en 1813 par Alexandre Brongniart à la faculté des sciences de Paris. *Trav. Comité fr. Hist. Géol.*, 3(13), 99–113.

Grandchamp, P. (2005). Le cours de géologie professé par Brochant de Villiers à l'École des mines dans les années 1810. *Trav. Comité fr. Hist. Géol.*, 3(19), 149–171.

Grandchamp, P. (2009a). Des leçons de géologie du Collège de France au *Discours sur les révolutions de la surface du Globe* : quatre synthèses successives de la pensée de Cuvier. *Trav. Comité fr. Hist. Géol.*, 3(23), 17–64.

Grandchamp, P. (2009b). Deux cours de géologie professés en France au début des années 1810, l'un à la Sorbonne, l'autre à l'École des mines. In *L'essor de la géologie française. Essais*, Gaudant, J. (ed.). Presses de l'École des mines de Paris, Paris.

Grandchamp, P. (2009c). Cuvier et la reconstitution de l'histoire du Globe : quatre étapes d'un cheminement intellectuel. In *L'essor de la géologie française. Essais*, Gaudant, J. (ed.). Presses de l'École des Mines de Paris, Paris.

Groessens, É. (2008). Quelques pionniers de la géologie officielle et la création d'un Service géologique de Belgique. *Trav. Comité fr. Hist. Géol.*, 3(22), 1–25.

Guntau, M. (1995). The beginnings of lithostratigraphic and biostratigraphic thinking in Germany. In *Rocks, Fossils and History*, Giglia, G., Maccagni, C., Moreno, N. (eds). Festina Lente, Florence.

Jourdy, É. (1917). Notice biographique sur Jules Marcou (1824-1898). *Bull. Soc. géol. Fr.*, 4(17), 3–19.

Laffitte, R. (1972). La notion stratigraphique d'étage. Colloque sur les méthodes et tendances de la stratigraphie, Orsay, September 1970. *Mém. BRGM*, 77(1), 17–23.

de Launay, L. (1940). *Une grande famille de savants. Les Brongniart.* Rapilly, Paris.

Marcou, J. (1875). *Explication d'une seconde édition de la carte géologique de la terre.* Wurster, Zurich.

Marcou, J. (1889). *Les géologues et la géologie du Jura jusqu'en 1870.* Declume, Lons-le-Saunier.

Marcou, J. (1897). Sur les équations personnelles et nationales dans les classifications stratigraphiques. *Bull. Soc. géol. Fr.*, 3(25), 803–830.

Mayer, C. (1864). *Tableau synchronistique des terrains jurassiques.* Orell Füssli and Co., Zurich.

Mayer, C. (1865). *Tableau synchronistique des terrains tertiaires de l'Europe.* Orell Füssli and Co., Zurich.

Mayer, C. (1868). *Tableau synchronistique des terrains tertiaires supérieurs.* Orell Füssli and Co., Zurich.

Munier-Chalmas, E. and de Lapparent, A. (1893). Note sur la nomenclature des terrains sédimentaires. *Bull. Soc. géol. Fr.*, 3(21), 438–488.

Omalius d'Halloy, J.-B.J. (1828). *Mémoires pour servir à la description géologique des PaysBas, de la France et de quelques contrées voisines.* Gérard, Namur.

Omalius d'Halloy, J.-B.J. (1858). Notice sur André Dumont. *Annu. Acad. r. Belg.*, 24, 91–100.

Oppel, A. (1856–1858). *Die Juraformation Englands, Frankreichs und des südwestlichn Deutschlands, nach ihren eintzelner Gliedern eingetheilt und verglichen.* Wurteinb. Naturwiss. Jahresh. Jahrg, Stuttgart.

d'Orbigny, A. (1842). *Paléontologie française. Description zoologique et géologique de tous les animaux mollusques et rayonnés fossiles de France, comprenant leur application à la reconnaissance des couches. Terrains oolitiques ou jurassiques.* Arthus Bertrand, Paris.

d'Orbigny, A. (1852). *Cours de paléontologie et de géologie stratigraphiques.* Masson, Paris.

Philips, J. (1841). Notices and inferences regarding the palaeozoic fossils of Devon and Cornwall. In *Figures and Descriptions of the Palaeozoic Fossils of Cornwell, Devon and West Somerset.* Longman, Brown, Green & Longmans, London, 155–182.

Rat, P. (2009). Regards sur deux siècles de stratigraphie. In *L'essor de la géologie française. Essais*, Gaudant, J. (ed.). Presses de l'École des mines de Paris, Paris.

Renevier, E. (1874). *Tableau des terrains sédimentaires formés pendant les époques de la phase organique du Globe terrestre.* Rouge & Dubois, Lausanne.

Renevier, E. (1897). *Chronographe géologique.* Bridel, Lausanne.

Rudwick, M. (1997). Smith, Cuvier et Brongniart, et la reconstitution de la géohistoire. In *De la géologie à son histoire*, Gohau, G. (ed.). Éditions du CTHS, Paris.

Rudwick, M. (2005). *Bursting the Limits of Time. The Reconstruction of Geohistory in the Age of Revolution.* University of Chicago Press, London.

Smith, W. (1816). *Strata Identified by Organized Fossils: Containing Prints on Colored Paper of the Most Characteristic Specimens in Each Stratum.* Arding, London.

Smith, W. (1817). *Stratigraphical System of Organized Fossils: With Reference to the Specimens of the Original Collection in the British Museum: Explaining Their State of Preservation and Their Use in Identifying the British Strata.* Cambridge University Press, Cambridge.

Sténon, N. (1669). *De solido intra solidum naturaliter contento dissertationis prodromus*. Ex typographia sub signo Stellae, Florence.

Sténon, N. (1916). *The Prodromus of Nicolaus Steno's Dissertation*. Macmillan, New York.

Taquet, P. (2009). Les contributions respectives de Georges Cuvier et d'Alexandre Brongniart à l'élaboration de l'*Essai sur la géographie minéralogique des environs de Paris* d'après les manuscrits retrouvés d'Alexandre Brongniart. *Trav. Comité fr. Hist. Géol.*, 3(23), 1–16.

Touret, J. (2006). Dans l'ombre de ses maîtres : Alexandre-Eugène Béguyer de Chancourtois (1820-1886). *Trav. Comité fr. Hist. Géol.*, 3(20), 61–73.

7

History of Mobilist Ideas: Wegener's Model

Philippe Le VIGOUROUX
Laboratoire CFV, Université de Nantes, France

> In the following, a first tentative attempt will be made to give a genetic interpretation of the principal features of the earth's surface, i.e. continents and ocean basins, by a single universal principle of horizontal mobility of the continents. Wherever we used to have ancient land connections sink into the depth of the oceans, we shall now assume rifting and drifting of continental rafts. The picture we obtain of the nature of the earth's "rind" is a new and somewhat paradoxical one, but as will be shown it does not lack a physical foundation. (Wegener 1912a, p. 185 in Jacoby 2001, p. 31)

It was with these words that in 1912 Alfred Wegener, a German meteorologist and specialist of the atmosphere and polar regions, introduced the article in which he set out his theory on the formation of the continents.

After presenting the theoretical context of geology in which the theory of continental drift emerged, we will examine its characteristics as well as the arguments advanced by its author to defend it and establish it within academic debates. We will finally address the reception it received among geologists, ranging from rejection to interest and support.

History of Earth Sciences,
coordinated by Pierre SAVATON. © ISTE Ltd 2026.

7.1. The Earth globe: a fixist conception under tension

The geology of the early 20th century was the heir to the theoretical constructions of the previous century. And since these had been founded, on either side of the Atlantic, on different methods, traditions and fields of study, geologists then had two competing visions to describe one and the same Earth globe. Each envisioned an Earth globe subjected to various constraints that explained the main features of the "face of the Earth", their modifications, and the processes at their origin, such as the formation of mountain ranges, volcanism and the distribution of sedimentary layers. However, one aspect remained unchanged over the course of the planet's history despite the events that could affect it: the distribution of continents and oceans remained, for the most part, a constant characteristic. In this respect, these geological conceptions of the globe can be described as "fixist".

7.1.1. *The American conception: permanentism*

In the United States, two concepts appeared at roughly the same time, in the mid-19th century, that made it possible to conceive the process of mountain-range formation: on the one hand, the theory of geosynclines, which evolved up to the synthesis proposed by Dana shortly before the end of the century, and, on the other hand, the theory of isostasy.

7.1.1.1. *Geosynclines*

Following the work on the Appalachian chain carried out by the two brothers Rodgers, Henry Darwin (1808–1866) and William Barton (1804–1882), in the 1830s and 1840s, James Hall (1811–1898) in 1859 and then James Dwight Dana (1813–1895) in 1866 and 1873 laid the foundations of the theory of geosynclines (the term itself was coined by Dana only in 1873). Sediments accumulated in a depression called the geosynclinal axis. Through subsidence (according to Hall), but also as a result of tangential stresses linked to the contraction of the Earth globe during cooling (for Dana), the sediments sank and the deepest layers lost their rigidity under the effect of an increase in their temperature. A contiguous zone, the geanticline, underwent a compensatory progressive uplift corresponding to the geosynclinal subsidence. For the Appalachians, Dana located the geanticline at the level of the Atlantic coastline (Dana 1888, pp. 163–166).

Cooling did not affect continents and oceanic domains in the same way, as their nature was assumed to be different: the former contracting before the latter, and then underwent subsidence. Under the effect of lateral stresses resulting from contraction, mountain ranges rose at the boundary between the two domains. It was then accepted that continents grew through the addition of such ranges, increasingly recent ones.

This concept was essentially developed in the context of studies of the Appalachian system and, with a few adjustments, it endured until the end of the 1960s (Aubouin 1959, 1961, 1965).

At the beginning of the 20th century, the notion of the geosyncline was adapted in France and Europe, in the Alpine context, by Émile Haug (1861–1927), professor of geology at the Sorbonne. Geosynclines were then defined as deep mobile troughs located between stable continental masses. For the Secondary period, Haug identified four continents – North Atlantic, Sino-Siberian, Africano-Brazilian and Australo-Indo-Malagasy – and postulated a hypothetical Pacific continent, now collapsed, whose existence was affirmed by some geologists and refuted by others, Haug (1900) himself judging it possible on the basis of a number of arguments.

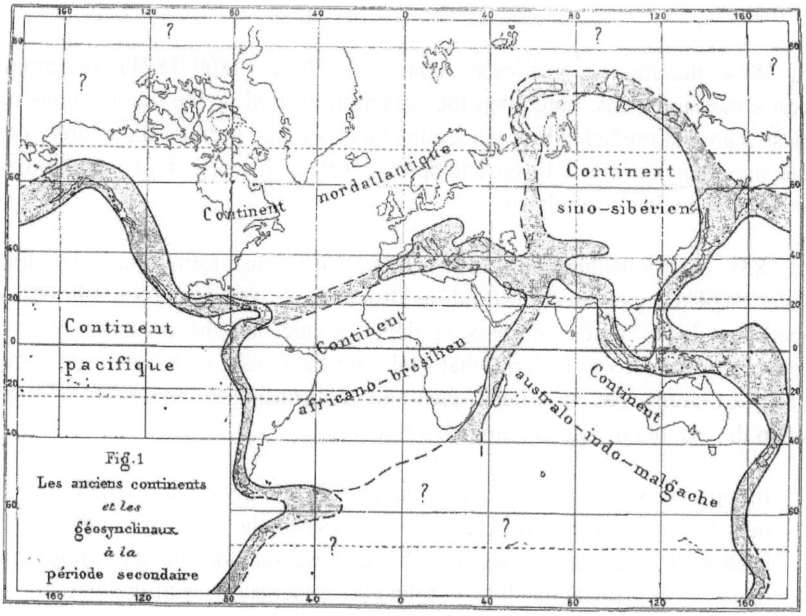

Figure 7.1. *The ancient continents and the geosynclines in the Secondary period (source: Haug 1900, p. 633)*

Over the course of geological periods, continents broke apart, some fragments subsided beneath the waters and others could merge. The geosynclines, as viewed by Haug, were filled with sediments and gradually deepened, until the moment when, at their location and under the action of orogenic movements, a new mountain range formed (Haug 1907, pp. 157–171).

7.1.1.2. *Isostasy*

Another fundamental aspect of the American conception of the Earth globe, the notion of isostasy was also developed in the mid-19th century, even though the term itself was coined only in 1889 by the American geophysicist Clarence Edward Dutton (1841–1912).

As early as the 18th century, Pierre Bouguer, during the expedition to Peru (1735–1744) undertaken to measure the length of an arc of a terrestrial meridian, observed that the gravitational attraction of the high relief of the Andes deflected the plumb line used for the measurements from the vertical (the Chimborazo experiment) (Goguel 1984). George Everest made similar observations in 1847 in the Himalayas.

In 1854, the English geodesist John Henry Pratt (1809–1871), reexamining Everest's measurements, noted that the attraction exerted by the mountainous mass was less than that predicted by calculations that took into account the entirety of the visible mountainous mass. In an attempt to explain this deficit of attraction, two hypotheses were then formulated.

In 1855, George Biddell Airy (1801–1892), an English astronomer, envisioning the interior of the globe as still fluid, held that the excess of mass at altitude was compensated by a deficit of mass at depth, suggesting the presence of a root composed of low-density material that sank into the denser underlying rocky layer. The level of compensation at depth then varied according to the density of surface rocks, in the manner of logs floating on a river:

> It appears to me that the state of the earth's crust lying upon the lava may be compared with perfect correctness to the state of a raft of timber floating upon water; in which, if we remark one log whose upper surface floats much higher than the upper surfaces of the others, we are certain that its lower surface lies deeper in the water than the lower surfaces of the others. (Airy 1855, p. 103)

A few years later, Pratt proposed to Airy a different explanation forged in the context of a globe that had by then undergone intense cooling and thus was essentially solidified, without an internal fluid mass. According to him, during this cooling, the rocks of the surface could have been compressed to a greater or lesser degree, or even undergone expansion, but in all cases without any variation in the mass of matter. Consequently, with a uniform level of compensation at depth, the higher a column of rock is, the lower its density is: as a result, the rocks of the highest summits are less dense than those of areas of lower altitude, and the pendulum is not deflected as one might expect if one assumes a homogeneous density for the mountainous mass (Pratt 1859).

Fig. 2.

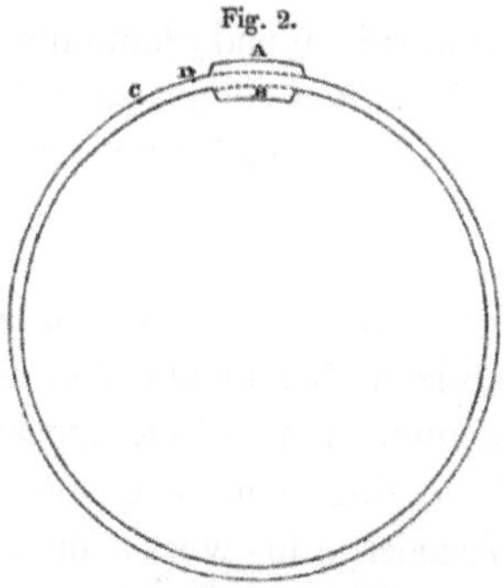

Figure 7.2. *Altitude compensated at depth (source: Airy 1855, p. 103)*

One consequence of the process of isostasy, whatever interpretation is adopted, is that continental and oceanic domains, formed of rocks of different densities, cannot be interchangeable over geological time: continents remain continents and can neither collapse nor become ocean floors; likewise, the ocean floor remains as such and cannot become continental material. Bailey Willis (1857–1949), a geologist at Stanford University, thus formalized this fixist and permanentist framework of thought: "The great ocean basins are permanent features of the earth's surface and they have existed, where they now are, with moderate changes of outline, since the waters first gathered" (Willis 1910, p. 243).

Thus expressed, the concept of isostasy was at the very least irreconcilable with that of European geologists.

7.1.2. *The European conception: contractionism*

The European conception of the globe at the beginning of the 20th century was dominated by the views of the Austrian geologist, professor at the University of

Vienna, Eduard Suess (1831–1914), author of a three-volume synthesis entitled *Das Antlitz der Erde*, published between 1883 and 1909. The work, translated and annotated by Emmanuel de Margerie (1862–1953), who secured the collaboration of several geologists, appeared in French under the title *La Face de la Terre* between 1897 and 1918 – the English translation was published between 1904 and 1924 – the first volume being prefaced by Marcel Bertrand, and the last receiving a postface by Pierre Termier (1859–1930).

The Earth was composed of three concentric layers: the superficial crust, composed of "Sal" (rocks rich in *s*ilicon and *al*uminum), covered the entire surface of the Earth. The intermediate layer was composed of "Sima" (rocks rich in *si*licon and *ma*gnesium), denser than Sal, and the central part was formed by "Nife" (or "Nifé"), short for *ni*ckel and iron (*Fe*).

This conception was part of a framework of secular cooling of the Earth globe, in line with the ideas of Léonce Élie de Beaumont (1798–1874) and James Dana, with, as a logical consequence, its contraction and the appearance of stresses leading to deformations of its surface. According to an image already old and often reused – it would also be taken up by Wegener in his work – the crust was to wrinkle like the skin of an apple which, as it dries, decreases in volume.

7.1.2.1. *Collapses*

The contraction of the globe caused a retraction of the Earth's crust that induced two kinds of stresses: radial ones, leading to vertical movements and to the collapse of continental masses at the origin of ocean basins, and tangential ones, which allowed, through horizontal movements, the development of folds and the formation of mountain chains:

> It is to subsidence and collapse that the Mediterranean seas and the largest oceans owe their origin and enlargement. The breaking up of the terrestrial globe, this it is we witness. […] It is not only in the great mountain ranges that the traces of this process are found. Great segments of the earth's crust have sunk hundreds, in some cases, even thousands, of feet deep, and not the slightest inequality of the surface remains to indicate the fracture… . (Suess 1897, pp. 823–824)

At different times in the past, regions that are currently above sea level had been reconnected to one another by other lands that are now submerged, forming continents. Thus, on the basis of the distribution of a fossil flora characterized by a particular genus, *Glossopteris*, whose leaves are found in sedimentary layers of the Carboniferous, the Permian and the Triassic, all around the Indian Ocean, Suess envisioned a continent, Gondwana, which grouped Africa, Madagascar and India, as well as the seabeds then supposed to have been emergent of the South Atlantic and the Indian Ocean:

> The first region includes the south and a good part of central Africa, then Madagascar, and finally the Indian peninsula. The high plateaus of this region have never been covered by the sea since an extremely remote time, since the end of the Carboniferous period, at least as far as we can judge in the present state of our knowledge; it is only at the foot of the plateaus that it deposited its sediments, insofar as the present Indian Ocean was forming by collapse within the tabular massif. We give it the name continent of Gondwana, after the ancient Gondwana flora, common to its different parts; it corresponds in large part to the Lemuria of zoologists. (Suess 1897, pp. 813–814)

Later, he specified: "This vast continent extended east of India as far as western Brazil and as far as the Argentine Cordilleras" (Suess 1918, p. 1678).

These ancient connections between present-day continents could sometimes have been narrow bands of emergent land, the "continental bridges", which species could use to migrate from one region to another. With terrestrial cooling and the retraction of superficial layers, these continents or these bridges collapsed. One aspect of the work of paleontologists therefore consisted of reconstructing suitable paleogeographies to explain the presence of biologically related species on different continents now separated by oceans.

7.1.2.2. *Thrust sheets and mountain chains*

From the first volume of his synthesis (1885), Suess referred to horizontal stresses linked to cooling, the folding that resulted from them, and the formation of mountain chains.

The American historian of science George Sarton noted that during the 20-year interval separating the publication of the second (1888) and third (1909) volumes of *La Face de la Terre*, numerous experimental and tectonic studies showed that horizontal movements could be of a much greater magnitude than what Suess had dared to imagine (Sarton 1919).

It was indeed in the Alps that the theory of overthrusts, overturning and the thrusting of sedimentary sheets was developed and then expanded at the end of the 19th century (Grandchamp 2015).

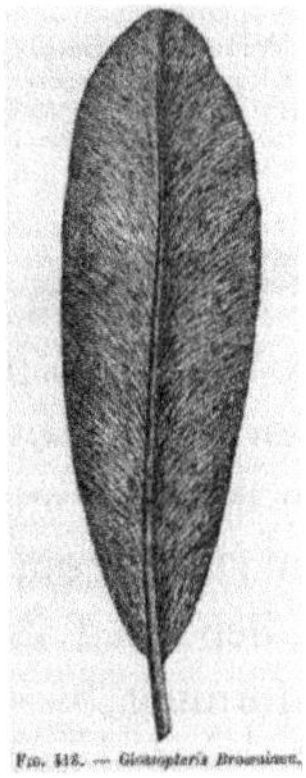

a)

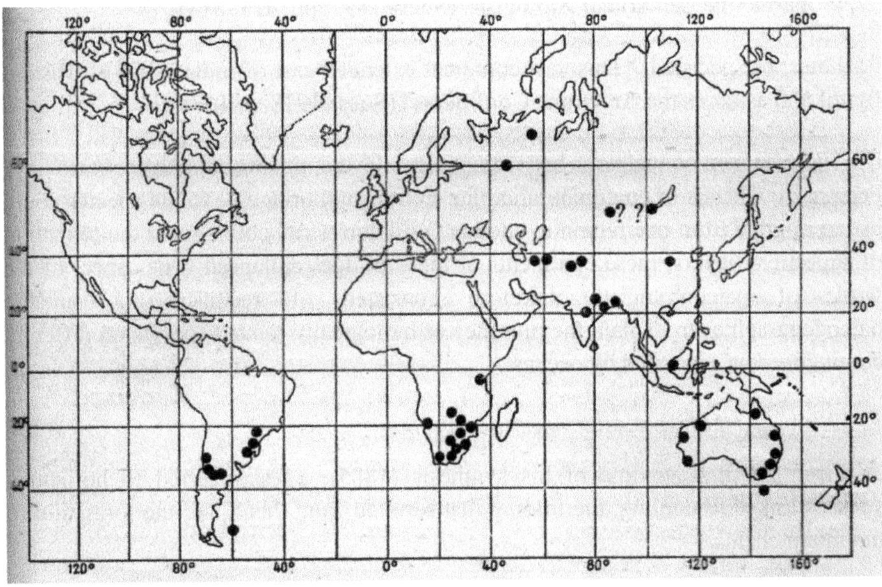

b)

Figure 7.3. *(a)* Glossopteris browniana*; (b) extension of the* Glossopteris *flora, according to David White (source: Suess 1918, p. 1679)*

In 1884, Marcel Bertrand, drawing from a comparison between the structure of the Glarus Alps and the Northern coal basin, brought out the idea of sedimentary overthrust sheets displaced from neighboring regions, sometimes over several hundreds, or even thousands, of kilometers. The thesis was developed and strengthened by the work of Bertrand himself, who extended it to the entire Alps (1887), by the Swiss geologists Hans Schardt (1858–1931) and Maurice Lugeon (1870–1953), and then by the French geologist Pierre Termier, his student, who identified the overthrust remnants as being of allochthonous origin. Finally, the Viennese master rallied to the theory of large nappes proposed by his disciple Marcel Bertrand and incorporated this new perspective into the chapter of the third volume of *La Face de la Terre* devoted to the Alps. Even though the generalization of nappe theory to other mountain chains such as the Pyrenees aroused controversy between "nappists" and "anti-nappists", a mountain chain such as the Alps could henceforth be understood as a substantial stacking of thrust terrains: "It took boldness to conceive in this way of an entire country moving over a neighboring region or inserting itself like a wedge between two layers" (Sarton 1919, p. 368).

In 1887, Marcel Bertrand undertook to demonstrate the unity of the Alpine chain from Switzerland to Romania, thereby adhering to Suess's view of an "Alpine System" that includes in the west the Betic chain of Andalusia and the Pyrenees, and that continues eastward to the Caucasus and the Himalayas. After examining the Alpine chain, Bertrand turned to the other European chains north of the Alps: the Hercynian chain, which crosses Europe from west to east, and then, farther north, the Caledonian chain, which, from Wales and Ireland, runs along the west of Norway. Above all, he established a correlation between these European chains and the mountain chains of the eastern part of the North American continent: the Alleghanis – for which we adopt M. Bertrand's spelling, 1887 – and the Green Mountains, which, from the standpoint of their structure and age, correspond to the Hercynian and Caledonian chains:

> If one examines on the map [...] the respective positions of these different chains, one sees that they face one another, as it were, on the two sides of the Atlantic Ocean, that the coal-bearing margin of the extended Hercynian chain goes to meet that of the Alleghanis, and that the Caledonian chain, likewise extended, would end near that of the Green Mountains. The possibility of an ancient continuous connection thus presents itself to the mind. (Bertrand 1887, p. 443)

A certain number of elements supported this hypothesis: the resemblance of floras and faunas at the extremities of each of the two chains. And this up to the Miocene, the time of the total disappearance in the Atlantic of a chain of islands or of a continent, remnants of the Hercynian chain. Bertrand nonetheless remained cautious, immediately adding: "No doubt one cannot consider it as proven" (Bertrand 1887, p. 443), thus forestalling any overly hasty interpretation. Indeed, on either side of the Atlantic, there faced each other, on one side, the two chains of the west of the Mediterranean, the Betic chain and the Atlas, and, on the other side, the Antilles: yet, according to Bertrand, the absence of prominences at the oceanic level testified to the absence of a submarine continuation. Indeed, since these were still recent, it seemed difficult to envisage that erosion had had time to reduce their highest summits.

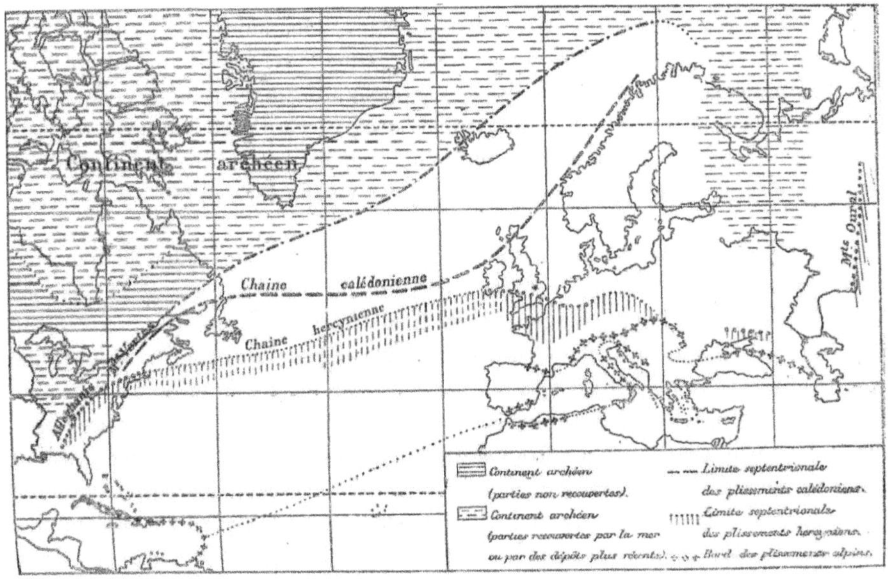

Figure 7.4. *Relative positions of the western European and eastern American mountain chains. The Archean continent with the Caledonian and Hercynian mountain chains (source: Bertrand 1887, p. 442)*

Beyond these conceptual innovations, the turn of the 20th century led, thanks to new discoveries, to the abandonment of certain explanations of the evolution of the globe.

7.1.3. *A renewal of ideas*

The beginning of the 20th century constituted "a critical moment in the evolution [of] ideas on the genesis of mountain chains, continents, and oceans, and, so to speak, [...] a turning point in the history of tectonics" (Gignoux 1925, p. 139). Indeed, the mechanisms envisaged up to that point were no longer very credible.

The demonstration of Alpine nappes raised the question of the forces at the origin of these gigantic thrustings: to explain overthrusts of several hundreds of kilometers for Alpine orogenesis, or even several thousands of kilometers for the Himalayan orogenesis (Termier 1924), it thus became necessary to envisage a very significant contraction of the globe, which was difficult to reconcile with the level of cooling that had until then been accepted.

Moreover, Suess's theory of subsidence, like Dana's theory of geosynclines, had precisely as its starting point the hypothesis of the gradual cooling of the globe. Yet the discovery of radium at the end of the 19th century suggested that the release of heat by radioactive decay might be sufficient, according to some authors, to prevent the cooling of the planet. It was therefore this very driving force behind orogeneses that the discovery of radioactivity led scholars to abandon.

Finally, the concept of isostasy developed in the United States had still scarcely penetrated the thinking and conceptions of European geologists at the beginning of the century. For example, it was only in 1900 that the word appeared in the *Comptes rendus hebdomadaires des séances de l'Académie des sciences de Paris* in a communication by Marcel Bertrand on the formation of mountain chains. The word then designated the principle, "universally accepted by American geologists", which

> immediately accounts for the geological significance of the law identified by Mr. Defforges[1] concerning variations in gravity: gravity is too strong above the seas, too weak above the continents; this merely indicates that the ocean basins are currently continuing to subside, and the continental salients to become more pronounced. (Bertrand 1900, p. 292).

It was only belatedly, over the course of the first three decades of the century, that European geologists, and French geologists in particular, began to integrate this new idea into their conceptions of the terrestrial globe. From that point on, it became difficult, within this renewed framework, to imagine the continental bridges required by biogeography that sometimes emerged temporarily and then sank into the depths.

It was in this context, still essentially one of tension between two competing and irreconcilable explanations of the form of the globe, and of the abandonment of a number of geophysical conceptions in step with scientific discoveries, that in 1912

1 Gilbert Étienne Defforges (1852–1915) was a graduate of the École Polytechnique (class of 1870) and a correspondent of the Bureau of Longitudes (1889). He published several studies on the measurement of gravitational intensity.

the theory of a German specialist in atmospheric science and polar meteorology, Alfred Wegener, emerged.

7.2. The invention of mobilism

"Fixism", whatever its version, American or European, rested on the idea of modifications of the Earth's crust being limited mainly to vertical movements: the uplift or subsidence of the crust. Displacement, limited, local and lateral, is considered to be caused by the contraction of a cooling Earth. Wegener's proposal, which he himself describes as a working hypothesis, has the ambition of reconciling the two opposing fixist conceptions by retaining only the elements he judges to be correct.

Before turning to the details of his theory, let us look at his scientific background: indeed, this would largely influence the reception of his theory among the specialists concerned.

7.2.1. *Alfred Wegener, a scientific background*

Alfred Lothar Wegener, the fifth and last child of Richard Wegener, a Protestant pastor and Anna Schwarz, was born on November 1, 1880, in Berlin. Drawn very early to the sciences, he pursued university studies in mathematics and natural sciences for several months in Heidelberg, Innsbruck and then in Berlin, where he defended, in 1904, a dissertation supervised by the astronomer Julius Bauschinger (1860–1934). This concerned the Alphonsine Tables compiled in the 12th century for Pope Alfonso X of Castile and intended to calculate the positions of the Sun, the Moon, and Mercury, Venus, Mars, Jupiter, and Saturn, the five planets known at that time (Wegener 1905).

As early as the following year, considering that "in astronomy, for the most part, everything has already been studied" and that this science "provides no opportunity for sporting activity" (these are the words of the Austrian physicist Hans Benndorf (1870–1953) in 1931 reported in Schwarzbach (1985, p. 22)), Wegener turned toward meteorology and joined his brother Kurt (1878–1964) as a technical assistant at the Royal Prussian Aeronautical Observatory in Lindenberg, southeast of Berlin, still under construction. The Observatory was intended to become the principal center for the study of the structure of the atmosphere (aerology), and, under the direction of the Observatory's director, the Wegener brothers were charged with launching instruments to record meteorological data into the upper atmosphere using balloons and tethered kites, as well as, for the highest altitudes, autonomous or manned free balloons. The objective was to learn about and understand the stratified organization of the atmosphere and the currents that affect it.

Informed of the preparation of a Danish mission to Greenland scheduled for 1906–1908 – the *Danmark* expedition led by the explorer Ludvig Mylius-Erichsen (1872-1907) – Alfred Wegener offered his services and was ultimately recruited as a physicist and meteorologist, also charged with carrying out geological observations and cartographic work (Greene 2015, p. 92). Despite the tragic death of Mylius-Erichsen in 1907, the expedition continued, as planned, over the following two years. Upon his return, in 1909, Wegener taught astronomy, physics and atmospheric optics at the University of Marburg. In 1911, he published *Thermodynamik der Atmosphäre*, a work that became a reference in the field of atmospheric physics (McCoy 2006, p. 10).

Wegener presented his theory of continental drift at the beginning of 1912 in the course of two lectures and in two articles, and then, from June onward, he took part in the Danish expedition of Captain Johan Peter Koch (1870–1928) to Greenland, until the autumn of 1913. When war was declared, he was mobilized first in Belgium and then on the French front in 1914. Wounded, he made use of a 6-month period of convalescence to work, among other things, on continental drift and to prepare the first edition of his book. In May 1915, he was called to Brussels to teach nocturnal aerial navigation to the captains of the zeppelins that were to bomb London. Remobilized for active service from September onward, he was assigned to various meteorological services until the end of the war. It was also in this scientific field that he worked in various universities, until his death in 1930 during a mission that he organized in Greenland.

Over the course of his career, Alfred Wegener took an interest in numerous scientific subjects more or less related to his specialty, polar meteorology and the upper atmosphere. One may cite, in addition to the theory of continental translations [continental drift], his interest in lunar craters, waterspouts, the Treysa meteorite and ice hair. In 1911, he hypothesized that the outermost layer of the atmosphere was composed of a new chemical element, geocoronium, so named in reference to the chemical element that, according to a 19th-century theory, constituted the solar corona, coronium. Neither the existence of coronium nor that of geocoronium was ultimately demonstrated.

He published with his father-in-law, Wladimir Köppen (1846–1940), a book that marked the science of past climates, *Die Klimate der geologischen Vorzeit* (1924), in which the explanatory power of the theory of continental drift was shown for understanding the distribution of traces of different climates over the surface of the globe:

> The large climatic variations which have caused the climate of Europe to be temperate to-day, frigid in the Pleistocene, sub-tropical in the Miocene and tropical in the Carboniferous, and have given sub-tropical if not tropical conditions in polar regions and polar conditions in equatorial regions, can only be explained by Wegener's hypothesis of the shift of the poles and the drift of the continents, the climatological consequences of which have been worked out in such great detail in Koppen and Wegener's book, *Die Klimate der geologischen Vorzeit*. (Simpson 1930, p. 302)

Yet it seems that, in the end, the book received rather little attention at the international level during the interwar period. It was published in 1924, the year in which the first translations of Wegener's book, *The Origin of Continents and Oceans*, appeared.

7.2.2. *Continental drift according to Wegener*

It was thus in January 1912, during two lectures entitled "Horizontalverschiebungen der Kontinente" ("Horizontal displacements of the continents"), that Wegener publicly set out for the first time his "working hypothesis" concerning the formation of continents and oceans: on January 6 before the Geological Society of Frankfurt am Main, and then on January 10 before the Society for the Advancement of Natural Sciences of Marburg. Articles were then published under the title "Die Entstehung der Kontinente" ("The formation of the continents") in two German scientific journals: one in *Petermanns Geographische Mitteilungen*, a German geography journal enjoying strong international recognition (Wegener 1912a), and the other in the third issue of *Geologische Rundschau*, a very young geology journal (Wegener 1912b).

Wegener took care to specify that this was "a first tentative attempt [...] made to give a genetic interpretation of the principal features of the earth's surface, i.e., continents and ocean basins, by a single universal principle of horizontal mobility of the continents," (Jacoby 2001, p. 31) emphasizing that it would certainly be necessary to modify various points of the hypothesis, since at that stage it was only a working hypothesis (Wegener 1912a, p. 185).

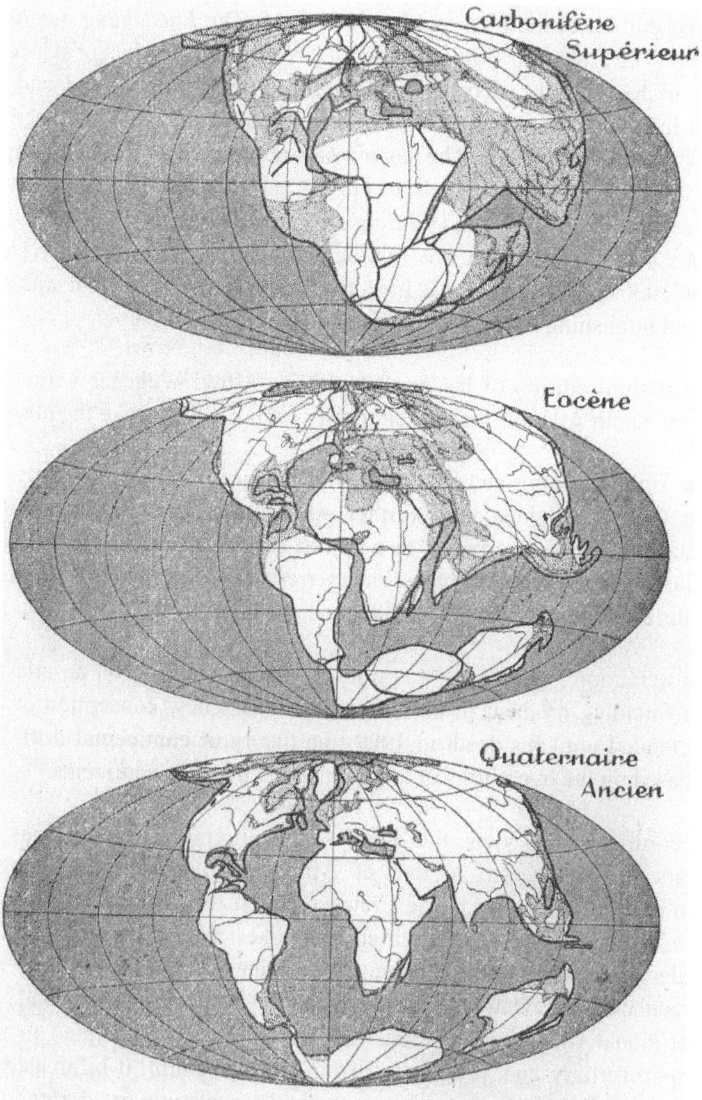

Fig. 1. — La carte du globe à 3 époques de l'histoire géologique selon la théorie des translations continentales.

En hachures : domaines abyssaux ; ponctué : mers épicontinentales. — Les contours actuels des continents et le tracé des fleuves ont été indiqués uniquement comme points de repère.

Figure 7.5. *Map of the globe at three geological periods according to the theory of continental drift. The Upper Carboniferous period, Eocene epoch and the early Quaternary period. The present outlines of the continents and the course of rivers are shown only as reference points. The cross-hatched areas are abyssal domains. The dotted areas are epicontinental seas (source: Wegener 1924, p. 4)*

The first edition of the book published in 1915, *Die Entstehung der Kontinente und Ozeane* (*The Origin of Continents and Oceans*), took up the substance of these articles, enriched by the debates that followed their publication and by new geological information. A second edition, revised and expanded, appeared in 1920, followed by a third in 1922. The latter was translated into French, English and Spanish (1924), and then into Russian (1925) and Swedish (1926). Finally, a last updated edition was published during Wegener's lifetime in 1929, a new posthumous printing of which was bibliographically supplemented by his brother Kurt in the 1930s. This version was translated into French in 1937. It was a genuine international publishing success for a scientific work.

From the third edition of his work in 1922[2], Alfred Wegener wrote about the shores of the South Atlantic, "whose similarity cannot fail to strike the observer":

> The right angle formed by Cape Saint Roques is repeated exactly in the Gulf of Guinea, and further south, each promontory of the Brazilian coast corresponds to a bay of similar shape on the African side. Measured with a compass on a terrestrial globe, the sizes of these features prove to be identical. (Wegener 1924, p. 1)

This observation – which he was not the first to make! – on an atlas in 1910, marks the founding moment in the development of a new conception of the Earth that he defended until his death in 1929: the theory of continental drift. This new vision is based on the hypothesis of lateral movement of the continents:

> According to this view, the South American craton was, millions of years ago, connected to that of Africa, forming with it a single continent. It was only in the Cretaceous that this landmass split into two distinct halves, which drifted apart like icebergs. North America underwent the same drift, having been joined to Europe, at least via Greenland and Newfoundland, which lay adjacent to Ireland. A meridional rift separated these continents. It reached Ireland in the Upper Tertiary and, progressing from south to north, it bifurcated at the latitude of Greenland, increasingly isolating it during the Quaternary. (Wegener 1924, pp. 1–2)

2 In this chapter, we will rely primarily on the first edition of Wegener's work that had wide international circulation, namely the third German edition (1922), translated into French (Wegener 1924).

Many features, often repeated on the Earth's surface, find an interpretation within the framework of continental mobility as envisaged by Wegener.

Thus, for example, the presence at the front of the American continents of a ridge formed by the Cordilleras (the Rocky Mountains and the Andes) results from compression of the leading edge of the Sial raft against the denser Pacific Sima ("bow effect").

Similarly, by a "stern effect", the raft sheds in its wake fragments of Sial that form strings of islands, such as New Zealand separated from Australia, the island chains off the eastern coast of Asia, the Antilles of Central America, or the southern Antilles between Tierra del Fuego and Antarctica. Wegener emphasized that the drift was most often directed westward with a tendency to move toward the equator (Wegener 1924).

7.2.3. *Two aspects of the theory*

In support of his interpretation, beyond the complementarity of the African and American coasts, Wegener multiplied observations of all kinds that suggested a former contact between shores now separated by an ocean: these observations constituted as many "proofs" of the correctness of his theory. The data he used came from the work of numerous authors across various Earth sciences disciplines, such as geology, geophysics, geodesy and paleoclimatology, but also from the life sciences, including paleontology, biology, and zoogeography and phytogeography.

However, to consider the mobility of continents, Wegener also had to rely on a revised conception of the globe. This is precisely the point he emphasized at the very beginning of his work: without an appropriate organization of the Earth that would make the movement of continents possible, it would be pointless to search for evidence across diverse disciplines.

It is therefore through a series of geophysical arguments that Wegener sought to reconcile the incompatible visions of the globe that had emerged on each side of the North Atlantic.

7.2.3.1. *A new conception of the Earth*

With the theory of continental drift, it was the organization of the Earth's crust, the outermost part of the globe, that had to be reconsidered in order to be compatible with the possibility of continental masses moving.

Borrowing Suess's terminology, Wegener proposed that continents, mobile on the Earth's surface, were composed of "Sial[3]" (for *si*licon, *al*uminum) and floated on a denser, more fluid underlying layer, the "Sima" (for *si*licon, *ma*gnesium). The ocean floor was composed of solidified Sima, not of Sial as European geologists assumed. Wegener therefore worked to defend this conception, relying on arguments drawn from geophysics. He highlighted a number of observations integrated into the conceptions then accepted by geologists and geophysicists.

First, there was the statistical distribution of land surfaces at two favored altitudes: Wegener cited the curve published by the Austrian meteorologist Wilhelm Trabert in his *Manuel de physique cosmique* (1911), which presented two elevation maxima, one around – 4,700 m and the other around +100 m, corresponding to the ocean floors and the continents.

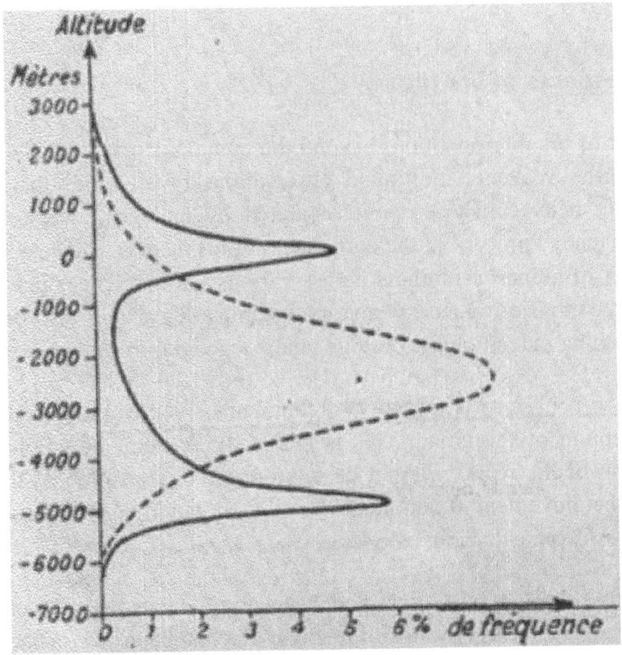

Figure 7.6. *The two elevation-frequency maxima. The frequency (%) of elevation in meters (source: Wegener 1924, p. 24)*

3 Wegener preferred to use the term Sial instead of Suess's Sal to designate the nature of continents, in order to avoid any risk of confusion with "sal", the Latin word for salt, and also to standardize the construction of these terms: Sial thus joins Sima and, in the deepest part of the globe, the Nife or Nifé. See Wegener (1924, p. 28).

This fact had long been known and led Wegener to consider that continental areas and abyssal plains constitute two distinct layers of the Earth's crust. Indeed, he argued against his opponents, if there had originally existed only a single level of equilibrium, one would have to explain why the random uplifts and subsidence of the crust over secular cooling do not produce an elevation distribution consistent with Gauss's law of errors.

In reality, these layers "behave with respect to each other like icebergs with the water in which they float" (Wegener 1924, p. 24). The explanation therefore relies on the principle of isostasy and assumes that continental masses are not of the same nature as the ocean floor; in particular, the density of the rocks composing each of the two domains is different: "This step seems so easy and so natural to us that our successors will be very surprised to learn that it took us so long to cross it" (Wegener 1924, p. 25).

The nature of the ocean floors still remained to be identified: little knowledge had been gathered until then, and the oceanic domain remained, for the most part, at the beginning of the 20th century, a *Terra incognita*.

However, Wegener found several types of data that corroborated his hypothesis:

– Despite uncertainties and incomplete data, some authors estimated that the ocean floors were composed of substances capable of being magnetized and were richer in iron than the continents.

– Even if the values tended to overlap, several studies suggested that surface seismic waves propagated faster through oceanic areas than across the continents: they were 0.1 km/s faster.

– The different data thus converged on the idea that the rocks of the oceanic crust differed from those of the continents by their higher density. A systematic sampling of the ocean floors could provide confirmation, but, Wegener acknowledged, "this will remain impossible for a long time" (Wegener 1924, p. 28). However, the few fragments recovered during dredging operations so far were high-density volcanic rocks with high iron content, rocks corresponding to Suess's Sima.

– Finally, the absence of folded mountain chains and the uniformity of the ocean floors, which exhibit relatively little relief, would indicate their plasticity:

> The reader has probably already concluded from what has been said that the rocks of the Sima group, of which we know only the intrusive

> representatives brought to the surface of the continental Sial by volcanism, where they appear foreign, are really in their proper place only beneath the continental crusts and simultaneously form the subsoil of the ocean basins. Basalt has all the properties that we demand of the hidden rock beneath the oceans. Its density is roughly that which, by a calculation based on the thickness of the continental crusts, we require of the mass that bathes them. (Wegener 1924, pp. 28–29)

By considering the dual composition of the Earth's surface – sialic at the continental level and simaic at the oceanic and subcontinental level – and relying on calculations of continental crust thickness estimated at about 100 km by other authors, Wegener was able to apply a consequence of the principle of isostasy:

> One must indeed admit that this thickness is much lower under continental shelves than under high plateaus such as Tibet. Its limits would be 50 to 300 kilometers. [...] From a quantitative point of view, in any case, these figures correspond well to our hypothesis. (Wegener 1924, pp. 29–30)

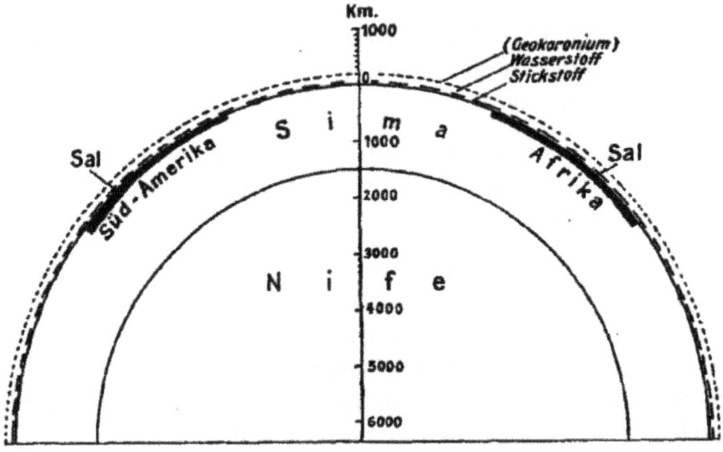

Figure 7.7. *Cross-section along the greatest circle through South America and Africa, at distinct (dimensions). At the most superficial level, Wegener identified the following layers: Stickstoff (Nitrogen), Wasserstoff (Hydrogen) and Geocoronium (source: Wegener 1912b, p. 279)*

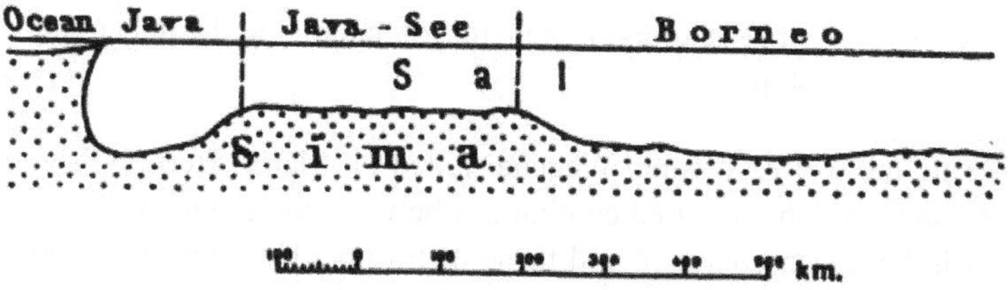

Figure 7.8. *Cross-section of the Java Sea plateau in true proportions (source: Wegener 1912a, p. 193)*

Despite the uncertainty of oceanographic data, Wegener noted the absence – unlike on continental surfaces – of mountain chains on the ocean floors, seeing in this an indication of their greater plasticity and higher degree of fluidity.

7.2.3.2. *Continental mobilism*

Once the organization of the Earth as he envisioned it was established and argued, Wegener proceeded to present a multitude of "proofs" in favor of the past convergence of continental masses now scattered across the surface of the globe.

7.2.3.2.1. Geological arguments

As the fit of South America into Africa is established based on the contours of the continental masses, Wegener considered that geological data could provide a criterion of primary importance. Indeed, the existence of continuities in folding, geological formations and their orientations on either side of the vastly widened rift now known as the Atlantic Ocean constitutes an independent criterion, separate from the exercise of reconstructing the fit.

From this perspective, Wegener emphasized, in particular, observations reported as early as 1913 before the International Geological Congress in Toronto by Hans Keidel, the German geologist in charge of the Geological Service of Argentina, who saw in the tectonic, stratigraphic and paleontological characteristics of old folds south of Buenos Aires on the South American continent similarities with the characteristics of the Cape chains at the southern tip of Africa.

Other points of comparison confirmed a continuity of several different petrological formations, both magmatic and sedimentary, between the two continents. According to some authors, Permo-Carboniferous erratic blocks identified in South America originated from Africa. Finally, the main structural directions of Brazil and those of West Africa coincide, provided that "a quite appreciable rotation is made to orient the Amazon parallel to the upper course of the Niger" (Wegener 1924, p. 40).

Regarding the North Atlantic, Wegener relied on Marcel Bertrand's conception (Bertrand 1887), which identified continuities between the chains of the eastern part of the North American continent and those of Europe. However, these continuities could concern only the oldest chains. Indeed, concerning a more recent fold, the Moroccan Atlas, whose deformation began at the end of the Cretaceous but whose main phase occurred during the Oligocene, Wegener considered that it does not continue toward the arc of the Lesser Antilles and the chains of Central America.

In any case, compared to the theory of subsidence, which explained the present configuration:

> The theory of translations, which allows the two apparent segments of the chain to be joined end to end, greatly simplifies the question. The portion that was assumed to have subsided is indeed longer than those belonging to the current continents. [...] Along the line passing through the two break points, the ocean floor presents some elevations which until now had been taken as peaks belonging to the submerged chain. In our view, these are fragments detached from the edges of the crusts. This fragmentation is all the more understandable because it occurs in a zone of tectonic disturbances. (Wegener 1924, p. 42)

7.2.3.2.2. Paleontological and biological arguments

Among the arguments Wegener employed, those from paleontology were fundamental. Indeed, after reading in the autumn of 1911 a scientific publication mentioning the necessity, for this discipline, of ancient terrestrial connections between Brazil and Africa, Wegener (1924, p. 6) began searching for data capable of confirming his intuition of continental drift.

The paleontological evidence advanced by Wegener relied on identifying fossils of identical or very closely related animal and plant species discovered on continents separated by an ocean, for example, Australian faunas closely related, depending on geological periods, to those of South Africa and India, or to those of South America or Southeast Asia, or the distribution of a *Glossopteris* flora around the Indian Ocean. Wegener's conception proposed a contiguous joining of the continents where this flora was found, with a relatively homogeneous climate, whereas the conventionally accepted view (at least in Europe) invoked the existence of a continent now submerged in the Indian Ocean, Gondwana, which brought together

Australia, the Indonesian archipelago, the Indian Peninsula, southern Africa and South America into a single vast emerged landmass.

Wegener summarized the problem:

> On the question of whether, between two isolated continents, a land connection once existed, scholars have generally answered in very diverse ways, each tending to generalize the data provided by their restricted field of investigation. (Wegener 1924, p. 58)

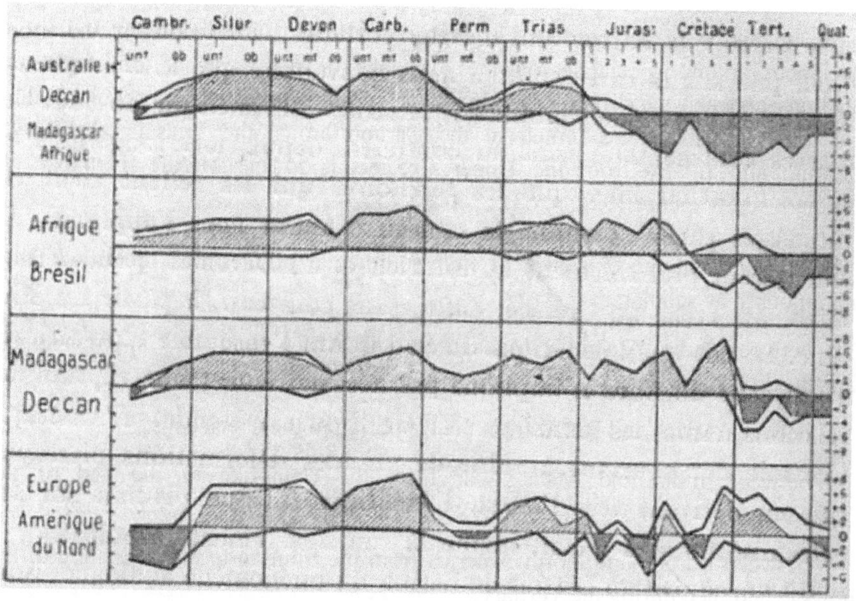

Figure 7.9. *Graph showing the number of votes in favor (upper curve) or against (lower curve) the existence of four continental bridges since the Cambrian period – Cambrian, Silurian, Devonian, Carboniferous, Permian, Triassic, Jurassic, Cretaceous, Tertiary and Quaternary periods. The bridges are Australia–Deccan–Madagascar–Africa, Africa–Brazil, Madagascar–Africa and Europe–North America. The difference (majority) is indicated by shading, cross-hatched in the case of opposition (source: Wegener 1924, p. 62)*

Wegener particularly drew on the work of Theodor Arldt (1878–1960), a German naturalist who, from 1907 onward, formalized paleogeography as a discipline with its own methodology and theoretical foundation (Ebach and Dowding 2017). Arldt had compiled data – "maps, in particular", Wegener emphasizes – from around 20 paleontologists and constructed a table presenting the number of authors in favor of or opposed to the existence of a continental bridge at different times from the Cambrian to the Quaternary, between (1) Australia, the Deccan (in India), and southern Africa (including Madagascar); (2) Africa and South America; (3) Peninsular India and Madagascar; (4) Europe and North America; (5) Tierra del Fuego and Antarctica; (6) Australia and Antarctica; (7) North America and South America; and (8) Alaska and Siberia.

From this, Wegener produced a graph that allowed visualization of the strong agreement among specialists on the existence or non-existence of continental junctions between the different continents at different times: "Most scholars admit that the continents were united in the Carboniferous, the Triassic, the Lower Jurassic, and finally from the Upper Cretaceous to the Middle Tertiary", he summarized (Wegener 1924, p. 63).

It appears that the existence of four continental bridges was identified fairly unanimously by scholars:

– between India, Australia, Madagascar and Africa from the Cambrian to the early Jurassic;

– between Africa and Brazil from the Cambrian to the Cretaceous;

– between India and Madagascar from the Cambrian to the end of the Cretaceous;

– between Europe and North America from the Silurian to the beginning of the Jurassic, and then during the Tertiary.

Wegener also drew on zoo- and phytogeographical "evidence" concerning the present distribution of species, for example, earthworms or snails on the shores of the North Atlantic: the ocean being an insurmountable obstacle, the presence of earthworms on either side of the Atlantic attests to the existence of ancient connections between the two opposite shores.

The similarities of faunas and floras between the two shores of the North Atlantic from the Carboniferous to the Eocene are explained by Wegener's cartographic representation of the two continents joined together.

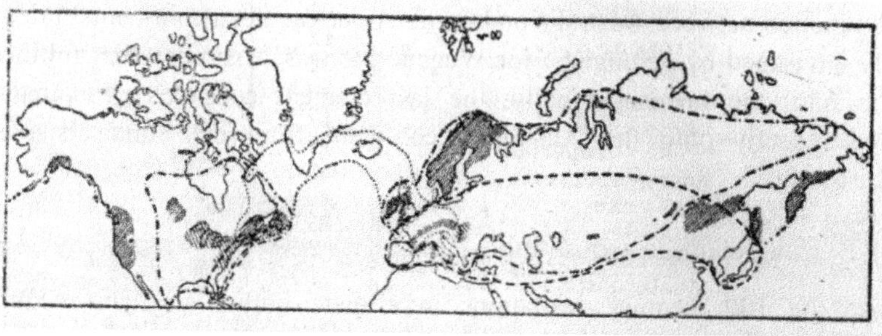

Figure 7.10. *Distribution of certain animals from regions bordering the North Atlantic (source: Wegener 1924, p. 64). Dotted: garden snails. Dashed: Lumbricines (earthworms). Dash-dot: perch. NE–SW hatching: pearl oyster. NW–SE hatching: mudminnows (Umbra)*

For the South Atlantic, in addition to the distribution of a *Glossopteris* flora, which will be discussed again a little further on, the presence of reptiles such as the Mesosaurids, as well as many other living beings, led widely to the acceptance of the existence of a continental mass uniting the southern continents. For example, the manatee is found both in western Africa and Brazil, and traits of relatedness are found between the South African flora and that of Australia.

Thus, Lemuria was imagined as early as 1864 by the English zoologist Philip Lutley Sclater (1829–1913) on the basis of a grouping of emerged lands bearing lemurs (Madagascar, India, Ceylon) (Sclater 1864). Lemuria was then interpreted as a "continental bridge" that had collapsed into the Indian Ocean. Popularized at the end of the 19th century by the German naturalist Ernst Haeckel (1834–1919), who made it the cradle of humanity from which the human species spread across the surface of the globe (Haeckel 1874), this vanished continent would constitute the link that made biological relations between the Deccan and Madagascar possible.

Regarding Australia, Wegener drew on a typology of faunal occupation of the island-continent proposed by Wallace. A first element, the oldest, made it possible to identify traits of relatedness between animals of southwestern Australia and the fauna of India, Madagascar and southern Africa. Australia was characterized by a distinctive mammalian fauna, composed of marsupials and monotremes, which differed markedly from that of the Sunda Islands and the Indo-Malayan archipelago, a fauna of so-called "placental" mammals separated by the "Wallace Line". This was the second element: a number of marsupial species – and even their parasites, Wegener emphasized – also existed on the American continent, suggesting an ancient junction between Australia and South America. Yet a continental bridge was scarcely envisaged by geologists: for Wegener, it was Antarctica that fulfilled this function until the Eocene. Finally, the last element consisted of animals that migrated recently onto the Australian continent from the Sunda Islands and Southeast Asia.

7.2.3.2.3. Paleoclimatic arguments

Since the 19th century, variations in climate within a given region over geological time had been fully accepted:

> The enormous quantity of facts [of a biological or geological nature] proves to us beyond doubt that most regions of the globe had, in the past, a climate very different from that which they enjoy today. (Wegener 1924, p. 73)

Markers of these ancient climates were identified through the study of geological formations and of fossil faunas and floras. Thus, luxuriant fossil floras attested to a hot and humid equatorial climate; sandstone and evaporite formations, as in present-day deserts, signaled a hot continental-type climate; finally, the flow of glaciers at high latitudes or at altitude left traces such as moraines or striations on the underlying rocks.

Over the course of the 19th century, the idea that a glacial episode occurred during the Pleistocene gradually became established among geologists based on the observation of various traces such as erratic blocks or the presence, in sedimentary layers of that period, of fossils of marine mollusks bearing witness to colder conditions than those prevailing during the Tertiary or more recently in the Northern Hemisphere.

Subsequently, several other phases of glaciation during earlier periods were documented and then accepted: first one in the Permo-Carboniferous, then another at the beginning of the Cambrian or at the end of the Precambrian.

At the beginning of the 20th century, to explain the existence of ancient climates differing from one point of the globe to another, the idea of a displacement of the Earth's axis of rotation – in other words, a migration of the poles and thus a displacement of the equator – was the subject of debate within the scholarly community: while many authors adhered to it, numerous opponents refused to accept it – "notably among geologists", Wegener (1924, p. 75) specified – proposing explanations linked to the composition of the atmosphere, variations in the energy released by the Sun, or regions of space that were more or less warm through which the Earth passed during its course (Macdougall 2013, p. 66). Around the explanation by migration of the polar axis, several aspects were also discussed: Was it a modification of the position of the polar axis, or a displacement of the entire Earth's crust over its substratum? Did the movement of the axis follow a strict oscillation within a determined circle? In any case, "all attempts to specify the locations [that the poles] would have occupied have resulted only in strange contradictions" (Wegener 1924, p. 75).

At the beginning of the 20th century, the general consensus broadly agreed, for recent times, on a migration of the North Pole from the Aleutian archipelago in the northern Pacific Ocean to its present position north of Greenland. The difficulty arose when attention turned to periods prior to the Cretaceous:

> We are in full cacophony. Not only do the opinions of the authors […] differ greatly from one another, but all the reconstructions they propose give rise to intractable contradictions and as such formally prevent determining, with any chance of success, the pattern of the pole's movements. (Wegener 1924, p. 76)

For the Carboniferous, various types of climate markers were identified:

– In today's Southern Hemisphere, glacial deposits such as moraines on polished and striated rocks produced by glacier flow were found in South Africa, Brazil, Argentina, India and Australia; to these geological indicators was added the presence of a flora considered characteristic of the polar climate of a tundra, the *Glossopteris* flora.

– In today's Northern Hemisphere, the numerous coal deposits of the United States, Europe, Russia and China, on the contrary, testified to a luxuriant vegetation characteristic of a tropical to subtropical climate.

– Finally, deposits of gypsum and other saline rocks, as well as sandstone formations characteristic of hot, dry climates, were found in present-day North Africa and in the Spitsbergen area.

This distribution of climates in the Carboniferous posed a problem: indeed, how could one explain that it did not follow the present distribution into parallel climatic zones, namely, a localization of the equatorial zone at 90° from the center of southern glaciation? How could one explain the large coal deposits in the United States or in China, in temperate regions, when the flora required for these formations was characteristic of the hot and humid climates of tropical zones?

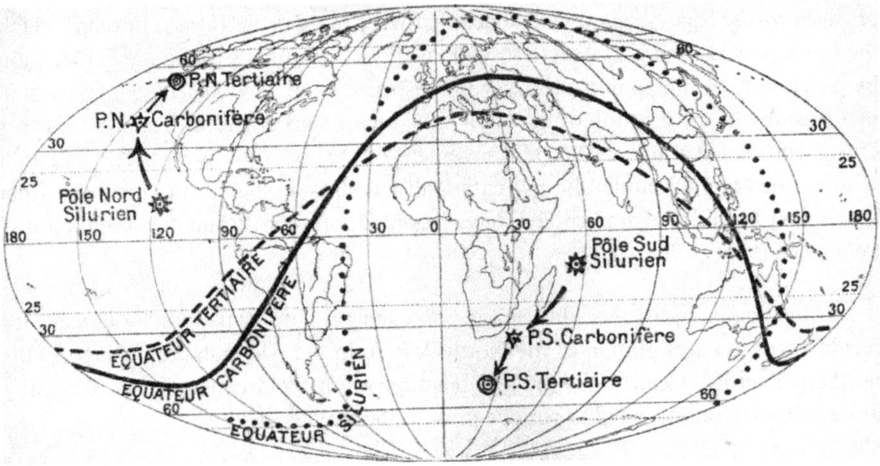

Figure 7.11. *Presumed migration of the poles between the Silurian and the Tertiary and visualization of the equators at different periods (source: Leuba 1938, p. 152). Top: Tertiary North Pole, Carboniferous North Pole and Silurian North Pole. Dashed line: Tertiary equator. Full line: Carboniferous equator. Dotted line: Silurian equator. Bottom: Silurian South Pole, Carboniferous South Pole and Tertiary South Pole*

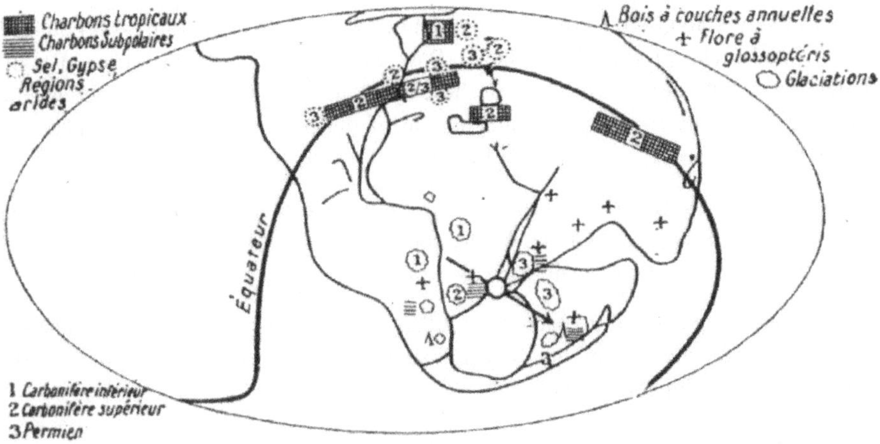

Figure 7.12. *Climatic indicators of the Permo-Carboniferous (source: Wegener 1924, p. 80). Dark cross-hatching: tropical coals. Light cross-hatching: subpolar coals. : salt, gypsum, arid regions. 1: Lower Carboniferous. 2: Upper Carboniferous. 3: Permian. Λ: trees with annual growth rings. +:* Glossopteris *flora. : Glaciations. Black line: equator*

Beyond the difficulty of properly explaining vertical movements of subsidence and uplift of the Earth's crust in certain places over geological time, the fixist conception – whether American or European – remains problematic in that it does not readily account for the presence, at the same time and at equivalent latitudes, of evidence for cold climates or, conversely, of tropical forests that gave rise to large coal deposits.

The theory of translations, therefore, has an advantage over the theory of continental bridges and, once the principle of continental displacement is accepted, provides a simple explanation for the ancient distribution of climates.

> The greatest success of Wegener's hypothesis has been in explaining the Mesozoic continent of Gondwanaland, with its widespread glaciation, its fresh water and temperate continental sediments, and its typical Glossopteris flora. (Tyrrell 1923, p. 552)

7.2.3.2.4. Geodetic arguments

Wegener initially attempted to determine the rate of past displacements on the basis of an estimate – still uncertain at the time and now known to be erroneous – of the duration of the various geological periods. Thus, estimating the time elapsed since the separation of Greenland and Scotland (between 50,000 and 100,000 years) and the distance between the two points (1,780 km), he calculated a westward drift rate of Greenland of between 18 and 36 m per year.

"The theory of continental translations has, among all those of such wide scope, the advantage of being capable of proof by exact astronomical measurements". (Wegener 1924, p. 89). Indeed, if one assumes that the continental drift that occurred in past ages is still continuing today, it should be possible to determine the position of a point on the Earth's surface in longitude and latitude by astronomical measurements and then, by repeating these measurements at regular intervals, to observe a change in these coordinates.

Such measurements had already been carried out during earlier missions in Greenland: in 1823 by the British astronomer Edward Sabine; in 1870 by the astronomers and geophysicists Börgen (German) and Copeland (British) during the *Germania* expedition; and finally in 1907–1908 during the Danish *Danmark* expedition by Koch (a mission in which Wegener himself also participated). Mobilized as early as his 1912 articles, these data allowed Wegener to estimate that Greenland moved away from Europe by 260 m between 1823 and 1870 and then by 690 m between 1870 and 1907, that is, 950 m over 84 years (Wegener 1912a). Wegener corrected these figures starting with the second edition of his book (1920), now presenting the following values: an increase in longitude of 260 m between 1823 and 1870 (i.e. 9 m per year) and 1,190 m between 1870 and 1907 (i.e. 32 m per year) (Wegener 1924, p. 92). Emphasizing that they may be affected by errors, he acknowledged that they needed to be confirmed by more accurate observations. It was planned that these measurements would be repeated every 10 or 15 years in order to detect, over a few decades, any possible change in the longitude of the two cable stations (Helmert 1915). The cutting of German submarine cables by the British, as early as the beginning of August, interrupted the measurements (Winkler 2008, pp. 5–33).

7.2.3.3. *In search of a driving force*

To explain the displacement of the sialic masses, Wegener initially considered a polar escape that he took from Eötvös (1848–1919) and the role of tidal forces influenced by the Moon and the Sun, then, in the last edition of *The Origin of Continents and Oceans* in 1929, he invoked fluid motions in the Sima. But he remained cautious, aware of the limits of his proposals regarding the forces ensuring the translation of continental masses, and acknowledged:

> The theory of translations does not yet have its Newton. One should not be concerned about his arrival: the theory is in fact only at its beginnings, many authors doubt it, and one should not blame theorists when they hesitate to lose their time and efforts in elucidating a law whose correctness is not yet agreed upon. It is nevertheless probable that some time will pass before the question of the forces involved in these movements can be completely resolved. (Wegener 1937, p. 161)

7.2.3.4. *Continental bridges and permanence at the same time: continental translations*

While geology was still imbued with the idea of global contraction, Wegener would not be mistaken about the essential characteristic of conceptions of the globe that must be challenged, when a few years later he targeted the idea summarized by this sentence of Suess: “It is the collapse of the earth’s body that we experience” (Jacoby 2001, p. 34).

To suppose the existence of land bridges between distant continents, separated by a deep ocean, for earlier geological periods, just as to attempt to preserve the idea of the permanence of the oceans by imagining chains of islands or detours through still unexplored regions, were ideas that were difficult to defend at the beginning of the 20th century. “The two schools have good impeccable arguments, but they draw unacceptable conclusions”. Wegener will “attempt to show that the correct postulates of both are easily satisfied by the hypothesis of rifting and horizontal drift of the continents" (Jacoby 2001, p. 34).

Some 15 years later, he summarized the advantageous position of his theory in relation to the two earlier theories in a synthetic formulation:

> If we take the theory of translations as our basis, we satisfy all justified requirements, both those of the law of former continental connections and those of permanence. We have only to state these laws as follows: Continental bridges? Yes, not through intermediate sunken continents, but through continental blocks that were once contiguous. Permanence? Yes, not of each continent or ocean taken individually, but permanence of the total oceanic surface and of the total continental surface taken as a whole. (Wegener 1937, p. 21)

Wegener's article in *Petermanns Mitteilungen* was the first conscious attempt to reconcile the disparate results of old geology with the new geophysics that was emerging at the beginning of the 20th century (Greene 2015). Before reviewing how Wegener's theory was received by specialists in the Earth sciences, one can note that the idea of a displacement of the continents was not original and that several earlier authors had already envisaged such a phenomenon, but most without developing a scientific argument.

7.2.4. *Other earlier drifts*

"It is curious to note, in the history of science, that as soon as a new hypothesis appears, one generally finds that it was, almost simultaneously, imagined by different scholars" (Bourcart 1924, p. 563). Indeed, although Wegener's theory was the most thoroughly substantiated and the one that had the greatest influence, in the sense that it would long occupy many scholars from various disciplines, it is useful to point out that the idea of the mobility of continental masses was not new and had been envisaged by predecessors belonging, or not, to the scientific world. Yet one should refrain from seeing in them, strictly speaking, "precursors", since the models then envisaged were only weakly supported by "evidence" and, above all, did not place these phenomena within a revision of the structure of the globe that would make it compatible with the alleged movements. It is not a matter here of taking inventory of, and discussing the relevance of, all the authors – and their conceptions – suspected of having anticipated Wegener and his theory, but a brief presentation of a few of them makes it possible to show the permeability of the scientific milieu when confronted with a new theory, with conceptions that were scarcely scientific or had until then been set aside.

Figure 7.13. *The South Pole a few billion years ago closely resembled the present-day North Pole. Left: continents of America, Africa and Australia. Right: continents of America, Asia and Africa with the Atlantic Ocean in the middle (source: Mantovani 1909, p. 597)*

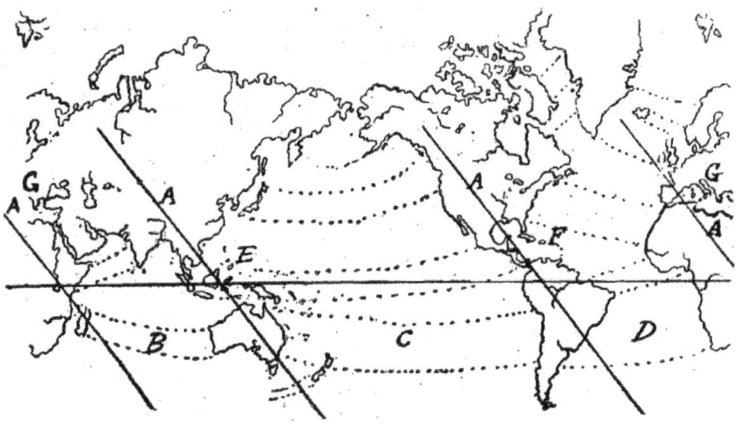

Figure 7.14. *Arrangement of spots on the surface of the terrestrial cell. Axis A highlights the tilt of the continental masses. The fan-shaped spread to the south of the three oceans B, C, and D corresponds to three rifts E, F, G. The dotted lines connect pairs of points that are now located on opposite shores separated by an ocean but were once merged into a single point, before the formation of the ocean due to the Earth's expansion and the widening of gigantic fractures. In the past, with all the oceans closed off, the Earth must have been smaller than it is today. (source: Mantovani 1909, p. 595)*

From the very first discussions of continental drift in the 1920s, various names were invoked as "precursors" (the term used at the time), not without a certain touch of chauvinism. If we confine ourselves to the French context, a few examples may be cited. Thus, during the special session of the Geological Society of France devoted to Wegener's theory in 1923, the geologist Gustave-Frédéric Dollfus (1850–1931) referred to Lamarck, who had envisaged that, under the influence of marine currents produced by the action of the Moon, the seas destroyed and invaded the eastern coasts of the continents, and that, conversely, the lowering of their level in the west left exposed land, continually modifying the distribution of seas and continents over the surface of the globe (Lamarck 1802; Joleaud 1923, p. 270). Then the geologist Jacques Bourcart, in 1924, identified the origins of the hypothesis of continental drift in Roberto Mantovani (1854–1933), the author of a theory of terrestrial expansion (Mantovani 1889, 1909), a proposal that would find a few defenders up to the present day (Scalera and Jacob 2003; Sudiro 2014) to explain the relative displacement of the continents. Be that as it may, Bourcart's bringing him to light led this violinist and music teacher to present his theory of "planetary expansion" before the Geological Society of France on two occasions (Mantovani 1924, 1927) and to exchange some correspondence with Wegener, who would include the 1909 article in the bibliography of the last edition of his book in 1929 (Scalera 2009).

Finally, when the theory of drift spread beyond the field of science, other authors also embarked on the quest for a precursor: this was, for example, the case of the Atlantologist Jean Gattefossé (1899–1960), author of a general bibliography of Atlantis (Gattefossé 1926), a continent whose disappearance might have had something to do with the processes described by Wegener:

> Permit us, in passing, to claim the theory of continental drift as an essentially French idea, taken as a whole, since it was first published, with explicit maps, by Snider-Pellegrini in his very curious work: *La création et les mystères dévoilés*, a work in which the nature of all beings, the origin of America, etc., are clearly set forth. (Gattefossé 1923, p. 598)

In reality, the Italo-American Antonio Snider-Pellegrini (1802–1885), often presented as a scientist or a geographer, was an eclectic figure, more a great traveler and businessman, who resided in France on several occasions (Le Vigouroux and Gohau 2016). He deserves a few lines to understand his irruption into the circles of geology and its history, for the lineage between the conceptions of these alleged predecessors and those of Wegener is often very tenuous.

Indeed, while in the "prehistory" of drift theory the historian of geology Albert Carozzi (Carozzi 1969, 1970) rejected authors invoked in the 1960s by some geophysicists as precursors of Wegener – such as Bacon (1620), Father Placet (1668), or even Humboldt (1801) – he retained Snider-Pellegrini (or Snider), whom he regarded "as the first precursor of the idea of continental drift, having assembled and dislocated the continents in roughly the same way as Taylor (1908) and Wegener (1910) did at the beginning of the 20th century" (Carozzi 1969, p. 178).

Yet, on closer inspection, one may question the relevance of introducing Snider into the field of Earth sciences. Indeed, although no credit was granted to his work *La création et les mystères dévoilés* (Snider 1858) within the French scholarly community at the time (Maury 1859), it did enjoy a certain posterity when an English popularizer, John Henry Pepper, reproduced in a book – *The Playbook of Metals* (1861) – the illustrations of the terrestrial hemispheres before and after the separation of America, illustrations to which Snider had given the following caption:

> We place here two engravings of the terrestrial globe. The first, marked 9, before the separation, shows our planet at the time of the interval, that is to say from Adam to Noah, when the Earth was still a single mass, furrowed at the surface by fissures. In this period appears the mass of Atlantis, which was later to give way to an ocean. The second engraving, marked 10, after the separation, represents this same globe with the separation of its parts at the surface after the Flood, that is to say since Noah, and as it exists today. If one takes in hand a planisphere on a sufficiently large scale, one will be better able to verify the fitting together of the terrains that nature offers to our examination in order to establish their primitive union. (Snider 1858, p. 314)

In Pepper's book, these engravings were intended to illustrate remarks by Lyell explaining that the Earth's surface has been reshaped again and again. After several further reproductions of these figures, the introduction of Snider's work into geological circles was completed by their publication in the treatise *Principles of Physical Geology* by the British geologist Arthur Holmes (1890–1965) in 1944. Thus, Snider acquired a notoriety that appears quite undeserved within the field of geology and its history.

During the 1970s, with the increasingly widespread adoption of the plate tectonics model by specialists in the Earth sciences, Wegener's continental drift experienced a renewed interest, and historians of science once again turned to the archives in search of the germ of the idea of continental drift. For example, an American geologist saw in the writings of the French geographer Élisée Reclus (Reclus 1868) the early stirrings of the idea of a continuity of the continents followed by their rupture, and even of many other characteristics discovered over the course of the 20th century (Berkland 1979).

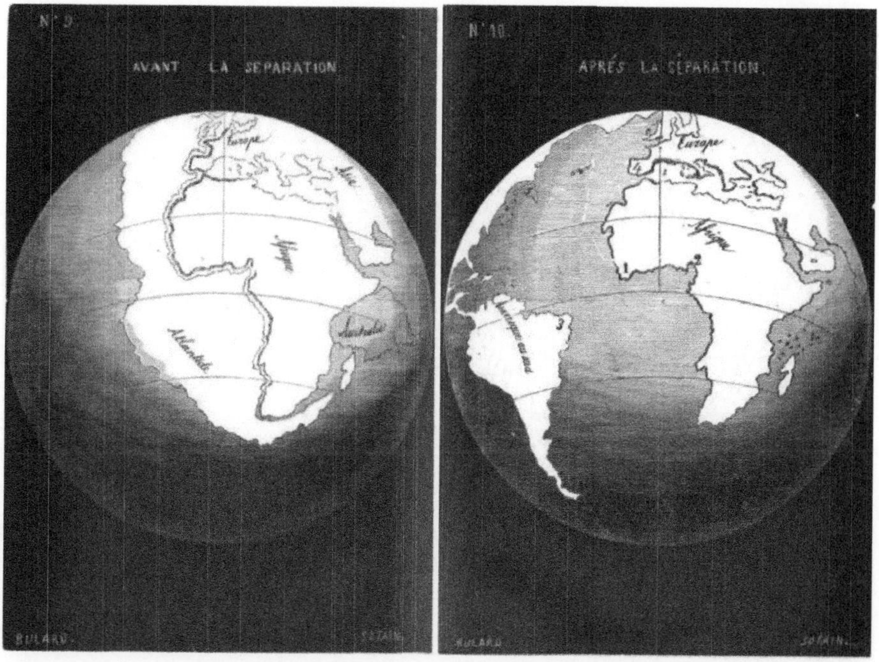

Figure 7.15. *Left: No. 9 before the separation. Right: No. 10 after the separation (source: Snider 1858, p. 314)*

The Anglo-Saxon world also has several forerunners of Wegener. Thus, shortly before the German scholar, an American geologist, Frank B. Taylor (1860–1938), proposed in 1910 a mobilist conception of the continents involving a migration of continental masses from the poles toward the equator. According to Taylor, the collision of continental masses coming from the north with those coming from the south led to the formation of the belt of Tertiary mountain chains at the Earth's surface (the Alps, the Himalayas, the Indonesian archipelago, the Rocky Mountains and the Andes).

Like Wegener, Taylor envisaged the separation of Greenland from North America (Labrador and Baffin Island) on the basis of the observed parallelism of the coastlines, and he considered a limited westward drift of the American continents – but only from the Mid-Atlantic Ridge. However, it is essentially on the interpretation of the equatorial positioning of the arcuate Tertiary mountain chains of the southern continent Eurasian continent that he based his reasoning: these chains were the result of the folding of the southern margin of Eurasia under the effect of a push from north to south, and the foldings of Southeast Asia were due to the northward movement of Australia toward the equator.

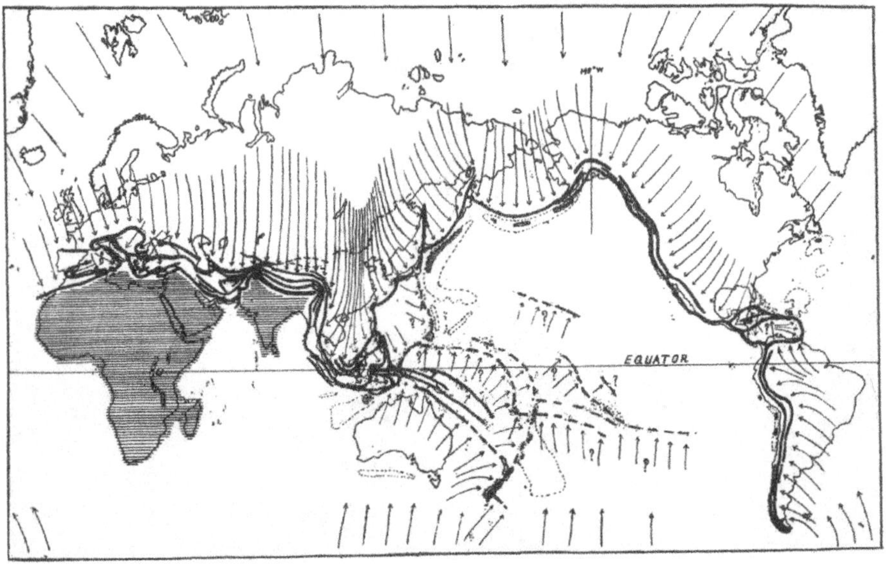

Figure 7.16. *The world in Mercator projection*

COMMENTS ON FIGURE 7.16.– *The map shows the distribution of Tertiary mountain chains in both hemispheres; the arrows indicate approximately the direction and the relative magnitude of crustal movement. The map also shows the position of oceanic trenches, the relationships between the Tertiary chains of Europe and western Asia as far as the Indo-African plateau, and the particular organization of the chains between Asia and Australia and in the Caribbean region (source: Taylor 1910, p. 211).*

Although situating himself within a contractionist framework, Taylor considered that the contractions linked to cooling advanced by Suess were not sufficient for his scenario; he therefore imagined complementary forces, linked to an acceleration of the Earth's rotation caused by the capture of the Moon by the Earth. This, in turn, led to a creep of continental masses from the poles toward the equator. In the United States, Taylor's ideas attracted little attention at the time (Oreskes 1999), and they were ultimately taken up only during the debates on continental drift according to Wegener in the 1920s.

Yet as early as 1912, Wegener cited Taylor's article and even regarded him as a precursor of his own work, while at the same time affirming that the two hypotheses of horizontal displacements of the continents were developed entirely independently (Wegener 1912a). Moreover, Wegener did not adhere to the explanations of the American geologist: since Taylor "did not investigate the implications despite their contradiction of traditional views, his suggestion has probably met mostly with skepticism" (Jacoby 2001, p. 32). Despite this distancing, in the years that followed, American geologists often referred to the theory of continental drift by calling it the "Taylor–Wegener theory".

The idea of a dispersal of emerged lands across the surface of the globe following the dislocation of a continent existed before Wegener, both among amateurs of scientific conceptions (Snider 1858; Mantovani 1909), who came to be incorporated into the narrative accompanying the debates and supporting the diffusion of the Wegenerian theory during the 1920s, and within the scholarly field of geology (Taylor), thereby marking the need for a renewal of conceptions.

7.3. The reception of Wegener's theory

Presented to the scholarly community through the two articles of 1912, Wegener's ideas met with little response in scholarly circles outside Germany at the time of their dissemination, although a few French- and English-language newspapers mentioned them discreetly in scientific columns during the years 1912–1914, in connection with their author's polar expeditions or with the longitude-measurement campaign undertaken by the Institute of Geophysics of Jena in July 1914, a campaign interrupted by the outbreak of the Great War.

Despite the wide range of data mobilized to argue for his conception, Wegener took care to specify:

> In spite of its broad foundation I call the new idea a working hypothesis and I wish it to be looked at as such, at least until it has been possible to prove by astronomical positioning with undoubtable accuracy that the horizontal displacements continue to the present day. Also, it is not superfluous to point out that this is a first outline of the hypothesis. In a more detailed elaboration, it will probably be necessary to modify the hypothesis here and there. (Jacoby 2001, p. 31)

This call to debate his proposal, so long as the measurements were not probative, did not seem to mobilize the geological community immediately: for that, it would therefore be necessary to wait for the publication of the book, in 1915. From that point on, two major phases of discussion of the theory in scientific circles can be distinguished.

The first, between 1912–1915 and 1922, concerned essentially German-speaking and Germanophile countries: Germany, but also Austria, the Netherlands, the Scandinavian countries and Switzerland. The invasion of Belgium in August 1914, the atrocities committed there by the occupying army, and the support given by German scholars and intellectuals to the belligerent enterprises of the German authorities led to the organization of a boycott of German scholars by the Allied countries. This boycott persisted in particular with respect to geologists of the Central Powers until the mid-1920s, after the organization of the International Geological Congress in Brussels in 1922, which marked the resumption of international exchanges within the geological community. It was, in fact, only from 1926 onward, at the Madrid Congress, that German and Austrian scholars were once again invited to participate by the host nation (Le Vigouroux 2018b, 2022).

Yet it was also in 1922 that the theory of continental translations was (re)placed at the forefront of the scientific scene in various countries, and that the second phase of discussion began. From the very next year, the debate over Wegenerian ideas reached more broadly the national communities of geologists in most of the Allied nations where it had until then been ignored: in addition to its critical presentation in scholarly journals, it became the subject of presentations and discussions at meetings of geologists, some of whom expressed their support or their reservations through various publications. Going beyond scientific circles, the theory of continental drift also found a wide resonance in popular works and journals (Le Vigouroux and Gohau 2016).

7.3.1. *A limited and mixed international reception*

Very early on, interest manifested itself within the community of German geologists, geographers and geophysicists: successive editions of *Die Entstehung der Kontinente und Ozeane* were enriched with references to German authors who confirmed or contested Wegener's interpretations. It was above all in publications in geography that the different aspects of the hypothesis were discussed, the 1912 article in *Geologische Rundschau* being the only one published by Wegener in a geology journal (Demhardt 2005).

In Austria, although facilitated by the earlier theses of Otto Ampferer (1875–1947) – an opponent of contractionism who as early as 1906 suggested the existence of sub-crustal currents allowing vertical movements of the Earth's crust and the erection of mountain chains – the acceptance of the Wegenerian theory remained moderate because of the strong influence of the fixist Leopold Kober (1883–1970) (Flügel 1984; Carozzi 1985). In the Netherlands, as early as 1916, geologists working in the Dutch East Indies, such as Gustaff Molengraaff (1860–1942) and Hendrik Albertus Brouwer (1886–1973) in particular, expressed their interest in the explanations given by Wegener for the characteristics of their fields of study (Carozzi 1985). The evolution of the thinking of the geophysicist Felix Andries Vening Meinesz (1887–1966) with regard to drift is representative of its reception among Dutch scholars. A specialist in gravity measurements, Vening Meinesz set as one of the objectives of his missions the verification of Wegener's hypothesis: gradually, results accumulated, but during the 1930s these came to contradict it, leading most Dutch geologists to reject continental translations (Barzilay 2010). Danish and Scandinavian geologists also welcomed the theory of continental drift, even though they had difficulty applying it to their own Precambrian and Paleozoic terrains (Carozzi 1985).

In Switzerland, Wegener's theory also enjoyed a certain success, notably in Neuchâtel, around the professor of geology Émile Argand (1879–1940) (Schaer 2003). As early as November 1916, and despite the strong anti-German sentiments then widespread in Switzerland, where it was strictly forbidden to read in public or in private any material printed in Germany (Carozzi 1985), Argand presented before the Neuchâtel Society of Natural Sciences his *Modern Views on Continents and Oceans*, explaining that the "new conception explains much better than the old one the similarities of faunas and floras between regions today separated by vast oceans; it is likewise supported by the study of the glaciations at the end of the Primary era" (Argand 1916, p. 115).

It was again a Swiss geologist, an assistant at the University of Lausanne, who in 1922 presented Wegener's theory in a French popular science journal, *La Revue générale des sciences pures et appliquées*. Élie Gagnebin (1881–1942) described it favorably there and expressed surprise that the program announced for the International Geological Congress to be held a few months later in Brussels made no mention whatsoever of Wegener's hypothesis:

> "This is a regrettable omission. For if Wegener's idea is open to criticism from more than one point of view, if reservations with regard to it are necessary, it can no longer be ignored today nor passed over in silence." And he regretted that "in France, where Einstein's theories and Freud's psychoanalysis are so furiously in fashion, Wegener's hypothesis is almost entirely ignored [and that] no specialized journal has yet spoken of it." (Gagnebin 1922, p. 293)

The first French translation of the 1922 edition of Wegener's book also came from Switzerland. It was carried out by Manfred Reichel (1896–1984), a young science graduate and assistant to Argand, who himself was involved in this translation. Finally, John Leuba (1884–1954), a student and assistant of Schardt at the Faculty of Science in Neuchâtel before the First World War, was the author in 1925 of *Introduction à la géologie*, a popular geology book published in the "Collection" series by Armand Colin, an encyclopedic pocket collection. In his book, reissued seven times until 1951, Leuba had the clever idea of presenting paleogeographies at different periods of the past, as envisioned both within Haug's geosynclinal theory and within the framework of continental drift theory. In his text, he compared the two theories, highlighting the hypothetical nature of continental boundaries in the first, while affirming that the theory of continental translations "is undoubtedly the one that accounts for the largest number of facts" (Leuba 1938, p. 107). This period also saw Argand finally publish his *Tectonique de l'Asie* (1924), long awaited by geologists since his opening lecture at the International Geological Congress in Brussels. While the first 26 chapters, covering the bulk of his work, followed classical tectonics, the last five chapters attempted to show to what extent classical tectonics and Wegener's ideas could be reconciled, and what compromises each required. This approach was guided by "The desire for objectivity and equally that of innovating on an established basis" (Schaer 1991, p. 534).

By the late 1920s, however, the Swiss geological community became increasingly critical of continental drift theory. Research directly inspired by it suffered from this negative climate, and references to Argand's work grew increasingly rare (Schaer 2010). Argand himself, after publishing his treatise, appears to have abandoned active research, though he continued teaching (Schaer 1991). Among other Swiss geologists, Léon W. Collet (1880–1957) of Geneva also adhered to the Wegener-Argand theory, while Rudolph Staub (1890–1961) incorporated the new concepts of Wegener and Argand. Staub confirmed Africa's northward migration, causing the compression of the Tethys into a mountain chain, and Europe's northward movement, which opened the Mediterranean Sea, but he considered a broader synthesis was still missing to explain the movements of other continental blocks. Later, he revised certain aspects proposed by Wegener and Argand. Albert Heim (1849–1937) initially rejected the theory as too speculative and insufficiently based on observational evidence – though he ultimately endorsed it in 1928 – as did Fritz Nussbaum (1880–1966) and Hans Schardt, the latter remaining hostile to the theory he judged romanesque, opposing it with stronger negative conviction than with precise arguments (Leuba 1931).

In Italy, reactions were later (from 1924 onwards) but fairly favorable among the younger generation of geologists (Carozzi 1985). As early as the late 19th century,

Italian geologists had already entertained the fragmentation of a large continent. In the 1920s, however, the theory was largely absent from studies focused on Italian geology (Romano et al. 2017). In Belgium, reactions were rather hostile, and the theory was generally rejected, whereas in Spain, which had remained neutral during the First World War, reactions were more open (Carozzi 1985).

In the United Kingdom, no reference to Wegener appeared in the major scientific journals between 1913 and 1922, prior to a brief unsigned presentation of the second German edition in *Nature* on 16 February 1922 (Marvin 1985). Two critical communications, emphasizing the revolutionary character of the theory and its potential, were published in March in *The Guardian* of Manchester and in the *Proceedings of the Manchester Literary and Philosophical Society*. In May, an article signed by Wegener occupied four and a half pages in *Discovery*, a popular monthly scientific magazine. The debates, initiated at the Manchester Literary and Philosophical Society, intensified in the following months and reached the columns of *Nature*, then those of the *Geographical Magazine*. Discussions were also organized at the British Association for the Advancement of Science and at the Royal Geographical Society. Many prominent geologists proved rather favorable, or at least open, to the new conception, even if certain details still remained to be clarified. However, two figures resolutely opposed Wegener's hypothesis: the geologist Philip Lake (1865–1949) and the geophysicist Harold Jeffreys (1891–1989). The English translation of Wegener's work was published in London in 1924, reviving debates over the relevance of his theory for several years.

The first mention of Wegener in an English-language scientific publication seems to have been an article by Alexander Du Toit in a South African journal in 1921. With the exception of this geologist, Wegener's ideas – which nevertheless regarded Australia and the Sunda Islands as key terrains for testing his theory – aroused little interest and few reactions among Australian scientists between 1920 and 1940. They largely remained silent or adopted a passive resistance, without publicly expressing support or opposition (Le Grand 1986). In the United States as well, Wegener's thesis was generally rejected, often in a more categorical and hostile manner than in Europe (Oreskes 1988).

7.3.2. *Reception in France*

In France, where Wegener's theory was introduced only in 1922 by Élie Gagnebin, scientific discussions during the 1920s remained very cautious. The theory was presented before the Geological Society of France in 1923 by the paleontologist Léonce Joleaud (1880–1938). The exchanges that followed showed that, while awaiting new data, scholars reserved/suspended their judgment. Faithful to the mechanism of uplifts and collapses of continental bridges to explain the distribution of faunas across the continents, Joleaud was not convinced by Wegener's views, which he judged too schematic, but he was nevertheless not entirely hostile to continental drift: for example, like other geologists, he viewed positively the way in which Wegener placed the South Pole at a point of coalescence of all the Gondwanan continents on the planisphere, positioning the climatic indicators of the Anthracolithique at the poles and on the equator: "Wegener's hypothesis is therefore in this case more plausible than the ways of thinking adopted up to the present day by geologists" (Joleaud 1923, p. 250). Moreover, drawing on the successive phases of the settlement of Australia, or of Madagascar, Joleaud proposed to amend Wegener's conception in an original way by imagining a process of accordion-like drift:

> The isolation of the Australian regions, in the zoological sense of the word, did not remain absolute since the Secondary era: both in New Guinea and in New Holland itself, Rodents and Artiodactyls penetrated during the Oligocene and the Quaternary, either by way of continental bridges, according to the old way of thinking, or thanks to an accordion-like movement between the Indian and Australian continental masses, in Wegener's theory modified according to my conception. (Joleaud 1923, pp. 237–238)

He remained alone in defending such a drift… .

Facing the waiting skeptics, an opponent rose up against the Wegenerian theory: the eminent geologist Pierre Termier, a member of the Academy of Sciences since 1909, a specialist in Alpine tectonics, where he had participated in establishing the existence of thrust nappes. It was the two aspects of the theory that Termier contested: both continental translations and the structure of the globe as Wegener conceived it. Moreover, two features of the Earth's surface had an almost permanent character: on the one hand, the great transverse scar, formerly occupied by a vast sea filling the intertropical zone between the continents of the Southern Hemisphere and those of the Northern Hemisphere between the Pacific and the Antilles, and today become the Alpine chain of Europe and Asia; and, on the other hand, the circum-Pacific trench, which was gradually turning into a mountain chain at the expense of the ocean.

> For me, Wegener's theory was a beautiful dream, a dream of a great poet. One tries to grasp it; and one realizes that one has in one's arms only a little vapor or smoke; it is at once seductive and elusive. (Termier 1924, p. 266)

Termier also reproached it for its "overly convenient" character, which in his eyes constituted a flaw: "with Wegener's theory, everything is arranged and there is no difficulty, it seems".

> It is admirable; and, in order not to be seduced, one must remember what I was saying only a moment ago, about the danger of hypotheses that are too convenient, which appeal to occult forces, perhaps nonexistent, and which give us the dreadful power to dispose the continents and the seas at our pleasure. (Termier 1924, p. 261)

Because he was the author of numerous articles and lectures, often taken up in works that met with a certain success among the public, Termier's opinions and grievances with regard to the Wegenerian theory were widely disseminated.

It is appropriate, finally, to underscore the support for the theory expressed later by the geologist Boris Choubert (1906–1983), born in Saint Petersburg and arrived in France in 1927. While attending geology courses at the Sorbonne, he undertook to verify Wegener's "brilliant synthesis" and in 1935 published his work in the journal of the Laboratoire de géographie physique et géologie dynamique directed by Léon Lutaud (Choubert 1935, 1981). He relied on bathymetric data to reconstruct a continental assemblage that differed from the one envisaged by Wegener: he chose to take the 1,000 m isobath to delimit the outline of the continental coasts (Wegener, for his part, had chosen 200–500 m). On the basis of the deep topography of the Bay of Biscay, he identified a rotation of the Iberian Peninsula and the Atlantic threshold, which he supposed to be composed of a mixture of Sial and Sima, and where the original rupture of the Atlantic would have occurred, from which the American continent and the African and European continents moved away in opposite directions.

Finally, judging it "illogical" to think that the Pangaea that preceded the last separation of the continents could have remained immutable throughout the entire Paleozoic era, he applied the principle of continental translations to periods prior to the Carboniferous-Permian, the temporal limit to which Wegener had restricted his hypothesis.

After this publication, Choubert no longer published on continental drift, and his work, which was not taken up again, was forgotten (Kornprobst et al. 2018). It nevertheless had an echo in the mid-1960s with the emergence of plate tectonics and the construction of a map of the fitting together of the American and African continents by the British geophysicist Edward C. Bullard (1907–1980), a map very similar to that of Choubert 30 years earlier (Bullard et al. 1965).

In the 1930s and 1940s, a few other French geologists found in the theory of continental translations an explanatory process for the structures observed in their fields of study or, at the very least, considered it useful to discuss it in their context. This was the case for the Afar region, the Cameroon Volcanic Line, or again the Kerguelen Archipelago.

Finally, in the 1940s and 1950s, biologists found it pertinent to mobilize the Wegenerian theory to explain the distribution of animal species across the different continents, thus extending the debates into the field of biogeography (Le Vigouroux 2018a).

7.4. Conclusion

The theory of continental drift held a particular status within 20th-century geology. It appeared at the beginning of the century, in 1912, but very quickly disappeared from the field of research and in geology from the end of the 1920s until the end of the 1950s – except for a few "daring" authors who took an interest in promoting and developing it – particularly because of the difficulty of supporting it with incontrovertible evidence.

In order to make his theory known, Wegener first set it out in a geography journal and in a geology journal as early as 1912. The latter was a young journal whose first volume had appeared only in 1910, and one may suppose that neither its existence nor its notoriety could at that time have contributed to a wide dissemination of the new theory. By contrast, the geography journal had been founded in the middle of the previous century and enjoyed wide international recognition, including among geologists concerned with physical geography, who could thus have access to Wegener's presentation of his conception. But very early on, the Great War constituted an obstacle to the possibility of submitting the hypothesis of continental translations to debate. The outbreak of the war caused the interruption of the mission to measure longitudes between Europe and North America initiated by the Institute of Geodesy of Potsdam, and with it the possibility of establishing the measurement of a possible present-day separation of the continents bordering the Atlantic Ocean. Moreover, the support of German scholars for the belligerent enterprises of the German authorities in 1914 prompted the organization of their boycott by the scientific institutions of the Allied countries within the International Research Council set up as early as 1919 (Le Vigouroux 2018b). As a result, discussion of Wegener's theory could emerge in the Allied countries only from 1922 onward, when the geologists gathered in Congress in Brussels lifted the boycott and invited German geologists to the following Congress, which was to meet in Spain in 1926 (Le Vigouroux 2022).

The absence of an identified driving force capable of ensuring the movement of the continents led many geologists to receive Wegener's mobilism with skepticism. Yet, a few decades earlier, the theory of thrust nappes and of major overthrusts in the formation of mountain chains, which implies the sliding of rock masses over several hundred, even several thousand kilometers, had been quite widely accepted despite the absence of an identified mechanism.

Geophysicists opposed the mechanisms invoked by Wegener, not failing to emphasize the weakness of the forces involved, like the physicist Henri Bouasse (1866–1953), for whom Wegener spoke only of ridiculously small forces that would be incapable "of crushing a very poor brick!" (Bouasse 1927, p. 360), and to highlight the difficulty of envisaging the mobility of sialic continents in a denser seismic substratum and considered rigid on the basis of seismic data.

The breadth of the scope of application of Wegener's theory means that it can hardly be burdened with details about measurements of the geophysical properties of the subsurface, about the precise nature and composition of rocks, about the particular arrangement of tectonic structures, or about the degree of resemblance of the fossils identified on either side of the oceans; yet the specificities of these matters are well known to specialists in each of these branches of the Earth sciences, as Joleaud points out, for example, with regard to the frequent analogies between living species on either side of an ocean, which "give a glimpse of the problem in all its complexity, a complexity infinitely greater than Wegener seems to imagine" (Joleaud 1923, pp. 212–213). As a non-specialist, Wegener was scarcely able to oppose, to challenges to the data used, a body of evidence comparable to those on which *The Face of the Earth* by Suess or the *Traité de géologie* by Émile Haug is based.

Very early on, this bias was highlighted by John Leuba:

> [...] if it has not won the immediate adherence of all geologists, this is no doubt due to that very commonplace psychological fact that one is inclined to accept only ideas put forward by specialists. Now Wegener is not a geologist, but a geophysicist: the resistance of certain geologists has no other reasons. (Leuba 1938, p. 208)

It can moreover be emphasized that while, for a few geologists, only a non-specialist could renew the way of looking at the problems that arose without burdening himself with concepts and a "hazy" vocabulary (Gignoux 1925), most specialists rejected in a fairly systematic way the inventor of continental drift by assigning him to another discipline: for geologists, he was a geophysicist or a meteorologist, whereas for geophysicists, he was a geologist (Marvin 1973; Le Vigouroux and Gohau 2016).

In the same way, the "overly convenient" character of the theory, which would make it possible to position the continents as one saw fit, could constitute a deterrent because of the lack of scientific rigor that would preside over the fate of the "sialiques" [silica- and alumina-rich continental crustal masses] (Termier 1924). But at the same time, Argand, seeing in it a quality, emphasizes its "extreme flexibility combined with great richness in operational maneuvers" and explains: "One thinks one has a decisive objection, one more blow and everything will crack, but nothing cracks: one has forgotten only one or several maneuvers. It is the protean resistance of a plastic universe" (Argand 1924, p. 292).

To explain the reserved attitude of Australian geologists, the historian of science Homer E. Le Grand puts forward specialization and "geological localism", the theory of translations not providing any added value to the problems that were occupying them at the time (Le Grand 1986, 1988).

The weight of the academic authority of a generation of scholars who built their conceptions of the globe within another paradigmatic framework, and whose questioning was hardly conceivable, blocked the evolution of ideas. The American geologist Rollin T. Chamberlin (1881–1948) summed up this stance by reporting a remark heard at a meeting of the *Geological Society of America*: "If we are to believe Wegener's hypothesis we must forget everything which has been learned in the last 70 years and start all over again" (Chamberlin 1928, p. 87). This situation can partly explain the timidity of the debates and the skepticism displayed by many geologists, younger ones in particular, in various countries. This is the case, for example, of the British geophysicist Harold Jeffreys, professor at Cambridge, who held an important tenured position and whose textbook, first published in 1924, was printed until 1976: "There is therefore not the slightest reason to believe that bodily displacements of continents through the lithosphere are possible. […] Secular drift of continents relative to the rest of the crust, such as have been maintained by A. Wegener and others, are out of the question" (Jeffreys 1929, pp. 304–305). Basing his position on seismological data, Jeffreys remained opposed to the movement of continents until his death despite the accumulation of evidence of its relevance since the 1960s.

According to the American historian of science Naomi Oreskes (1988), the resistance of American geologists to Wegener's thesis rested above all on the rejection of a new way of doing science that continental drift would imply, in contradiction with the foundations of the method that these geologists usually employed. Indeed, following the formalization written by the American geologist Thomas C. Chamberlin (1843–1928), Americans were attached to the method of multiple hypotheses, a method that gives primacy to field observations rather than to theory:

> Too often a theory is promptly born and evidence hunted up to fit in afterward. Laudable as the effort at explanation is in its proper place, it is an almost certain source of confusion and error when it runs before a serious inquiry into the phenomenon itself. (Chamberlin 1897, p. 838)

Moreover, geological reasoning remained based on the principle of uniformitarianism given by Lyell in the 19th century, which considers geological processes of the past only by analogy with what is observable in the present time. Oreskes also notes the strong adherence of American geologists to two conceptions of the terrestrial globe.

On the one hand, shortly before the formulation of the mobilist conceptions of Taylor and Wegener, the theory of the permanence and fixism of continents and oceans – independently of the geosynclinal tectonic activity that can occur at their margins – was reaffirmed by Willis. On the other hand, American geologists, following John Hayford (1868–1925), who regarded it as more convenient than the Airy model as a first approximation, gave their preference to the Pratt isostatic model when Wegener took the Airy model into consideration.

While the theory of continental translations failed to convince geologists, it nonetheless very quickly went beyond the sole circle of Earth science specialists to be offered up to the discussions of "learned amateurs", detractors as well as defenders, popularizers, or even a circle of "parascientists": Wegener's science, unlike that of academic specialists, had the advantage of being easily translatable into simple terms and striking images that captured the imagination. Its encompassing explanatory power gave it the capacity to arouse admiration, even wonder. For similar reasons, secondary-school geology textbooks retained it from the late 1930s to the 1950s to explain the formation of mountain chains, even when most geologists had abandoned it (Le Vigouroux and Gohau 2016; Le Vigouroux 2018c).

Wegener's theory, like the other conceptions it competed with, was definitively replaced by plate tectonics theory in the 1960s and 1970s. Yet, however fleeting its period of discussion, Wegener's theory remained in the collective mind and even became the emblematic example used, by analogy, to explain complex processes and mechanisms in all kinds of contexts. This scientific theory, like a few others, left its original restricted field to spread within a population that, without mastering its ins and outs, used it and integrated it into its culture.

7.5. References

Airy, G.B. (1855). On the computation of the effect of the attraction of mountain masses, as disturbing the apparent latitude of stations in geodetic surveys. *Philosophical Transactions of the Royal Society of London*, 145, 101–104.

Argand, E. (1916). Compte-rendu de la séance du 3 novembre. *Bulletin de la Société neuchâteloise des sciences naturelles*, 42, 115.

Aubouin, J. (1959). À propos d'un centenaire : les aventures de la notion de géosynclinal. *Revue de géographie physique et de géologie dynamique*, 2(3), 135–188.

Aubouin, J. (1961). Propos sur les géosynclinaux. *Bulletin de la Société géologique de France*, S7-III(7), 629–711.

Aubouin, J. (1965). *Geosynclines*. Elsevier, Amsterdam.

Barzilay, W.F. (2010). Vening Meinesz and the theory of Wegener. *Earth Sciences History*, 29(2), 213–231.

Berkland, J.O. (1979). Élisée Reclus: Neglected geologic pioneer and first (?) continental drift advocate. *Geology*, 7, 189–192.

Bertrand, M. (1887). La chaîne des Alpes et la formation du continent européen. *Bulletin de la Société géologique de France*, 15(3), 423–447.

Bertrand, M. (1900). Essai d'une théorie mécanique de la formation des montagnes. Déplacement progressif de l'axe terrestre. *Comptes rendus hebdomadaire des séances de l'Académie des sciences*, 130, 291–298.

Bouasse, H. (1927). *Séismes et sismographes*. Delagrave, Paris.

Bourcart, J. (1924). Les origines de l'hypothèse de la dérive des continents. *Revue scientifique*, 62(18), 563–564.

Bullard, E., Everett, J.E., Smith, A.G. (1965). The fit of the continents around the Atlantic. *The Philosophical Transactions of the Royal Society London A*, 258(1088), 41–51.

Carozzi, A.V. (1969). À propos de la théorie des dérives continentales : Francis Bacon (1620), François Placet (1668), A. von Humboldt (1801) et A. Snider (1858). *Compte rendu des séances de la Société de physique et d'histoire naturelle de Genève*, 4(3), 171–179.

Carozzi, A.V. (1970). New historical data on the origin of the theory of continental drift. *Geological Society of America Bulletin*, 81, 283–286.

Carozzi, A.V. (1985). The reaction in continental Europe to Wegener's theory of continental drift. *Earth Sciences History*, 4(2), 122–137.

Chamberlin, T.C. (1897). Studies for students: The method of multiple working hypotheses. *Journal of Geology*, 5(8), 837–848.

Chamberlin, R.T. (1928). Some of the objection to Wegener's theory. In *Theory of Continental Drift: A Symposium on the Origin and Movement of Land Masses Both Inter-continental and Intra-continental, as Proposed by Alfred Wegener*, van Waterschoot van der Gracht, W.A.J.M., Willis, B., Chamberlin, R.T., Joly, J., Molengraaff, G.A.F., Gregory, J.W., Wegener, A., Schuchert, C., Longwell, C.R.L., Taylor, F.B. et al. (eds). American Association of the Petroleum Geologists, Tulsa.

Choubert, B. (1935). Recherche sur la genèse des chaînes paléozoïques et antécambriennes. *Revue de géographie physique et géologie dynamique*, 8(1), 5–50.

Choubert, B. (1981). Aperçu de l'évolution des idées concernant la formation de la croûte terrestre d'Alfred Wegener à nos jours. *Geologische Rundschau*, 70(1), 40–49.

Dana, J.D. (1888). *New Text-book of Geology*. American Book Company, New York.

Demhardt, I.J. (2005). Alfred Wegener's hypothesis on continental drift and its discussion in Petermanns Geographische Mitteilungen (1912-1942). *Polarforschung*, 75(1), 29–35.

Du Toit, A.L. (1921). Land connections between the other continents and South Africa in the past. *South Africa Journal of Science*, 18, 120–140.

Ebach, M.C. and Dowding, E.M. (2017). Theodor Arldt (1878-1960): Parochial pauker and pioneering palaeobiogeographer. *Zootaxa*, 4319(1), 157–168.

Flügel, H.W. (1984). A. Wegener, O. Ampferer, R. Schwinner: The first chapter of the New Global Tectonic. *Earth Sciences History*, 3(2), 178–186.

Gagnebin, E. (1922). La dérive des continents selon la théorie d'Alfred Wegener. *Revue générale des sciences pures et appliquées*, 33(10), 293–304.

Gattefossé, J. (1923). À propos de l'Atlantide. *La Géographie*, 40(5), 594–598.

Gattefossé, J. (1926). *Bibliographie de l'Atlantide et des questions connexes*. Imprimerie Bosc frères & Riou, Lyon.

Gignoux, M. (1925). Les théories de Wegener. À propos de la traduction française de son ouvrage. *Revue générale des sciences pures et appliquées*, 36(5), 139–142.

Goguel, J. (1984). Bouguer au Chimborazo, et l'acceptation de la théorie newtonienne de l'attraction. *Travaux du Comité français d'histoire de la géologie*, 2(II), 31–48.

Grandchamp, P. (2015). L'histoire de la découverte de l'allochtonie des Préalpes revue et corrigée. *Travaux du Comité français d'histoire de la géologie*, 3(XXIX), 79–95.

Greene, M.T. (2015). *Alfred Wegener. Science, Exploration, and the Theory of Continental Drift*. Johns Hopkins University Press, Baltimore.

Haeckel, E. (1874). *Histoire de la création des êtres organisés d'après les lois naturelles*. C. Reinwald et Cie Libraires-Éditeurs, Paris.

Haug, É. (1900). Les géosynclinaux et les aires continentales. Contribution à l'étude des transgressions et des régressions marines. *Bulletin de la Société géologique de France*, 3(28), 617–711.

Haug, É. (1907). *Traité de géologie 1 : les phénomènes géologiques*. Armand Colin, Paris.

Helmert, F.R. (1915). *Jahresbericht des Direktors des Königlichen Geodätischen Instituts für die Zeit von April 1914 bis April 1915*. Königl. Preußischen Geodätischen Instituts, Potsdam.

Holmes, A. (1944). *Principles of Physical Geology*. Thomas Nelson and Son Ltd, London.

Jacoby, W.R. (2001). Translation of Die Entstehung der Kontinente, Dr Alfred Wegener. Petermanns Geographische Mitteilungen, 58 I, 185–195, 253–256, 305–309, 1912. *Journal of Geodynamics*, 32, 29–63.

Jeffreys, H. (1929). *The Earth: Its Origin, History and Physical Constitution*. Cambridge University Press, Cambridge.

Joleaud, L. (1923). Essai sur l'évolution des milieux géophysiques et biogéographiques. *Bulletin de la Société géologique France*, 23(4), 205–270.

Köppen, W. and Wegener, A. (1924). *Die Klimate der geologischen Vorzeit*. Borntraeger, Berlin.

Kornprobst, J., Àbalos, B., Barbey, P., Boullier, A.-M., Burg, J.-P., Capdevila, R., Claesson, S., Cordani, U. G., Corrigan, D., Gabrielsen, R. H. et al. (2018). Boris Choubert: Unrecognized visionary geologist, pioneer of the global tectonics. *BSGF – Earth Sciences Bulletin*, 189(2), 7. doi: 10.1051/bsgf/2018006.

Lamarck, J.B. (1802). *Hydrogéologie*. Agasse et Maillard, Paris.

Le Grand, H.E. (1986). Specialities, problems and localism. The reception of continental drift in Australia 1920-1940. *Earth Sciences History*, 5(1), 84–95.

Le Grand, H.E. (1988). *Drifting Continents and Shifting Theories: The Modern Revolution in Geology and Scientific Change*. Cambridge University Press, Cambridge.

Le Vigouroux, P. (2018a). La théorie de Wegener sur la dérive des continents en France jusqu'aux années 1950. Critiques, soutiens, oubli. PhD Thesis, Université de Nantes, Nantes.

Le Vigouroux, P. (2018b). De la préparation de l'après-guerre au boycott des géologues allemands. In *14-18, la Terre et le Feu. Géologie et géologues sur le front occidental*, Bergerat, F. (ed.). AGBP-COFRHIGÉO-SGN, Paris.

Le Vigouroux, P. (2018c). L'introduction de la théorie des translations continentales de Wegener dans l'enseignement des sciences naturelles en France. *Travaux du Comité français d'histoire de la géologie (COFRHIGEO)*, 3(XXXII), 77–102.

Le Vigouroux, P. (2022). The XIII International Geological Congress of 1922 in Brussels (Belgium): Between wounds of the Great War and theory passed under silence. *Episodes Journal of International Geoscience*, 45(4), 457–467.

Le Vigouroux, P. and Gohau, G. (2016). De l'amateurisme en géologie : la réception française de la théorie de la dérive des continents (1920-1950). *Gesnerus*, 73(2), 337–359.

Leuba, J. (1931). Le professeur Hans Schardt 1858–1931. *Bulletin de la Société neuchâteloise des sciences naturelles*, 56, 103–109.

Leuba, J. (1938). *Introduction à la géologie*. Armand Colin, Paris.

Macdougall, D. (2013). *Frozen Earth: The Once and Future Story of Ice Ages*. University of California Press, Berkeley.

Mantovani, R. (1889). Les fractures de l'écorce terrestre et la théorie de Laplace. *Bulletin de la Société des arts et sciences de la Réunion*, 41–53.

Mantovani, R. (1909). L'Antarctide. *Je m'instruis. La science pour tous*, 38, 595–597.

Mantovani, R. (1924). Les points de contact entre la théorie de la dilatation planétaire et la théorie de la dérive des continents de Wegener. *Compte rendu sommaire des séances de la Société géologique de France*, 24(16), 167–168.

Mantovani, R. (1927). L'Atlantide et la découverte de la dilatation planétaire. *Compte rendu sommaire des séances de la Société géologique de France*, 27(4), 153–155.

Marvin, B.U. (1973). *Continental Drift: The Evolution of a Concept*. Smithsonian Institution Press, Washington, D.C.

Marvin, B.U. (1985). The British reception of Alfred Wegener's continental drift hypothesis. *Earth Sciences History*, 4(2), 138–159.

Maury, A. (1859). Les premiers âges de notre planète. *Revue des deux mondes*, 21(2), 329–348.

McCoy, R. (2006). *Ending in Ice: The Revolutionary Idea and Tragic Expedition of Alfred Wegener*. Oxford University Press, Oxford.

Oreskes, N. (1988). The rejection of continental drift. *Historical Studies in the Physical and Biological Sciences*, 18(2), 311–348.

Oreskes, N. (1999). *The Rejection of Continental Drift. Theory and Method in American Earth Science*. Oxford University Press, New York.

Pepper, J.H. (1861). *The Playbook of Metals: Including Narratives of Visits to Coal, Lead, Copper, and Tin Mines*. Routledge, London.

Pratt, J.H. (1859). On the deflection of the plumb-line in India, caused by the attraction of the Himalaya mountains and of the elevated regions beyond; and its modification by the compensating effect of a deficiency of matter below the mountain mass. *Philosophical Transactions of the Royal Society of London*, 149, 745–796.

Reclus, E. (1868). *La Terre. Description des phénomènes de la vie du globe 1 : les continents*. Hachette, Paris.

Romano, M., Console, F., Pantaloni, M., Fröbisch, J. (2017). One hundred years of continental drift: The early Italian reaction to Wegener's "visionary" theory. *Historical Biology: An International Journal of Paleobiology*, 29(2), 266–287.

Sarton, G. (1919). La synthèse géologique de 1775 à 1918. *Isis*, 2(2), 357–394.

Scalera, G. (2009). Roberto Mantovani (1854-1933) and his ideas on the expanding Earth as revealed by his correspondence and manuscripts. *Annals of Geophysics*, 52(6), 615–649.

Scalera, G. and Jacob, K.-H. (eds) (2003). *Why Expanding Earth? – A Book in Honour of O.C. Hilgenberg*. INGV, Rome.

Schaer, J.-P. (1991). Émile Argand 1879-1940. Life and portrait of an inspired geologist. *Eclogae Geologicae Helvetiae*, 84(3), 511–534.

Schaer, J.-P. (2003). La dérive des continents, son accueil à Neuchâtel et parmi les géologues alpins. *Travaux du Comité français d'histoire de la géologie*, 3(17), 159–173.

Schaer, J.-P. (2010). Swiss and Alpine geologists between two tectonic revolutions. Part 1: From the discovery of nappes to the hypothesis of continental drift. *Swiss Journal of Geosciences*, 103, 503–522.

Schwarzbach, M. (1985). *Wegener : le père de la dérive des continents*. Belin, Paris.

Sclater, P.L. (1864). The mammals of Madagascar. *The Quarterly Journal of Science*, 1(2), 212–219.

Simpson, G.C. (1930). Discussion on geological climates. *Proc. Roy. Soc. London, B*, 106, 299–302.

Snider, A. (1858). *La création et ses mystères dévoilés*. Franck et Dentu, Paris.

Sudiro, P. (2014). The Earth expansion theory and its transition from scientific hypothesis to pseudoscientific belief. *History of Geo- and Space Sciences*, 5, 135–148.

Suess, E. (1897). *La face de la terre (Das antlitz der erde)*. Armand Colin, Paris.

Suess, E. (1918). *La face de la terre (Das antlitz der erde)*. Armand Colin, Paris.

Taylor, F.B. (1910). Bearing of the tertiary mountain belt on the origin of the Earth's plan. *Bulletin of the Geological Society of America*, 21(2), 179–226.

Termier, P. (1924). La dérive des continents. *Revue scientifique*, 62(9), 257–267.

Tyrrell, G.W. (1923). Geology. *Science Progress in the Twentieth Century (1919-1933)*, 17(68), 552–556.

Wegener, A.L. (1905). Die astronomischen Werke Alfons X. *Bibliotheca Mathematica*, 3(6), 129–185.

Wegener, A.L. (1911). *Thermodynamik der Atmosphäre*. J.A. Barth, Leipzig.

Wegener, A.L. (1912a). Die Entstehung der Kontinente. *Petermanns Geographische Mitteilungen*, 58(I), 185–195, 253–256, 305–309.

Wegener, A.L. (1912b). Die Entstehung der Kontinente. *Geologische Rundschau*, 3, 276–292.

Wegener, A.L. (1924). *La genèse des continents et des océans*. Blanchard, Paris.

Wegener, A.L. (1937). *La genèse des continents et des océans. Théorie des translations continentales*. Bastard, Paris.

Wegener, A.L. (2002). The origins of continents. *International Journal of Earth Sciences (Geol Rundsch)*, 91, S4–S17.

Willis, B. (1910). Principles of paleogeography. *Science*, 31(790), 241–260.

Winkler, J.R. (2008). *Nexus: Strategic Communications and American Security in World War I*. Harvard University Press, Cambridge.

List of Authors

Joëlle DUCOS
STIH
Sorbonne Université
and
SAPRAT
EPHE
PSL
Paris
France

Jean-Claude DUPONT
CHSSC
Université de Picardie Jules Verne
Amiens
France

Gaston GODARD
IPGP
Université Paris-Cité
France

Philippe GRANDCHAMP
COFRHIGEO
Paris
France

Frédéric LE BLAY
Laboratoire CFV
Université de Nantes
France

Philippe LE VIGOUROUX
Laboratoire CFV
Université de Nantes
France

Bernard MAITTE
Université de Lille
Villeneuve d'Ascq
France

Pierre SAVATON
Université Caen Normandie
and
Laboratoire CFV
Université de Nantes
France

Fleur VIGNERON
Université Grenoble Alpes
France

Index of Terms

G, H

I, L

M, N

O, P

R, S

T, U

V, W

Index of Names

L, M

N O

P, R

S, V

W

Printed and bound by CPI Group (UK) Ltd, Croydon, CR0 4YY

13/09/2026

14961010-0006